Target Organ Toxicity

Volume I

Editor

Gerald M. Cohen
Reader in Toxicology
Head of Toxicology Unit
School of Pharmacy
University of London
London, England

CRC Press
Taylor & Francis Group
Boca Raton London New York

CRC Press is an imprint of the
Taylor & Francis Group, an **informa** business

CRC Press
Taylor & Francis Group
6000 Broken Sound Parkway NW, Suite 300
Boca Raton, FL 33487-2742

Reissued 2019 by CRC Press

A Library of Congress record exists under LC control number:

Publisher's Note
The publisher has gone to great lengths to ensure the quality of this reprint but points out that some imperfections in the original copies may be apparent.

Disclaimer
The publisher has made every effort to trace copyright holders and welcomes correspondence from those they have been unable to contact.

ISBN 13: 978-0-367-25645-6 (hbk)
ISBN 13: 978-0-429-28884-5 (ebk)

**Visit the Taylor & Francis Web site at http://www.taylorandfrancis.com and the
CRC Press Web site at http://www.crcpress.com**

INTRODUCTION

The specificity of certain chemicals to induce a wide variety of differing toxicities, including hepatotoxicity, neurotoxicity, pulmonary toxicity, nephrotoxicity, ototoxicity, and cancer in particular organs is a fascinating problem currently under active investigation in many laboratories.

The purpose of these volumes is to provide a valuable reference for established investigators and postgraduate students in toxicology. Essential information on the general principles of target organ toxicity is provided in the first seven chapters in Volume I. Included are the importance of pharmacokinetics, metabolic activation and key defense mechanisms, excretion, species variation, and tissue-specific biochemistry. Next these general principles are illustrated by dealing with specific examples of toxicity to different target organs and systems. Volume II, Chapter 9, deals with DNA modifications and repair in tumor induction, and the final chapter discusses organ specificity in tumor initiation and complete carcinogenesis.

THE EDITOR

Gerald M. Cohen Ph.D. is Head of the Toxicology Unit, Department of Pharmacology, The School of Pharmacy, University of London.

Dr. Cohen graduated in 1966 from Leeds University, Leeds, U.K., with a B.Sc. Honours degree in pharmacy and obtained his Ph.D. degree in pharmacology in 1973 from the University of Minnesota, Minneapolis. He is a member of the British Toxicology Society, British Association for Cancer Research, American Association for Cancer Research, Society for Free Radical Research and the Biochemical Society.

Dr. Cohen has presented many invited lectures at both International and National Meetings. He has published more than 60 research papers. His current major research interests include mechanisms of selective toxicity and target organ toxicity.

CONTRIBUTORS

Th. P. M. Akerboom
University of Düsseldorf
Düsseldorf, F.R.G.

Rebecca J. Anderson
Warner Lambert/Parke-Davis
Ann Arbor, Michigan

Steven D. Aust
Michigan State University
East Lansing, Michigan

Tibor Balazs
Division of Drug Biology
Food and Drug Administration
Washington, D.C.

J. W. Bridges
Robens Institute
University of Surrey
Guildford, Surrey, U.K.

M. H. Briggs
University of Cambridge
Cambridge, U.K.

John Caldwell
St. Mary's Hospital Medical School
London, U.K.

John C. Connelly
Robens Institute
University of Surrey
Guildford, Surrey, U.K.

Maurice M. Coombs
Imperial Cancer Research Fund
London, U.K.

V. Ferrans
National Institutes of Health
Bethesda, Maryland

J. Hanig
Division of Drug Biology
Food and Drug Administration
Washington, D.C.

Ernest S. Harpur
Aston University
Birmingham, U.K.

E. Herman
Division of Drug Biology
Food and Drug Administration
Washington, D.C.

Ralph Heywood
Huntingdon Research Centre
Huntingdon, Cambs, U.K.

Jerry B. Hook
Smith Kline & French Laboratories
Philadelphia, Pennsylvania

J. Brian Houston
University of Manchester
Manchester, U.K.

Toshihisa Ishikawa
University of Düsseldorf
Düsseldorf, F.R.G.

Paul Kleihues
University of Zurich
Zurich, Switzerland

Walter G. Levine
Albert Einstein College of Medicine
Bronx, New York

W. J. Malaisse
Brussels Free University
Brussels, Belgium

Harry D. M. Moore
Institute of Zoology
MRC/AFRC Comparative Physiology Group
Zoological Society of London
London, U.K.

Benoit Nemery
MRC Toxicology Unit, Carshalton
Surrey, U.K.

Rudy J. Richardson
University of Michigan
Ann Arbor, Michigan

Glenn F. Rush
Smith Kline & French Laboratories
Philadelphia, Pennsylvania

Susan A. Sangster
Tate and Lyle, Ltd.
Reading, U.K.

Bradley W. Schwab
University of Michigan
Ann Arbor, Michigan

Helmut Sies
University of Düsseldorf
Düsseldorf, F.R.G.

Lewis L. Smith
Imperial Chemical Industries, PLC
Macclesfield, Cheshire, U.K.

J. David Sutton
St. Mary's Hospital Medical School
London, U.K.

J. A. Timbrell
University of London
London, U.K.

Otmar Wiestler
University of Freiburg
Freiburg, F.R.G.

Dedicated
To Judith, Gidon, Raffi, and Aron
For Their Encouragement and Understanding

TARGET ORGAN TOXICITY

Volume I

Volume II

TABLE OF CONTENTS

Volume I

Chapter 1

BASIC PRINCIPLES OF TARGET ORGAN TOXICITY

Gerald M. Cohen

TABLE OF CONTENTS

I. INTRODUCTION

A baffling array of different toxicities may be induced in various organs of the body by thousands of chemicals to which we are exposed. All tissues are susceptible in varying degrees to these toxic effects, but many chemicals exhibit a marked propensity to damage specific organs; these are known as the *target organs of toxicity*. An understanding of the remarkable ability of some chemicals to cause these organ-specific lesions is the subject of this book. Some examples of specific compounds and their target organs of toxicity are given in Table 1.

In this chapter, I shall discuss the basic principles which determine the susceptibility of particular organs to the toxicity of chemicals. Several factors predispose particular organs to toxicity. Portals of entry and exit from the body will generally be exposed to high concentrations of chemicals and thus are more likely to be susceptible to toxicity. For example, the lungs are exposed to a wide variety of inhaled chemicals including atmospheric pollutants, irritants, and cigarette smoke, as well as medicinal and household aerosols. Other portals of entry include the skin and GI tract, which will be exposed to chemicals either from dermal contact or by ingestion, respectively. Similarly, in the process of excretion of chemicals from the body, the kidney and the biliary tree and intestines may be exposed to high concentrations of drugs and/or metabolites. Many xenobiotics (foreign chemicals) are lipid soluble compounds which, if not metabolized to more polar and more readily excretable metabolites, would remain in the body for long periods.[1,2] Quantitatively the major site of xenobiotic metabolism is the liver and as many chemicals are metabolized to toxic metabolites, the liver is often a target organ for toxicity. Often it is the balance in a particular tissue between activating and deactivating enzymes which determines the susceptibility of different organs to toxicity.

A. Factors Affecting Toxicity

Many factors may influence the susceptibility of specific organs to toxicity and particular consideration should be given to the distribution of toxic compounds in the body and their mechanisms of toxicity. It is self-evident that for any chemical to exert a deleterious effect on a specific organ it must first gain access to "its site of action" (Figure 1). The quantitation of the time course of distribution of a toxic chemical and its metabolites in the body is known as toxicokinetics and is dealt with in detail in Chapter 2. The response of an organ is affected by:

1. Factors affecting the disposition of the toxic compound to an organ (Figure 1 and Table 2). These include physicochemical properties of the compound, such as pKa and lipid solubility, as well as absorption, plasma protein binding, and excretion of the compound and its metabolites.
2. Metabolic fate of a chemical and its metabolites within the organ and the body, e.g., the generation and detoxication of reactive metabolites in the target tissue.
3. The ability of the organ to respond to the effects of the chemical, e.g., its ability to repair chemically induced damage.

The importance of these three factors will now be illustrated with appropriate examples.

II. DISPOSITION OF THE TOXIC CHEMICAL

The concentration of a xenobiotic or its reactive metabolite and the duration of exposure of a target organ will generally determine the severity of the toxicological response. However,

Table 1
SOME TARGET ORGANS OF TOXICITY OF CERTAIN CHEMICALS

Organ or tissue	Chemical	Toxicity
Nervous system	Acrylamide	Axonopathy
Liver	Paracetamol	Liver necrosis
Kidney	Cephaloridine	Nephrotoxicity
Respiratory system	Bleomycin	Pulmonary fibrosis
Eye	Chloroquine	Retinopathy
Ear	Streptomycin	Ototoxicity
Blood	Primaquine	Hemolytic anemia
Bone marrow	Chloramphenicol	Aplastic anaemia
Reproductive system	1,2-Dibromo-3-chloropropane	Male sterility
Heart	Adriamycin	Cardiomyopathy
Skin	Phenylbutazone	Exfoliative dermatitis
GI tract	Methotrexate	Ulceration
Fetus	Phenytoin	Congenital abnormalities (cleft palate)
Bone	Anticonvulsants	Osteomalacia
Pancreas	Alloxan	Damage to β cells

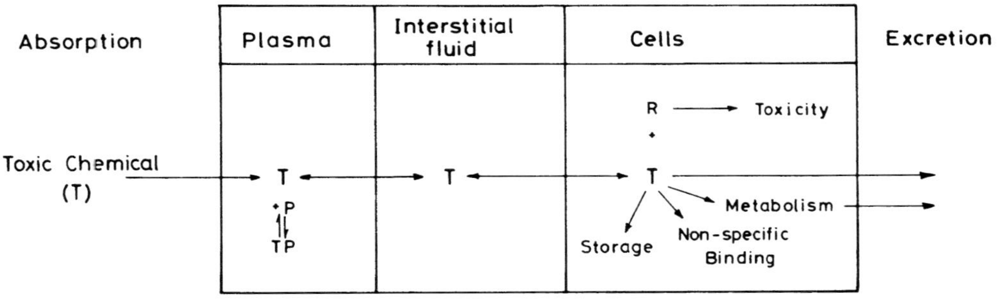

FIGURE 1. Distribution of a toxic chemical (T) in the body. Many factors such as binding to plasma proteins (P) and storage may affect the concentration of T available to interact with critical receptors (R) and elicit toxicity in the target organ.

Table 2
MAJOR FACTORS AFFECTING THE SUSCEPTIBILITY OF ORGANS TO CHEMICAL-INDUCED TOXICITY

Factor	
Tissue distribution of the compound	Physicochemical properties; pKa, lipid solubility, absorption, distribution
	Plasma protein binding and excretion of compound
	Possibility of active transport or secretion
	Dose-dependent kinetics
Metabolism	Depends whether the parent compound or a metabolite is toxic
	Tissue distribution of drug-metabolizing enzymes
	Qualitative and quantitative nature of enzymes present
	Balance of activating and deactivating enzymes in liver and/or target organ(s). Different enzyme inducers or inhibitors may affect these tissues differently
Individual biochemistry of different tissues	Presence of a particular biochemical pathway
	Absence or low levels of key defensive mechanisms, e.g., glutathione
Host	Ability to repair a particular damage or lesion
	Presence of impaired host function, e.g., liver or kidney disease

Table 3
TARGET ORGAN TOXICITY OF CHEMICALS WHERE TISSUE-SPECIFIC ACCUMULATION IS OF MAJOR IMPORTANCE

Chemical	Target organ of toxicity	Comments	Ref.
Paraquat	Lung	Active uptake system into type I and II alveolar epithelial cells due to structural similarity to endogenous polyamines	3 and Vol. II, Chap. 3
Streptozotocin	Pancreas	High specificity for pancreatic β cells possibly due to glucose moiety which causes specific binding to cells with glucose receptors	4
Chloroquine, chlorpromazine, and other N-substituted phenothiazines	Eye	High persistent concentrations in the pigmented eye due to avid binding to melanin	5
Kanamycin, chloroquine	Ear	Binding to melanin in ear	6
Cephaloridine	Kidney	Uptake into the kidney by anion transport system and subsequent accumulation due to presence of cationic grouping in cephaloridine preventing its excretion by the anion transport system	7
6-Hydroxydopamine	Adrenergic nerves	Taken up into adrenergic nerve terminals because of structural similarity to endogenous substrate	8

high concentrations of a toxic chemical in a particular tissue do not necessarily lead to toxicity in that tissue. For example, DDT and related insecticides accumulate in adipose tissue without exerting any apparent deleterious effect on it. However, selective accumulation of a chemical in a tissue may be a critical factor in determining the target organ of toxicity and several examples of this are given in Table 3. Further details of the nature and importance of the accumulation of paraquat and streptozotocin are given in Volume II, Chapters 3 and 8, respectively.

The normal physiological function of an organ may also predispose it to toxicity. Many chemicals are nephrotoxic due in part to their accumulation in the kidney as a result of its large blood supply but also because of particular aspects of its normal physiological functions.[9,10] Any potentially toxic chemical in the systemic circulation will be presented to the kidney in significant quantities, where it will be filtered at the glomerulus. Thus, toxic concentrations that may be reached in the tubular fluid as salt and water are reabsorbed from the glomerular filtrate. Compounds may also become particularly concentrated in tubular cells because of active secretion or reabsorption. Any chemical which is actively secreted will first be concentrated, in cells of the proximal tubule, to levels greater than those found in plasma. These cells may, therefore, be exposed to potentially toxic concentrations. High concentrations of chemicals may also be attained in the renal medulla, although this tissue only receives about 10% of total renal blood flow; medullary cells will be exposed to high concentrations of chemicals in the tubular urine as they pass through the loop of Henle and the medullary collecting duct. The countercurrent effect in the medulla may also act to concentrate chemicals. Due to normal physiological functions, certain areas of the kidney may, therefore, be exposed to excessively high concentrations of potentially toxic compounds resulting in nephrotoxicity.[9,10] These aspects are discussed in more detail in Volume II, Chapter 1.

A. Binding to a Specific Macromolecule: Melanin

Binding to specific macromolecules either intracellularly or to plasma proteins[2] may profoundly affect the body distribution of xenobiotics. Many drugs including chloroquine, phenothiazine derivatives, and aminoglycoside antibiotics bind avidly to melanin.[5] Melanin is found in the eye, the inner ear, the skin, nuclei of the brain stem, and some other tissues. Following long-term administration, these drugs accumulate in melanin-containing tissues, often resulting in toxicity. This has been best documented for the antimalarial chloroquine, which on prolonged high-dose therapy, may damage both the eye and the inner ear. The chloroquine-induced retinopathy has been related to very high concentrations in the pigmented eye and these high levels persist long after other tissues are depleted. A similar accumulation and retention of chloroquine with the melanin of the inner ear has also been observed. A high retention was found 1 year after a single administration of radiolabeled chloroquine.[6] Following administration of kanamycin, the stria vascularis, the melanin-containing region in the cochlea, was damaged to a greater extent in pigmented guinea pigs than albinos, supporting the involvement of melanin in kanamycin-induced ototoxicity.[6,11]

III. ROLE OF METABOLIC ACTIVATION IN TARGET ORGAN TOXICITY

Toxicity of a chemical may be due to the parent compound, a chemically reactive metabolite or a chemically stable metabolite. Many types of toxicity such as carcinogenesis, mutagenesis, teratogenesis, cell necrosis, and hypersensitivity reactions may be mediated by reactive metabolites.[12,13] Such toxicities often result following the metabolism of xenobiotics to reactive electrophiles which combine covalently with critical cellular macromolecules such as DNA, RNA, or proteins.[12-14]

When toxicity is mediated by a metabolite, consideration must be given to whether (1) the formation of the toxic metabolites occurs in the liver followed by transport to the target organs, (2) metabolic activation occurs in the target organ(s), or (3) the process requires a combination of both. These concepts have been recently reviewed[15-17] and are also discussed in Chapter 3. Thus, the metabolic activation of a compound in a particular organ may be a critical factor in determining its susceptibility to toxicity.

A. Toxicity Mediated by the Parent Compound

The effects of metabolism are self-evident when the parent compound is itself toxic (Figure 2). Any change in the rate of metabolism of the compound, either in the liver or in any extrahepatic tissue, will alter its effective concentration in the target organ(s) and cause a corresponding effect on toxicity.[15,17] It may also be readily appreciated from Figure 2 that any other factors that serve to increase or decrease the concentration of the toxic compound in the target tissue will have a correspondingly predictable effect on toxicity, provided, of course, that the toxicity is a dose-related phenomenon. Thus, induction of drug-metabolizing enzymes with inducing agents such as phenobarbitone or 3-methylcholanthrene results in increased metabolism of the active parent compound and therefore in decreased amounts available for pharmacological or toxicological activity.

B. Toxicity Mediated by Reactive Metabolites

The situation is more complex when toxicity is due to a metabolite. If the metabolite formed is chemically stable, then it may be formed in one organ and transported by the systemic circulation, as illustrated in Figure 3.[15,17-20] In contrast, some metabolites may be chemically reactive, i.e., their half-lives are very short.[20] Such reactive metabolites would most likely react either intracellularly, within the cell in which they were formed, or if they escaped from the cells, with constituents or blood. The extrahepatic toxicity of these metabolites will most likely be mediated by reactive intermediates generated *in situ* within their

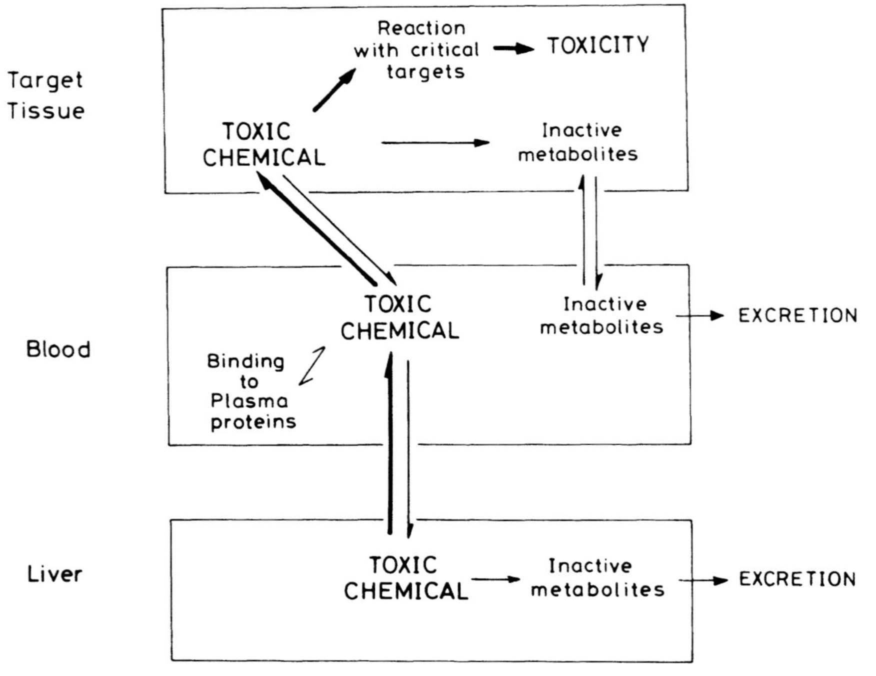

FIGURE 2. Toxicity mediated by the parent chemical. Any factor, such as plama protein binding, which affects the concentration of the toxic chemical available to interact in the target organ, will affect the toxicity of the chemical.

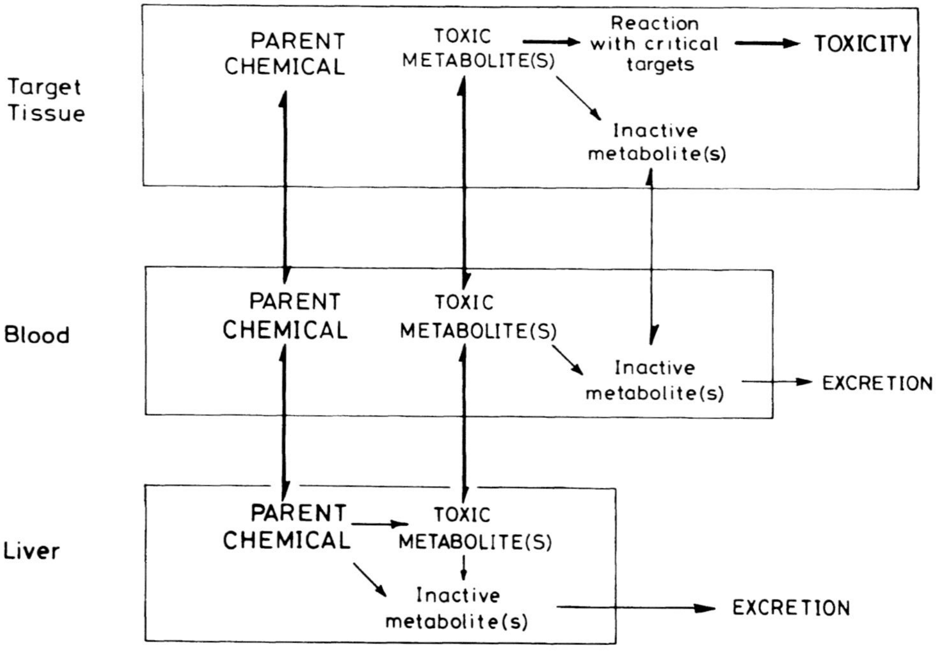

FIGURE 3. Toxicity mediated by a toxic metabolite generated in the liver and transported to the target organ. The toxic metabolite is formed in the liver and is sufficiently stable that it may then be transported via the circulation to the target organ of toxicity.

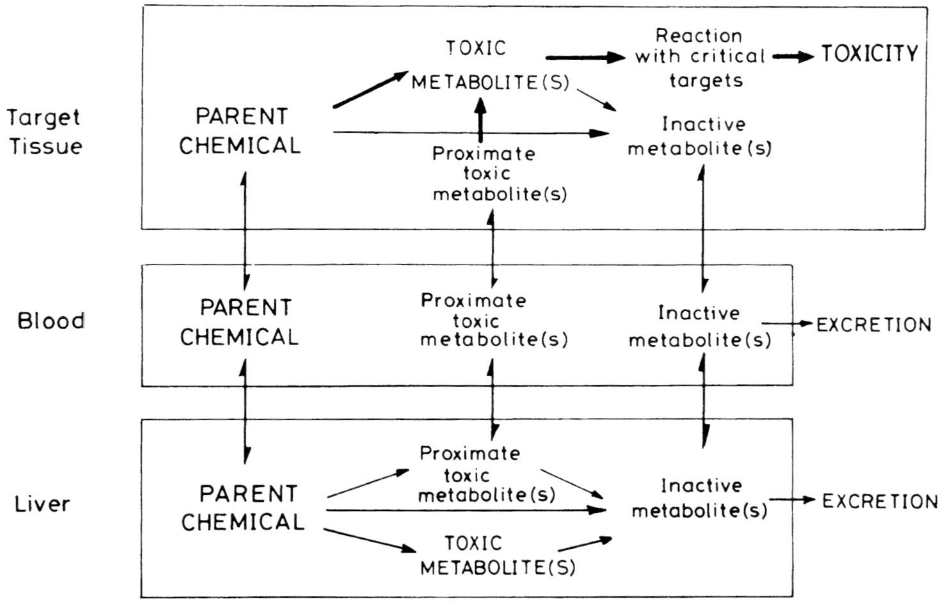

FIGURE 4. Toxicity mediated by reactive metabolites formed in the target organ either from the parent chemical or from a proximate toxic metabolite. If the toxic metabolite is chemically very reactive, it may have to be generated *in situ* in the target organ in order to exert toxicity. Depending on the metabolic capability of the tissue, the metabolic activation from the parent chemical may be completely carried out in the target organ. Alternatively, part of the metabolic activation may be carried out in other tissues, such as the liver, but the ultimate toxic metabolites are formed *in situ* in the target organ.

target organs,[15,17,20] as illustrated in Figure 4. This latter case is particularly well exemplified by the work of Boyd[21] and colleagues with 4-ipomeanol which causes specific damage to the Clara cells in a number of species (Volume I, Chapter 5 and Volume II, Chapter 3). In order to establish whether a compound is activated by either one or more of the previously discussed mechanisms, it is necessary to consider other factors, including the tissue distribution of the xenobiotic metabolizing enzymes, as illustrated in the following section.

C. Tissue Distribution of Xenobiotic Metabolizing Enzymes

While it is generally recognized that the liver is the major site of most xenobiotic metabolism, a significant amount of extrahepatic metabolism also takes place.[22] The ability of extrahepatic tissues to metabolize xenobiotics has recently received more attention. Although quantitatively the contribution of a particular extrahepatic tissue may represent only a small percentage of the overall metabolism, it may be of particular toxicological significance. The enzymes responsible for the activation of xenobiotics will often be the phase I oxidative enzymes but may in some cases be the phase II conjugating enzymes, such as the activation of *N*-hydroxyacetylaminofluorene following sulfate conjugation.[14,23] The tissue distribution of both phase I and II drug metabolizing enzymes and the relative availabilities of their various co-factors are, therefore, vital factors in determining the relative susceptibilities of different tissues to toxic chemicals (see Chapter 3).

The wide tissue distribution of some of these xenobiotic metabolizing enzymes[22,24] may be illustrated by the large number of tissues capable of metabolizing benzo(*a*)pyrene, a ubiquitous environmental carcinogen and mutagen (Table 4). This particular reaction has been extensively studied because benzo(*a*)pyrene requires metabolic activation before exerting its carcinogenic and mutagenic effects and also because some of the major metabolites (i.e., 3- and 9-hydroxybenzo(*a*)pyrene) may be measured by an extremely sensitive fluo-

Table 4
TISSUE DISTRIBUTION OF BENZO (*a*)PYRENE 3-MONOOXYGENASE

Liver	Testes
GI tract	Thyroid
Lung	Adrenal
Trachea	Spleen
Skin	Thymus
Kidney	Lymph nodes
Placenta	Bone marrow
Muscle	Lymphocytes
Aorta	Monocytes
Mammary gland	Leukocytes
Salivary gland	Ovary
Prostate	Kidney
Pancreas	Eye

rometric assay. This enzyme activity is known as benzo(*a*)pyrene 3-monooxygenase, benzo(*a*)pyrene hydroxylase, aryl hydrocarbon hydroxylase, or AHH (E.C. 1.14.14.2). A large amount of data has been accumulated over the last few years strongly implicating the formation of "bay region" diol epoxides as the ultimate carcinogens of polycyclic aromatic hydrocarbons (PAH).[25] It is readily apparent that a large number of tissues may, under appropriate conditions, metabolize foreign chemicals[22] (Table 4 and Chapter 3). It should be stressed that although such metabolism may be demonstrated in vitro, it does not mean that these tissues will necessarily metabolize these compounds in vivo; it will depend on many factors such as blood flow, the nature of the substrate and its availability, and a supply of the necessary co-factors.

Important differences in the distribution of the drug-metabolizing enzymes in any particular tissue may also predispose particular cell types in a tissue to toxicity. Even in the liver, considered a relatively homogeneous tissue, large differences in the tissue distribution of cytochrome P-450 and glutathione have been observed using quantitative cytochemistry.[26,27] Therefore, it is not surprising that in heterogeneous tissues such as the lung, with more than 40 cell types present,[28] one observes an increased concentration of drug-metabolizing enzymes such as cytochrome P-450 in certain cell types.[21,29,30] In addition to quantitative differences in the distribution of these enzymes, important differences in isoenzymes, with different substrate specificities, may be observed which may also contribute to susceptibility to target organ toxicity.[30]

Thus the Clara (nonciliated bronchiolar epithelial) cells, which constitute only 1% of the cells of the lung, contain the majority of the pulmonary cytochrome P-450.[21,29] This high intracellular enzyme concentration predisposes the Clara cells to damage from a number of agents including 4-ipomeanol, 3-methylfuran, and carbon tetrachloride.[21,30] Similarly, a number of studies suggest that the highest activities of microsomal cytochrome P-450 mixed function oxidases in the kidney are found in the S_3 cells of the proximal tubules.[31,32] The highest concentration of S_3 cells is found in the pars recta, that portion of the proximal tubule most susceptible to toxic damage.

D. Balance of Activating and Deactivating Enzymes

It is vitally important when considering the generation of reactive metabolites in any tissue, either hepatic or extrahepatic, that due attention is given to both the activating and deactivating enzymes. It is the balance of the activities of these enzymes and the availability of their respective co-factors which will ultimately determine both quantitatively and qual-

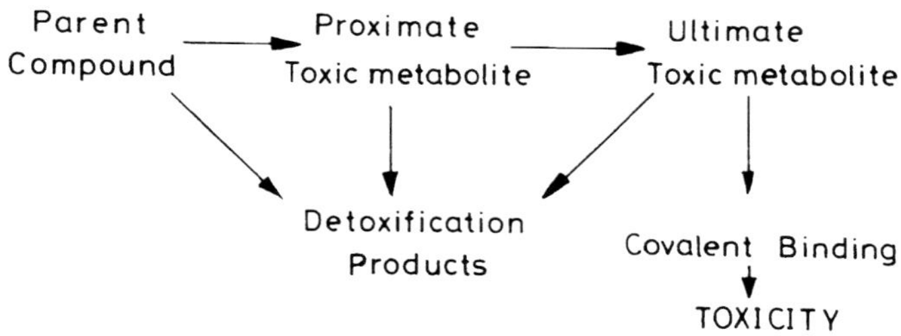

FIGURE 5. Multistep metabolic activation to generate ultimate toxic metabolite.

Table 5
CHEMICALS REQUIRING MULTISTEP METABOLIC ACTIVATION TO GENERATE THEIR ULTIMATE TOXIC METABOLITES

Benzene	Bromobenzene
Naphthalene	7,12-Dimethylbenz(*a*)anthracene
Benzo(*a*)pyrene	Hexachlorobutadiene
Benz(*a*)anthracene	2-Naphthylamine
3-Methylcholanthrene	4-Aminobiphenyl
Acetylaminofluorene	Isoniazid
Chlorobiphenyl	

itatively how much of a reactive metabolite(s) is generated in a particular tissue and, therefore, is subsequently available for interaction with critical cellular targets ultimately leading to toxicity. This has been well illustrated in studies with the carcinogenic PAHs, benzo(*a*)pyrene[33] and 7,12-dimethylbenz(*a*)anthracene,[34] when the hydrocarbon-deoxyribonucleoside adducts formed following microsomal activation of the carcinogen are markedly different to those formed either in vivo or in intact cells. This is primarily due to the fact that with microsomes there is a disruption of the normal cellular integrity, resulting in the loss of the normal balance of activating and deactivating enzymes and the formation of reactive metabolites not normally formed in vivo. Such problems would appear to be of more importance with compounds which require a multistep metabolic activation such as the PAHs, rather than with those requiring a single step activation.

E. Formation of Proximate Toxic Metabolites in the Liver and Ultimate Reactive Metabolites in the Target Tissue

Several years ago James and Elizabeth Miller,[14,35] at the McArdle Laboratory in Wisconsin, proposed the importance of proximate and ultimate carcinogens in their unifying concept of the role of metabolism in the generation of ultimate reactive electrophiles in chemical carcinogenesis (Figure 5). It is realized that several toxic chemicals, in addition to many carcinogens, require multistep metabolic activation before generating their ultimate toxic metabolites (Table 5). It is, therefore, conceivable that the initial activation to a stable proximate toxic metabolite may take place in the liver or any other tissue and this metabolite could be transported to the target organ where it is metabolized to the ultimate toxic metabolite (Figure 4). The revelation that many chemicals require such multistep activation suggests that this mechanism may be far more important than previously realized. Different enzymes may also be involved in different stages of the activation and this may be of importance in target organ toxicity. For example, a target tissue may not be able to metabolize the parent chemical to its proximate toxic metabolite but may be able to convert this to the ultimate reactive species. Much further work will be needed to verify these suggestions. Indirect support comes from recent findings that several extrahepatic tissues, while low in the activities of cytochrome P-450 dependent mixed function oxidases, may activate xenobiotics to reactive metabolites by a variety of different enzymes and mechanisms including prostaglandin endoperoxide synthetase,[36] peroxidase,[37] xanthine oxidase,[38] tyrosinase,[39] and even by-products derived during lipid peroxidation.[40]

IV. ROLE OF THE SPECIFIC BIOCHEMISTRY OR PHYSIOLOGY OF THE TISSUE IN DETERMINING TARGET ORGAN TOXICITY

Major differences in the basic biochemistry or physiology of different tissues and cells is obviously of prime importance in determining the inherent susceptibility of different organs to toxicity. The presence of vital metabolic pathways and specific receptors in certain tissues may predispose these tissues to the toxicity of certain chemicals (Chapter 6). Undoubtedly this is one of the least documented areas of target organ toxicity because our basic knowledge of the mechanisms of toxicity of the majority of chemicals is very poor.[41] Much more is known about the factors governing both the disposition of toxic chemicals and their subsequent metabolic activation, if required, than about their interaction with cellular targets and the biochemical pathways then involved before clinical signs of toxicity are manifested.[41]

Although the factors governing the inherent susceptibility of a particular organ to toxicity are less well understood it may be useful to subdivide them into those related either to (A) organ function or (B) biochemistry of the tissue.

A. Organ Function

The normal physiology and function of the tissue may predispose it to toxicity. Thus the nephrotoxicity of many chemicals is related to their concentration in the kidney due to its normal physiological function[10] (see Volume II, Chapter 1). Similarly, the respiratory tract will be exposed to high concentrations of inhaled pollutants (Volume II, Chapter 3).

The normal function of a tissue may be impaired in certain diseased states such as liver or kidney disease. This may lead to impaired metabolism and/or excretion of compounds followed by an accumulation of toxic concentrations of the chemicals[42] and may be one of the reasons for the high incidence of adverse drug reactions observed in patients with impaired renal function. In addition to effects on the parent compound, impaired renal excretion of metabolites may also be observed leading to increased concentrations of metabolites.[42,43] The metabolites may be toxicologically active per se or they may interfere with the metabolism and/or distribution of the parent compound. This area and its clinical implications has been recently reviewed.[42,43]

Tissues with a high proportion of dividing cells such as bone marrow, GI tract, and hair follicles are more sensitive to the toxic effects of many anticancer drugs than tissues with a lower proportion of such cells.[44] Thus, treatment with anticancer drugs such as the antimetabolites or alkylating agents often results in damage to the bone marrow. Since the marrow contains stem cells, i.e., the immature precursors of the red cells, the platelets, and the white cells, this damage may result in pancytopenia, a decrease in the numbers of the three major groups of formed elements, which are circulating. If the damage is very severe the marrow may no longer proliferate, a condition known as aplastic anemia. The damage to the platelets can lead to thrombocytopenia and bleeding while damage to the leukocytes leads to an increased risk of infection in patients treated with many anticancer drugs.

A particular problem associated with the inhibition of rapidly dividing cells occurs with the fetus. Drugs which inhibit rapidly dividing fetal cells may produce developmental abnormalities and some anticancer drugs have been shown to be teratogenic.[45]

B. Organ-Specific Biochemistry

The differentiated functions of different cells endow them with specialized biochemical pathways and enzymes. Organ-specific toxicity may result from the interaction of chemicals with enzymes or other macromolecules connected with these specialized functions. In Table 6, a number of examples are given of biochemical differences, which predispose particular tissues or organs to toxicity. These examples do not include differences due either to the presence of specific uptake mechanisms or metabolic activation by the mixed function oxidase system which have already been discussed. Our current understanding of the role of tissue-specific biochemistry in target organ toxicity is relatively poor but will increase as our understanding of the mechanism of toxicity of chemicals improves (Chapter 6).

V. FACTORS AFFECTING TARGET ORGAN TOXICITY

Numerous factors are known to affect the toxicity of chemicals.[2,54,60] In a short review one cannot consider many of these factors in detail; however, it is obvious that in many cases such factors may also affect the organ-specific toxicity of a chemical. Such variables include age, route of administration, diet, species (Chapter 5), and environment. A number of examples are given in Table 7.

VI. SUMMARY

Some basic principles of organ-specific toxicity have been discussed. For convenience these were considered together with a number of examples in three distinct, but interrelated, areas: (1) the pharmacokinetics or toxicokinetics of the compound which determines the tissue distribution of the chemical, (2) the metabolism of the compound to either toxic or inert metabolites, and (3) the individual biochemistry of different organs and their response to chemically induced damage.

Induction of toxicity in a specific organ in one species does not necessarily mean that the same organ will be affected in another species.[65] Only by understanding more about the mechanisms of toxicity and target organ toxicity will it be possible to extrapolate scientifically the meaning of toxicity data obtained in animals for man.

ACKNOWLEDGMENT

I should like to thank Mrs. M. Fagg for typing the manuscript.

Table 6
ORGAN-SPECIFIC BIOCHEMICAL DIFFERENCES PREDISPOSING PARTICULAR ORGANS TO TOXICITY

Biochemical difference	Examples	Ref.
DNA repair	Individuals with defective DNA repair, e.g., xeroderma pigmentosum, are very susceptible to skin cancer; in animals, differences in DNA repair correlate with organ-specific carcinogenicity of certain nitrosamines and nitrosamides	46, 47, and Vol. 2, Chapter 10
Neurotoxic esterase	Toxicity of certain organophosphorous esters related to their ability to combine with a specific neurotoxic esterase	48, 49
Ah locus	There are large differences in the susceptibility of different strains of mice to drug toxicity, chemical carcinogenesis, and teratogenesis, much of which is due to the presence, absence, or defects in a cytosolic receptor controlled by the Ah locus; this is responsible for the activities of a battery of different enzymes including monooxygenase activities; alterations in the receptor have been related to differing toxicities to mice of either 2,3,7,8-tetrachlorodibenz-*p*-dioxin (TCDD) or the induction of cataracts by paracetamol or naphthalene	50—52
Glucose 6-phosphate dehydrogenase deficiency	Individuals with a deficiency of this enzyme in their erythrocytes are susceptible to drug-induced hemolysis for a variety of compounds including phenylhydrazine, vitamin K, and fava beans	2, 53
Galactose metabolism	Cataracts commonly occur in congenital galactosemia associated with a deficiency of galactose-1-phosphate-uridyl transferase; rats fed a diet high in galactose develop cataracts due to reduction of galactose by aldose reductase in the lens, to galactitol; this sugar alcohol does not readily permeate membranes so the lens is subject to osmotic stress	54, 55
Dihydrodiol dehydrogenase	Naphthalene induces experimental cataracts in animals, particularly rabbits; it is metabolized in the liver to 1,2-dihydro-1,2-dihydroxynaphthalene and then conjugated with UDP-glucuronic acid; in the lens, the dihydrodiol is readily metabolized by the lens dihydrodiol dehydrogenase to 1,2-dihydroxynaphthalene; this immediately autooxidizes to the toxic 1,2-naphthoquinone	54, 56
Tyrosinase	Cells, such as melanoma cells, may be selectively destroyed by compounds activated by tyrosinase; thus, γ-L-glutaminyl-4-hydroxybenzene is metabolized in melanomas to a cytotoxic quinone	39
C-S or β-lyase	This enzyme cleaves cysteine but not glutathione of *N*-acetyl cysteine conjugates to a reactive product; high concentration of this enzyme in the kidney may be involved in the nephrotoxicity of *S*-(1,2-dichlorovinyl) cysteine and hexachloro-1-butadiene	57
Alcohol dehydrogenase	Highest concentrations in periportal area of liver results in periportal necrosis following activation of allyl alcohol to acrolein	Chapter 8
Amine oxidase	Allylamine is toxic to the cardiovascular system, most probably due to acrolein which is formed by the action of benzylamine oxidase; the activity of this enzyme is much higher in cardiovascular tissue than liver, brain, or plasma	58

Table 6 (continued)
ORGAN-SPECIFIC BIOCHEMICAL DIFFERENCES PREDISPOSING PARTICULAR ORGANS TO TOXICITY

Biochemical difference	Examples	Ref.
Oxidative defenses	Adriamycin causes a life-threatening cardiotoxicity possibly due to a decreased ability of cardiac tissue to cope with oxidative stress because of low levels of catalase and superoxide dismutase; adriamycin also decreased cardiac glutathione peroxidase	59

Table 7
FACTORS AFFECTING TARGET ORGAN TOXICITY

		Ref.
Age	CNS of children affected by organic Pb because blood brain barrier not fully developed; in adults the peripheral nervous system affected	54
	Calcium disodium ethylene diaminetetracetic acid causes liver necrosis in very young rats and severe kidney lesions in 12-week-old animals	61
Developmental stage	Teratogenic agents act selectively on developing cells	45
Route of administration and duration of exposure	Benzene on inhalation of high concentrations for a short time depress the CNS whereas chronic exposure to low levels affects the hemopoietic system	
Diet	In rats fed a normal diet, paraquat causes lung damage, but in selenium deficient animals, liver damage is observed	62
	Toxicity due to cassava is only observed in individuals with a dietary deficiency	63
	Oral administration of dimethylnitrosamine induces liver tumors to rats on a normal diet; it causes kidney tumors in animals fed a low-protein diet	64
Species	Dithizone and ethambutol cause a retinopathy in dogs and rabbits, which possess a tapetum, but not in rats and monkeys	54 and Vol. 2, Chapter 6
	Dihydrostreptomycin causes cochlear toxicity in man and patas monkey, but not in macaque monkey	
Environment	Various factors in environment, including enzyme inducing agents or inhibitors, may alter the metabolism of a chemical and thereby change its target organ of toxicity	
Chemical structure	Small variations in chemical structure can profoundly influence the target organ of toxicity; paraquat, but not structurally related diquat, damages the lung	3 and Vol. 2, Chapter 3

REFERENCES

1. **Williams, R. T.,** *The Metabolism and Detoxication of Drugs, Toxic Substances and Other Organic Compounds*, Chapman & Hall, London, 1959.
2. **Goldstein, A., Aronow, L., and Kalman, S. M.,** *The Principles of Drug Action: The Basis of Pharmacology*, 2nd ed., John Wiley & Sons, New York, 1974.
3. **Smith, L. L.,** The identification of an accumulation system for diamines and polyamines into the lung and its relevance to paraquat toxicity, *Arch. Toxicol. Suppl.*, 5, 1, 1982.
4. **Srivastava, L. M., Bora, P. S., and Bhatt, S. D.,** Diabetogenic action of streptozotocin, *Trends Pharmacol. Sci.*, 3, 376, 1982.
5. **Potts, A. M.,** The reaction of uveal pigment *in vitro* with polycyclic compounds, *Invest. Ophthalmol.*, 3, 405, 1964.
6. **Denckner, L., Lindquist, N. G., and Ullberg, S.,** Mechanism of drug-induced chronic otic lesions. Role of drug accumulation on the melanin of the inner ear, *Experientia*, 29, 1362, 1973.
7. **Wold, J. S.,** Cephalosporin nephrotoxicity, in *Toxicology of the Kidney*, Hook, J. B., Ed., Raven Press, New York, 1981.
8. **Thoenen, H. and Tranzer, J. P.,** Chemical sympathectomy by selective destruction of adrenergic nerve endings with 6-hydroxydopamine, *Naunyn-Schmiedebergs Arch. Pharmacol. Exp. Pathol.*, 261, 271, 1968.
9. **Hook, J. B., Ed.,** *Toxicology of the Kidney*, Raven Press, New York, 1981.
10. **Hook, J. B., McCormack, K. M., and Kluwe, W. M.,** Biochemical mechanisms of nephrotoxicity, in *Reviews in Biochemical Toxicology*, Vol. 1, Hodgson, E., Bend, J. R., and Philpot, R. M., Eds., Elsevier/North Holland, Amsterdam, 1979, 53.
11. **Comis, S. D. and Leng, G.,** Kanamycin ototoxicity and pigmentation in the guinea pig, *Hearing Res.*, 3, 249, 1980.
12. **Gillette, J. R., Mitchell, J. R., and Brodie, B. B.,** Biochemical mechanisms of drug toxicity, *Ann. Rev. Pharmacol.*, 14, 271, 1974.
13. **Jollow, D. J., Kocsis, J. J., Snyder, R., and Vainio, H., Eds.,** *Biological Reactive Intermediates, Formation, Toxicity and Inactivation*, Plenum Press, New York, 1977.
14. **Miller, E. C.,** Some current perspectives on chemical carcinogenesis in humans and experimental animals, *Cancer Res.*, 38, 1479, 1978.
15. **Boyd, M. R.,** Biochemical mechanisms in chemical-induced lung injury: roles of metabolic activation, *CRC Crit. Rev. Toxicol.*, 7, 103, 1980.
16. **Boyd, M. R., Grygiel, J. J., and Minchin, R. F.,** Metabolic activation as a basis for organ-selective toxicity, *Clin. Exp. Pharmacol. Physiol.*, 10, 87, 1983.
17. **Cohen, G. M.,** Metabolic basis of target organ toxicity, in *Biological Basis of Detoxication*, Caldwell, J. and Jakoby, W. B., Eds., Academic Press, New York, 1983, 325.
18. **Gillette, J. R.,** A perspective on the role of chemically reactive metabolites of foreign compounds in toxicity. I. Correlation of changes in covalent binding of reactive metabolites with changes in the incidence and severity of toxicity, *Biochem. Pharmacol.*, 23, 2785, 1974.
19. **Gillette, J. R.,** A perspective on the role of chemically reactive metabolites of foreign compounds in toxicity. II. Alterations in the kinetics of covalent binding, *Biochem. Pharmacol.*, 23, 2927, 1974.
20. **Gillette, J. R.,** The use of theoretical pharmacokinetic concepts in studies of the mechanisms of formation of chemically reactive metabolites *in vitro* and *in vivo*, *Drug Metab. Rev.*, 14, 9, 1983.
21. **Boyd, M. R.,** Evidence for the Clara cell as a site of cytochrome P-450-dependent mixed-function oxidase activity in lung, *Nature (London)*, 269, 713, 1977.
22. **Gram, T. E., Ed.,** *Extrahepatic Metabolism of Drugs and Other Foreign Compounds*, MTP Press, Lancaster, U.K., 1980.
23. **Caldwell, J.,** The significance of phase II (conjugation) reactions in drug disposition and toxicity, *Life Sci.*, 24, 571, 1979.
24. **Baron, J. and Kawabata, T. T.,** Intratissue distribution of activating and detoxicating enzymes, in *Biological Basis of Detoxication*, Caldwell, J. and Jakoby, W. B., Eds., Academic Press, New York, 1983, 105.
25. **Jerina, D. M., Lehr, R., Schaeffer-Ridder, M., Yagi, H., Karle, J. M., Thakker, D. R., Wood, A. W., Lu, A. Y. H., Ryan, D., West, S., Levin, W., and Conney, A. H.,** Bay region dihydrodiols: a concept explaining the mutagenic and carcinogenic activity of benzo(a)anthracene, in *Origins of Human Cancer*, Book B, Hiatt, H. H., Watson, J. D., and Winsten, J. A., Eds., Cold Spring Harbor Laboratory, Cold Spring Harbor, N.Y., 1977, 639.
26. **Gooding, P. E., Chayen, J., Sawyer, B., and Slater, T. F.,** Cytochrome P-450 distribution in rat liver and the effect of sodium phenobarbitone administration, *Chem. Biol. Interact.*, 20, 299, 1978.
27. **Smith, M. T., Loveridge, N., Wills, E. D., and Chayen, J.,** The distribution of glutathione in the rat liver lobule, *Biochem. J.*, 182, 103, 1979.

28. **Sorokin, S. P.,** The cells of the lung, in *Morphology of Experimental Respiratory Carcinogenesis,* Nettesheim, P., Hanna, M. G., and Deatherage, J. W., U.S. Atomic Energy Commission, Washington, D.C., 1970, 3.
29. **Serabjit-Singh, C. J., Wolf, C. R., Philpot, R. M., and Plopper, C. G.,** Cytochrome P-450: localization in rabbit lung, *Science,* 207, 1469, 1979.
30. **Minchin, R. F. and Boyd, M. R.,** Localization of metabolic activation and deactivation systems in the lung: significance to the pulmonary toxicity of xenobiotics, *Ann. Rev. Pharmacol. Toxicol.,* 23, 217, 1983.
31. **Zenser, T. V., Mattammal, M. B., and Davis, B. B.,** Differential distribution of the mixed function oxidase activities in rabbit kidney, *J. Pharmacol. Exp. Ther.,* 207, 719, 1978.
32. **Dees, J. H., Masters, B. S. S., Muller-Eberhard, U., and Johnson, E. F.,** Effects of 2,3,7,8-tetrachlorodibenzo-p-dioxin and phenobarbital on the occurrence and distribution of four cytochrome P-450 isozymes in rabbit kidney, lung and liver, *Cancer Res.,* 42, 1423, 1982.
33. **Ashurst, S. W. and Cohen, G. M.,** The formation of benzo(a)pyrene-deoxyribonucleoside adducts *in vivo* and *in vitro, Carcinogenesis,* 3, 267, 1982.
34. **Bigger, C. A. H., Tomaszewski, J. E., Dipple, A., and Lake, R. S.,** Limitations of metabolic activation systems used with *in vitro* tests for carcinogens, *Science,* 209, 503, 1980.
35. **Miller, E. C. and Miller, J. A.,** Mechanisms of chemical carcinogenesis: nature of proximate carcinogens and interactions with macromolecules, *Pharmacol. Rev.,* 18, 805, 1966.
36. **Marnett, L. J.,** Polycyclic aromatic hydrocarbon oxidation during prostaglandin biosynthesis, *Life Sci.,* 29, 531, 1981.
37. **McLachlan, J. A., Metzler, M., and Lamb, J. C.,** Possible role of peroxidase in the diethylstilboestrol-induced lesions of the Syrian hamster kidney, *Life Sci.,* 23, 2521, 1978.
38. **Pan, S.-S. and Bachur, N. R.,** Xanthine oxidase catalysed reductive cleavage of anthracycline antibiotics and free radical formation, *Mol. Pharmacol.,* 17, 95, 1980.
39. **Vogel, F. S., Kemper, A. K., Boekeheide, K., Graham, D. G., and Jeffs, P. W.,** Intracellular activation of γ-L-glutaminyl-4-hydroxybenzene by tyrosinase, a mechanism for selective cytotoxicity against melanocarcinoma, *Cancer Res.,* 39, 1490, 1979.
40. **Dix, T. A. and Marnett, L. J.,** Metabolism of polycyclic aromatic hydrocarbon derivatives to ultimate carcinogens during lipid peroxidation, *Science,* 221, 77, 1983.
41. **Aldridge, W. N.,** Mechanisms of toxicity: new concepts are required in toxicology, *Trends Pharmacol. Sci.,* 2, 228, 1981.
42. **Drayer, D. E.,** Pharmacologically active drug metabolites: therapeutic and toxic activities, plasma and urine data in man, accumulation in renal failure, *Clin. Pharmacokinet.,* 1, 426, 1976.
43. **Verbeeck, R. K., Branch, R. A., and Wilkinson, G. R.,** Drug metabolites in renal failure: pharmacokinetic and clinical implications, *Clin. Pharmacokinet.,* 6, 329, 1981.
44. **Pratt, W. B. and Ruddon, R. W., Eds.,** *The Anticancer Drugs,* Oxford University Press, New York, 1979.
45. **Sieber, S. M. and Adamson, R. H.,** The clastogenic, mutgenic, teratogenic and carcinogenic effects of various antineoplastic agents, in *Pharmacological Basis of Cancer Chemotherapy,* Williams & Wilkins, Baltimore, 1975, 401.
46. **Setlow, R. B., Ahmed, F. E., and Grist, E.,** Xeroderma pigmentosum: damage to DNA is involved in carcinogenesis, in *Origins of Human Cancer,* Book B, Hiatt, H. H., Watson, J. D., and Winsten, J. A., Eds., Cold Spring Harbor Laboratory, Cold Spring Harbor, N.Y., 1977, 889.
47. **Pegg, A. E.,** Formation and metabolism of alkylated nucleosides: possible role in carcinogenesis by nitroso compounds and alkylating agents, *Adv. Cancer Res.,* 25, 195, 1977.
48. **Johnson, M. K.,** The delayed neuropathy caused by some organophosphorous esters: mechanism and challenge, *CRC Crit. Rev. Toxicol.,* 3, 289, 1975.
49. **Johnson, M. K.,** Delayed neurotoxicity induced by organophosphorous compounds — areas of understanding and ignorance, in *Mechanisms of Toxicity and Hazard Evaluation,* Holmstedt, B., Lauwerys, R., Mercier, M., and Roberfroid, M., Eds., Elsevier/North-Holland, Amsterdam, 1980, 27.
50. **Nebert, D. W. and Jensen, N. M.,** The Ah locus: genetic regulation of the metabolism of carcinogens, drugs, and other environmental chemicals by cytochrome P-450-mediated monooxygenases, *CRC Crit. Rev. Biochem.,* 6, 401, 1979.
51. **Poland, A. and Glover, E.,** 2,3,7,8-Tetrachlorodibenzo-*p*-dioxin: segregation of toxicity with the Ah locus, *Mol. Pharmacol.,* 17, 86, 1980.
52. **Shichi, H., Tanaka, M., Jensen, N. M., and Nebert, D. W.,** Genetic differences in cataract and other ocular abnormalities induced by paracetamol and naphthalene, *Pharmacology,* 20, 229, 1980.
53. **Beutler, E.,** Drug induced hemolytic anemia, *Pharmacol. Rev.,* 21, 73, 1969.
54. **Doull, J., Klaassen, C. D., and Amdur, M. O.,** *Cassarett and Doull's Toxicology. The Basic Science of Poisons,* 2nd ed., Macmillan, New York, 1980.
55. **Gehring, P. J.,** The cataractogenic activity of chemical agents, *CRC Crit. Rev. Toxicol.,* 1, 93, 1971.

56. **Van Heyningen, R. and Pirie, A.,** The metabolism of naphthalene and its toxic effects on the eye, *Biochem. J.*, 102, 842, 1967.
57. **Nash, J. A., King, L. J., Lock, E. A., and Green, T.,** The metabolism and disposition of hexachloro-1:3-butadiene in the rat and its relevance to nephrotoxicity, *Toxicol. Appl. Pharmacol.*, 73, 124, 1984.
58. **Nelson, T. J. and Boor, P. J.,** Allylamine cardiotoxicity. IV. Metabolism to acrolein by cardiovascular tissue, *Biochem. Pharmacol.*, 31, 509, 1982.
59. **Doroshow, J. H., Locker, G. Y., and Myers, C. E.,** Enzymatic defenses of the mouse heart against reactive oxygen metabolites: alterations produced by doxorubicin, *J. Clin. Invest.*, 65, 128, 1980.
60. **Brown, V. K.,** *Acute Toxicity in Theory and Practice*, John Wiley & Sons, Chichester, U.K., 1980.
61. **Reuber, M. D.,** Hepatic lesions in young rats given calcium disodium edetate, *Toxicol. Appl. Pharmacol.*, 11, 321, 1967.
62. **Bus, J. S. and Gibson, J. E.,** Lipid peroxidation and its role in toxicology, *Rev. Biochem. Toxicol.*, 1, 125, 1979.
63. **Cock, J. H.,** Cassava: a basic energy source in the tropics, *Science*, 218, 755, 1982.
64. **McLean, A. E. M. and Magee, P.,** Increased renal carcinogenesis by dimethylnitrosamine in protein deficient rats, *Br. J. Exp. Pathol.*, 51, 587, 1970.
65. **Heywood, R.,** Target organ toxicity, *Toxicol. Lett.*, 8, 349, 1981.

Chapter 2

ROLE OF PHARMACOKINETICS IN RATIONALIZING TISSUE DISTRIBUTION

J. Brian Houston

TABLE OF CONTENTS

I. INTRODUCTION

It is well known that many pharmacological and toxicological responses are dependent on the concentration of xenobiotic within the target organ and/or the length of exposure time. A variety of models have been explored which attempt to relate degree of effect to concentration and time.[1,2] For a number of reasons, including practical, ethical, and economic constraints, tissue concentrations of xenobiotics are not readily measurable. Therefore, the focal point of the majority of models is the circulating blood or plasma concentrations of xenobiotics and this necessitates the use of pharmacokinetics.

A major objective of pharmacokinetic studies is to describe, in a quantitative fashion, the distribution of a xenobiotic to the various body tissues. With refinements in the physiological approach to pharmacokinetics which have become evident in the last decade,[3,4] the prediction of specific tissue concentrations based on blood concentrations of xenobiotics can be readily achieved. Physiological models, in contrast to the traditional compartmental models, are based on an understanding of the actual anatomical and physiological factors which dictate xenobiotic disposition.[5] Hence, blood flow, tissue size, rates of transport and elimination, etc., together with the physicochemical properties of the xenobiotic provide the basis for such models.

Physiological parameters vary in a predictable fashion in different mammalian species[6] and this provides the opportunity to correlate and scale pharmacokinetic characteristics across species.[7] This may circumvent the lack of experimental data available in the human situation which nevertheless may be delineated in other mammalian systems. This approach, which has been successfully applied for a number of drugs,[8] may be applicable to target organ toxicity providing the biochemical factors may be extrapolated. The potential of a physiological pharmacokinetic approach to elucidate certain features of target organ toxicity contrasts sharply with the classical compartmental approach to pharmacokinetics. Although the latter approach is quite adequate for many situations,[9] it has little value in the context of target organ toxicity. The compartmental model parameters have no physiological reality and hence whether these parameters are a valid indicator of the transport processes and dilution spaces involved in xenobiotic tissue distribution is highly questionable.

A. The Constant Exposure Model

One of the first applications of pharmacokinetics was in the rationalization of drug plasma concentration-time profiles during chronic drug therapy. Two common situations prevail[10] and are illustrated in Figure 1.

In the controlled constant input case (for example, constant zero order input via an intravenous infusion pump or implant) plasma concentrations rise to a steady-state value. By definition, at steady-state, rate of input equals rate of elimination and all tissues are in equilibrium in terms of xenobiotic concentrations. When discrete multiple dosing occurs the controlled multiple unit input case is applicable (for example, multiple oral or parenteral dosing). A build up in plasma concentrations to steady-state is again evident, but as input is not constant, fluctuations during each dosing intervals occurs. The extent of fluctuations will be dependent on the respective rates of absorption and elimination of drug. Since the dosage regimen is controlled, the fluctuations are regular. If the average plasma concentrations for each dosing interval are plotted then the curve is identical to the controlled constant input case. Thus, the latter case approximates the controlled multiple unit input case.

In environmental toxicology, neither of the previous cases may accurately reflect the method of exposure to xenobiotics. Multiple unit input may occur but at varying doses and at irregular time intervals. In Figure 1, this situation is illustrated as the random multiple unit input case. The fluctuations within the dosing interval are extremely variable and the concept of a steady-state plasma or tissue concentration somewhat tenuous. However, since

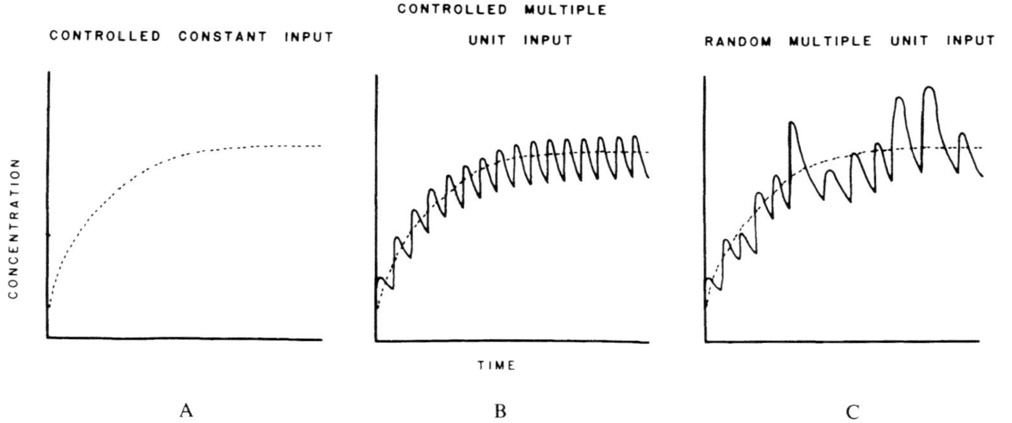

FIGURE 1. Xenobiotic concentration-time profiles following various modes of input. (A) Dotted line is controlled constant input; (B) solid line is controlled multiple unit input, dotted line is controlled constant input; (C) solid line is random multiple unit input, dotted line is controlled constant input.

in many situations there is no information on the extent and frequency of exposure, the approximation of this case by the controlled constant input case may be of value if only to provide a basis for more accurate modeling.

For the purposes of this chapter, the role of pharmacokinetics in the rationalization of target organ toxicity will be illustrated using the constant exposure model. The factors influencing the tissue and blood concentration-time profiles will be discussed within this context.

II. FACTORS GOVERNING XENOBIOTIC CONCENTRATIONS IN SPECIFIC TISSUES

The distribution of xenobiotics to particular tissues occurs at various rates and to different extents. The factors governing both the rate and extent of distribution of a xenobiotic to an idealized tissue are summarized in Figure 2. The idealized tissue represents a region of the body at some uniform concentration of xenobiotic. It may be a single capillary with surrounding tissue, a discrete organ like the kidney or liver, or a widely distributed anatomical region such as muscle or fat.

The idealized tissue is composed of three well-mixed phases of known volumes — the capillary (V_1), interstitial (V_2), and intracellular (V_3) fluids. Within each phase, the xenobiotic may bind to soluble proteins, cell membranes, or to subcellular organelles. Hence, both bound (Cb) and unbound (Cu) concentrations are shown. Since only relatively small molecules (mol wt <1000) cross membranes readily, it is assumed that only the unbound xenobiotic is available for transmembrane passage.

Mass balance equations may be readily formulated to describe the time course for the xenobiotic within each tissue fluid. The compound enters the capillary phase according to the arterial blood flow (Q) and the arterial concentration (C_A). It moves from the plasma to the interstitial fluid at some net flux (flux 1) and is transferred from the interstitial fluid to the intracellular compartment at another net rate (flux 2). Although transport between adjacent tissues is possible, it is considered to occur too slowly to be of importance. The xenobiotic leaves the capillary compartment at the venous concentration (C_V) which is equal to the average capillary concentration within that compartment. The differential mass balance equations for each phase are shown in the legend to Figure 2.

In addition to physiological and anatomical parameters, certain physicochemical properties of the xenobiotic dictate tissue uptake. Information on binding within the tissue and transport

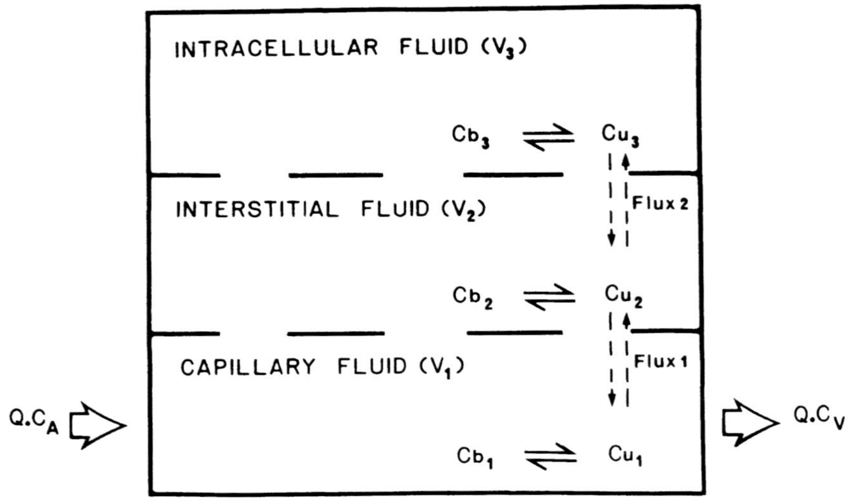

FIGURE 2. Schematic representation of the capillary, interstitial, and intracellular phases within an idealized tissue. Perfusion to and from the tissue is shown by the unidirectional arrows, equilibrium between bound (Cb), and unbound (Cu) concentrations by solid arrows and flux across the membrane by stippled arrows. The differential equations to describe the xenobiotic kinetics in each phase are — Rate of change within the capillary phase = $Q.C._A - Q.C_V -$ Flux 1; Rate of change within the interstitial phase = Flux 1 − Flux 2; Rate of change within the intracellular phase = Flux 2.

properties for both the capillary and cellular membranes are required to solve the equations in Figure 2. In particular, whether these processes are of a saturable (nonlinear) nature must be defined. In certain tissues, the presence of metabolizing enzyme systems or irreversible binding to macromolecules may be additional complexities to be quantified and incorporated into the system illustrated in Figure 2. Anatomical and physiological parameters (tissue volumes and tissue perfusion rates) can often be derived from previous literature studies (see Table 1).

The mathematical description of tissue uptake can often be simplified by appropriate assumptions regarding intercompartmental transport. In many cases the capillary membrane transport (flux 1) is so rapid that the plasma and interstitial fluid are essentially at equilibrium. Therefore, the fluid volumes V_1 and V_2 can be combined to give one extracellular volume and only two mass balance equations are required to describe the tissue uptake process. Further simplification may be possible dependent on the relative rates of tissue perfusion and cellular membrane transport. For very permeable xenobiotics, cellular membrane transfer is very rapid and hence extra- and intracellular fluids are in equilibrium and uptake is governed solely by the rate at which blood perfuses the tissue. For more polar molecules, the flux across the cellular membrane is slow with respect to perfusion rate and this latter simplification may not be applied.

In the case of the constant exposure model, when the concentration of xenobiotic in the blood is constant and hence rate of presentation ($Q.C._A$) is also constant, a steady-state between the tissue and arterial concentrations will eventually be achieved. At steady-state, the net uptake will be zero and the amount of xenobiotic in the tissue will be dependent on the tissue to blood partition coefficient (K_P) and the volume of the tissue (V_T) as shown in Equation 1

$$\text{Amount of xenobiotic in tissue at steady-state} = K_P \cdot V_T \cdot Css_A \tag{1}$$

Table 1
PHYSIOLOGICAL PARAMETERS FOR SPECIFIC TISSUES IN MOUSE, RAT, DOG, AND MAN

	Mouse[11]	Rat[12]	Dog[13]	Man[14]
Body weight (g)	**22**	**500**	**12,000**	**70,000**
		Tissue volume (mℓ)		
Plasma	1.0	19.6	500	3,000
Muscle	10.0	245	5,530	35,000
Kidney	0.34	3.6	60	280
Liver	1.3	20.0	480	1,350
Gut	1.5	11.2	480	2,100
Gut lumen	1.5	8.8	—	2,100
Heart	0.10	1.1	120	300
Lungs	0.12	2.1	120	—
Spleen	0.10	1.3	36	160
Fat	—	35.0	—	10,000
		Plasma flow rate (mℓ/min)[a]		
Plasma	4.4	85	512	3,670
Muscle	0.5	22.4	138	420 (0.012)
Kidney	0.8	12.8	90	700 (2.5)
Liver	1.1	4.7	60	800 (0.6)
Gut	0.9	14.6	81.5	700 (0.33)
Heart	0.3	1.6	60	150 (0.5)
Lungs	4.4	2.2	512	—
Spleen	0.05	0.95	13.5	240 (0.67)
Fat	—	3.6	—	200 (0.02)

[a] Values in parentheses indicate plasma flow per unit volume of tissue, mℓ/min/mℓ of tissue.

Css_A is the arterial concentration of xenobiotic at steady-state and V_T is the sum of the volumes of the three tissue phases. K_P is defined under steady-state conditions when arterial and venous (Css_V) concentrations are equal

$$K_P = \frac{Css_T}{Css_A} = \frac{Css_T}{Css_V} \qquad (2)$$

The remainder of this section is concerned with further discussion of tissue characteristics and xenobiotic properties which determine the rate and extent of tissue distribution.

A. Blood Flow Limitation

When neither the capillary nor the cellular membrane present a barrier to xenobiotic transport, the rate of uptake is directly proportional to the rate of presentation, i.e., tissue blood flow. Such a situation commonly prevails for many compounds for the slowly perfused tissues but only for lipophilic compounds in the rapidly perfused tissues.

Table 1 lists the blood flows to various tissues of several animal species employed in toxicology. In addition the same information is supplied for man together with the flow rates expressed per unit volume of tissue. The extremes in this table are the kidney and muscle which display a variation of 200-fold. All other factors remaining equal, a well-perfused tissue will take up a xenobiotic more rapidly than poorly perfused tissues.

The net rate of tissue uptake will be the difference between the rates of presentation and exit, hence

$$\text{Net rate of uptake} = Q(C_A - C_V) \tag{3}$$

The time course for the approach of tissue concentration to a steady-state value will be an exponential process since the rate of uptake will decrease as the tissue concentrations rises. The exponent governing this process is the rate constant for tissue uptake (k_T)[11] which is defined as

$$k_T = \frac{Q}{V_T \cdot K_P} \tag{4}$$

From k_T the time to reach 50% of steady state can be obtained

$$T_{50} = \frac{0.693 \cdot K_P \cdot V_T}{Q} \tag{5}$$

which serves to illustrate the interrelationship between these three parameters.

The lower the blood flow, the longer it takes to reach steady-state. This steady-state is achieved in different tissues at different times. The product of K_P and V_T may be regarded as the effective tissue space available to the xenobiotic. The greater this number the longer is T_{50}. This latter point enforces the limitation of perfusion rate. The tissue affinity is so great that the time taken to occupy this effective tissue space is appreciable.

The importance of k_T in defining the time course for the approach to steady-state for a given tissue may be appreciated by considering Equation 6[15]

$$C_T = K_P \cdot C_A(1 - e^{-k_T \cdot t}) \tag{6}$$

where C_T is the tissue concentration at any time t.

B. Diffusion Rate Limitation

Polar compounds do not traverse membranes readily and hence their tissue uptake is frequently diffusion rate limited. This situation is very common for well-perfused tissues where the rate of presentation is efficient. The diffusion barrier may occur at the capillary membranes (for example, in CNS uptake) or at the cellular membrane (for example, uptake into the kidney and muscle).

Figure 3 illustrates the time course of entry of various drugs from the plasma into the cerebrospinal fluid during a constant intravenous infusion of drug.[16] Antipyrine reaches steady-state conditions within a short period of time, exemplifying a blood flow limitation. However, the other drugs studied display a diffusion rate limitation. The differences between the rates of entry being a function of the lipophilicity and the extent of ionization of the compound. This behavior is in accord with the classic pH-partition theory.

The flux across a membrane may be described by numerous mathematical forms depending on the nature of the transport process. One of the simplest descriptions involves Fickian diffusion, a first order process in which the net flux is proportional to the concentration difference across the membrane (ΔC)

$$\text{Flux} = k_P \cdot A \cdot C \tag{7}$$

k_P is the permeability constant and A the area of membrane, hence the product (k_P A) has units of mℓ/min.

When active transport processes are available for the xenobiotic, the net flux may be described in terms of a saturable transport system.

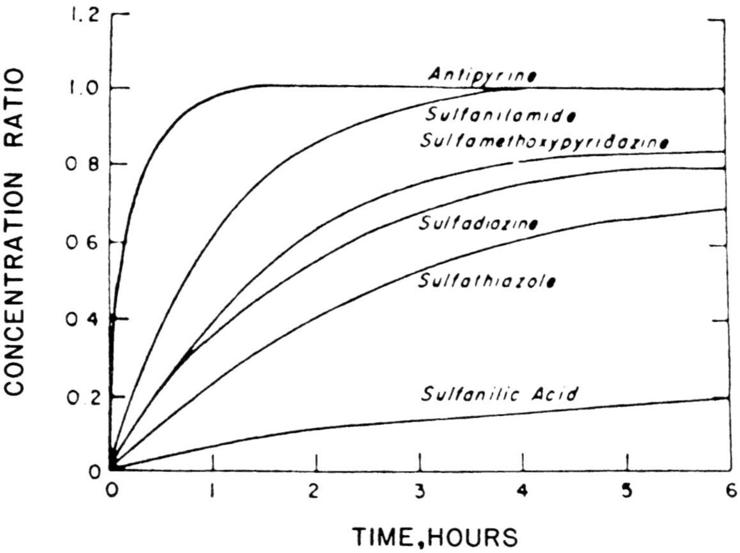

FIGURE 3. Time course for entry of various drugs into the cerebrospinal fluid in dog while plasma concentrations are maintained constant. The concentration ratio on the ordinate relates to cerebrospinal fluid concentration divided by plasma unbound concentration. (From Rall, D. P., Stabenau, J. R., and Zubrod, C. G., *J. Pharmacol,* 125, 185, 1959. American Society for Pharmacology and Experimental Therapeutics. Williams & Wilkens. With permission.)

$$\text{Flux} = \frac{T_{max_2}\, C_2}{K_2 + C_2} - \frac{T_{max_3}\, C_3}{K_3 + C_3} \tag{8}$$

The first term on the right side of Equation 8 is the transmembrane influx and the second term the transmembrane efflux. T_{max} represents the maximum transport rate and K represents the Michaelis constant for transport (i.e., the concentration at which the transport rate is half its maximum). The subscripts identify the different tissue fluids (for example, interstitial and intracellular fluid).

Figure 4 illustrates the consequences of active transport processes on the intracellular concentration time profiles for the constant exposure model at three different arterial concentrations. Panel A relates to the situation where the transport is concentration independent and Panel B depicts the saturable transport case. In the latter case the Michaelis constant for transport is 1 μg/mℓ. At the lowest input concentration, 0.1 μg/mℓ, which is well below the K, both panels show essentially the same time profile. At the medium input concentration, 1 μg/mℓ, which equals K, the two cases start to deviate and the approach to steady-state is slower for the saturable transport case. At the highest input concentrations (10 μg/mℓ), the difference between the two transport systems is very pronounced. Although the rate of entry is much slower for the saturable system, the intracellular concentration will eventually equal that of blood.

C. Binding within Blood

There are many macromolecules within the blood which are capable of reversibly binding xenobiotics. Also many compounds may interact with the red blood cells. In tissue distribution it is often appropriate to consider the unbound concentration of xenobiotic in the blood rather than the total concentration.

The rates of association and dissociation between compounds of low molecular weight and macromolecules are usually rapid (milliseconds). Therefore there exists an equilibrium

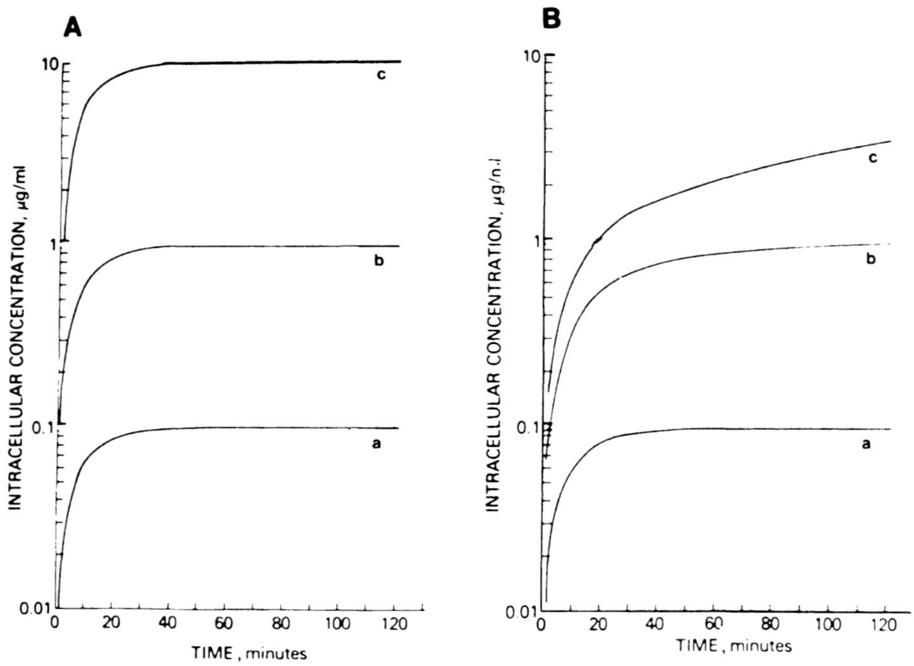

FIGURE 4. Linear and nonlinear tissue uptake: simulations of intracellular concentrations of xenobiotic while blood concentrations are maintained constant at 0.1 (a), 1 (b), and 10 (c) μg/mℓ. Panel A illustrates the time courses when the uptake is first order (according to Equation 7 where $k_P.A = 0.1$). Panel B illustrates the corresponding time courses when saturable transport occurs (according to Equation 8 where $K_1 = K_2 = 1$ μg/mℓ and $T_{max1} = T_{max2} = 0.1$ μg/min). (From Lutz, R. J., Dedrick, R. L., Zaharko, D. S., *Pharmacol. Ther.*, 11, 559, 1980. Pergamon Press. With permission.)

between bound and unbound xenobiotic at all times under virtually all circumstances. Only the unbound xenobiotic is regarded as capable of diffusion due to the large molecular size of the xenobiotic-macromolecular complex. Accordingly in Figure 3, which illustrates a diffusion limitation for plasma to cerebrospinal fluid transfer, the abscissa is the ratio of the cerebrospinal fluid concentration to the unbound plasma concentration.

Although binding to blood may decrease the extent to which a xenobiotic distributes into a given tissue, it is unlikely to influence the rate of distribution. In the diffusion rate limiting case, where the concentration gradient across the membrane is of major importance, the large tissue volume and binding within the tissue itself ensures that a gradient exists despite extensive blood binding. In the perfusion rate limiting case, binding within the blood may augument the rate of uptake as binding provides a mechanism by which a greater amount of xenobiotic may be delivered to the tissue during each transit time.

The unbound concentrations of a xenobiotic may be readily calculated from the total concentration when the fraction of unbound xenobiotic (fu) is known

$$Cu = fu \cdot C \tag{9}$$

The parameter fu may be determined in vitro and extensive investigations into the factors influencing binding have been carried out with plasma proteins, in particular albumin.[18,19] The determinants of fu may be summarized as follows

$$fu = \frac{1 + Ka \cdot Cu}{1 + Ka \cdot Cu + n \cdot P \cdot Ka} \tag{10}$$

The affinity constant for the interaction is Ka, P is the protein concentration, and n the number of binding sites per protein molecule.

Equation 10 summarizes the curvilinear relationship between fu and Cu. However, at low concentrations of a xenobiotic which displays avid binding (low Ka), this equation reduces to

$$fu = \frac{1}{1 + Ka \cdot n \cdot P} \tag{11}$$

and fu is a constant. The nonlinear behavior in binding is only apparent when the concentration of bound drug approaches the total concentration of binding sites (n.P). When there is a knowledge of the particular protein involved and its molar concentration this may be predicted. Many acidic compounds bind primarily to albumin which has a plasma concentration of 600 μM. An acidic xenobiotic with a mol wt of 250 would be predicted to show nonlinear binding above concentrations of 50 mg/ℓ. On the other hand, for basic compounds which bind primarily to α_1-acid glycoprotein (15 μM plasma concentration) the critical concentration would be in the region 1 mg/ℓ (if the mol wt of the base was 250), above which nonlinearity in binding would be expected.

D. Tissue to Blood Partition Coefficient

A K_P value of unity would occur for a xenobiotic which shows no binding either within the tissue or within the blood. However, in general, interactions occur and K_P represents a relative measure of the extent of tissue and blood binding. This parameter may be defined as

$$K_P = \frac{fu_B}{fu_T} \tag{12}$$

where fu_B is the unbound blood fraction and fu_T is the fraction of compound within the tissue which is not bound. In practice, it is very difficult to resolve the binding interactions within each of the tissue fluids unless the binding is very specific and the particular macromolecular concentrations are known. However, the use of Equations 2 and 12 allows the fraction unbound for the overall tissue (fu_T) to be calculated from tissue and blood concentrations when fu_B is known.

When K_P exceeds unity then tissue binding is more substantial than blood binding. In contrast, when K_P is less than unity, binding to blood is very strong, thus restricting tissue uptake. Other mechanisms by which large K_P values may result include active uptake processes and dissolution in fat.

Determination of K_P may be carried out following administration of xenobiotics in vivo prior to tissue removal and assay or by directly adding xenobiotic to a tissue homogenate. The advantage of the latter is that a concentration range may be investigated simultaneously and hence any nonlinearity detected. However, the loss of cellular integrity following homogenization may produce artefactual results.[20]

Figure 5 illustrates the effect of K_P on the time curve of tissue concentrations in the constant exposure model. Panel A illustrates the proportional increase in Css_T as the K_P is increased (in accord with Equation 1). In contrast, panel B relates to the nonlinear situation, where K_P is not concentration independent, and saturation of tissue binding is evident. K_P may be defined in such circumstances as

$$K_P = \frac{A_{max}}{K + Cu_B} \tag{13}$$

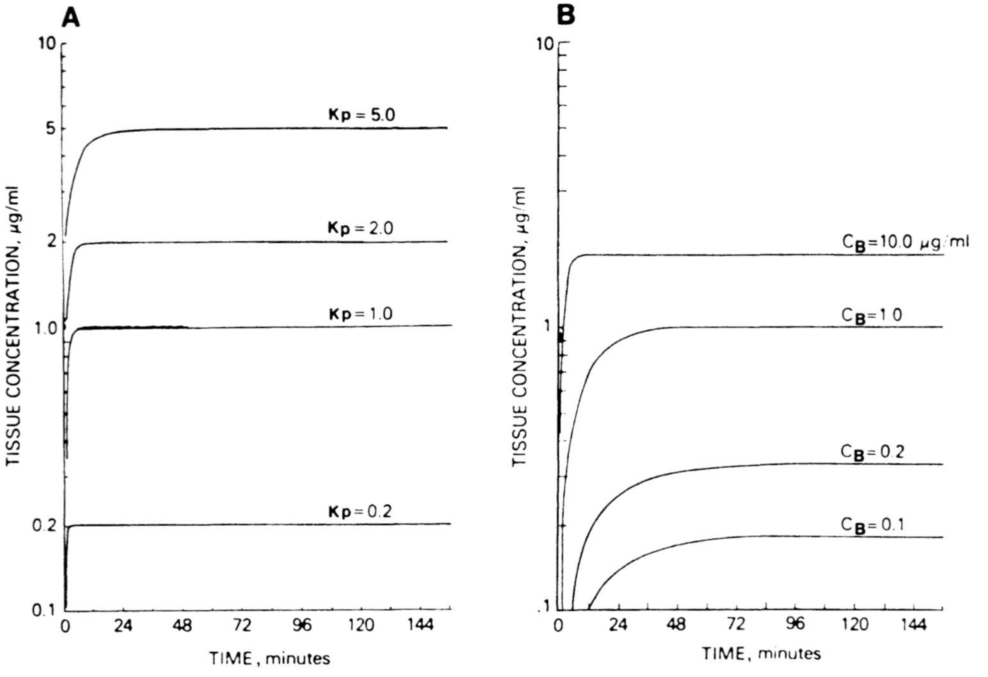

FIGURE 5. Linear and nonlinear tissue partitioning: simulations of tissue concentrations of xenobiotic while blood concentrations are maintained constant and uptake is blood flow limited. Panel A illustrates the time courses for various K_P values when partitioning is linear and blood flow and blood concentrations are 0.5 mℓ/min/mℓ of tissue and 1 μg/mℓ, respectively. Panel B illustrates the consequences of nonlinear partitioning where A_{max} and K are 2 μg/mℓ and 1 μg/mℓ, respectively, blood concentrations are maintained as shown (K_P being defined as in Equation 13), and blood flow is 0.12 mℓ/min/mℓ of tissue. (From Lutz, R. J., Dedrick, R. L., Zaharko, D. S., *Pharmacol. Ther.*, 11, 559, 1980. Pergamon Press. With permission.)

where A_{max} is the maximum binding capacity, K is the dissociation constant, and Cu_B is the unbound concentration in blood. At the lower input concentrations of 0.1 and 0.2 μg/mℓ (well below K), K_P is approximately 2 and Css_T increases proportionally with blood concentration. Higher input concentrations either equal to or exceeding K result in small K_P values and disproportionately small increases in Css_T.

The changes in K_P in Figure 5 also influence the time taken to reach steady-state within the tissue. In the linear case shown in panel A this time increases as K_P increases. In panel B when K_P shows concentration dependence, the rise in arterial concentration, and hence the decrease in K_P, shortens the time necessary to reach the steady-state condition. Both of these trends are in accord with Equation 5.

So far in this section binding within the tissue has been reversible and such interactions have increased the capacity of the tissue for xenobiotic. Certain interactions are not reversible, such as those resulting in covalent binding or metabolism. Under such circumstances the loss of xenobiotic either by turnover of macromolecular adduct or production of metabolite must be considered. The meaning of K_P becomes ambiguous unless a correction for the extent of irreversible loss is made.[21]

$$K_P = K_P' (1 - E) \tag{14}$$

where K'_P is the apparent K_P and E is the fraction of xenobiotic concentration irreversibly lost or extracted by the tissue. The parameter, extraction ratio, will be discussed in some detail later in the text.

E. Eliminating Tissues

An additional complication in the assessment of xenobiotic uptake into a particular tissue exists when that tissue can eliminate xenobiotic from the body. For example, elimination of xenobiotic can occur via the lungs or the kidney by various mechanisms and metabolism of the xenobiotic can occur to a varying extent in different tissues according to their enzymic complement (Chapters 1 and 3). The following discussion will focus on hepatic metabolism although the general principles are equally applicable to other eliminating tissues.

For an eliminating tissue, steady-state effluent venous concentration is less than the arterial steady-state concentration. The efficiency of xenobiotic removal by an eliminating tissue may be expressed in terms of extraction ratio across the tissue

$$E = \frac{Css_A - Css_V}{Css_A} \tag{15}$$

or by the parameter clearance (CL) which relates the rate of xenobiotic elimination to the arterial presentation rate

$$CL = \frac{Q \cdot Css_A - Q \cdot Css_V}{Q \cdot Css_A} \tag{16}$$

Substitution of Equation 15 in Equation 16 yields

$$CL = Q \cdot E \tag{17}$$

and the equation to describe the rate of tissue metabolism may be written

$$\text{Rate of metabolism by tissue} = CL \cdot C_V \tag{18}$$

The mass balance equation for describing rate of uptake for eliminating tissues is therefore

$$\text{Rate of uptake} = Q(C_A - C_V) - CL \cdot C_V \tag{19}$$

Classifying hepatic metabolism of drugs in terms of their extraction ratio has proved useful.[22] From Equation 17 two extreme cases are evident. For compounds with a high extraction ratio (approaching unity), hepatic clearance is approximated by hepatic blood flow. Under this perfusion rate limitation, all other possible factors contributing to hepatic clearance are secondary to hepatic blood flow. In contrast, the clearance of drugs of low extraction are not limited by perfusion but by some other factor(s), such as enzyme activity, cofactor availability, or degree of binding within the blood.

Hepatic clearance is an apparent term to express the rate of metabolism based on circulating concentrations of xenobiotic and is a complementary parameter to intrinsic clearance. The latter term reflects the full potential of the enzyme system to metabolize a compound in the absence of all other limitations. Thus, intrinsic clearance has no upper limit of hepatic blood flow, unlike hepatic clearance. Intrinsic clearance may also be expressed in terms of unbound drug within the liver. Hence, this parameter is independent of both the extent of binding within the blood and the rate at which blood perfuses the liver; it is a pure measure of hepatocellular activity. In terms of enzyme kinetics, intrinsic clearance (CLu_{INT}) may be viewed as[23]

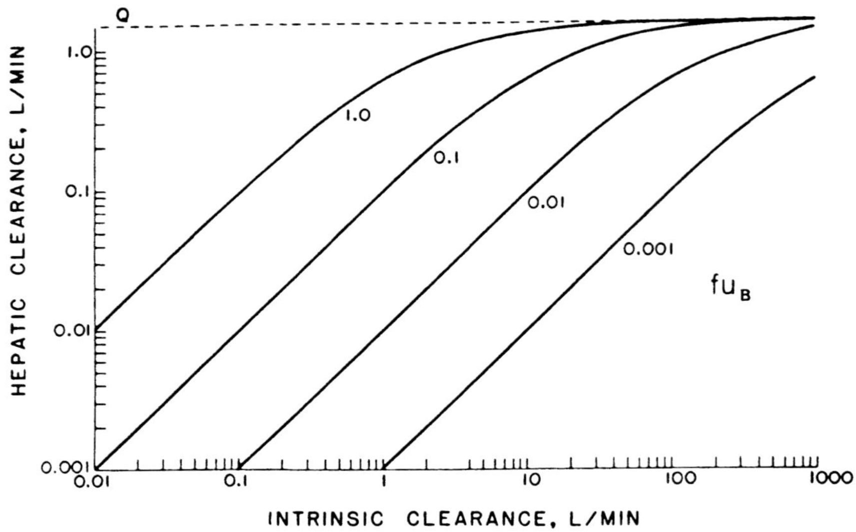

FIGURE 6. Relationship between hepatic clearance and intrinsic clearance of xenobiotics when blood binding (fu_B) occurs to differing extents. (From Tozer, T. N., *Pharmacol. Ther.*, 12, 109, 1981. Pergamon Press. With permission.)

$$CLu_{INT} = \frac{V_{max}}{K_m} \tag{20}$$

where V_{max} is the maximal rate of metabolism, K_m the Michaelis constant, and the unbound liver concentration is much lower than the K_m. Thus, Equation 18 may be written in terms of unbound concentration and intrinsic clearance

$$\text{Rate of metabolism by tissue} = CLu_{INT} \cdot Cu_V \tag{21}$$

The interrelationships between hepatic clearance, hepatic blood flow, blood binding, and intrinsic clearance have been explored in some detail. Figure 6 summarizes some of the important elements.[24]

The curvilinear relationship between hepatic clearance and CLu_{INT} is readily apparent. Linearity is observed below a hepatic clearance of 0.3 ℓ/min (which corresponds to a hepatic extraction ratio below 0.2). The assymptotic value in this figure is governed by hepatic blood flow (1.5 ℓ/min). In the situation where no binding occurs within the blood, hepatic clearance will be essentially flow limited when CLu_{INT} exceeds hepatic clearance by a factor of five. Any further increase in the enzyme activity will not be manifest in the hepatic clearance term.

Figure 6 also illustrates the consequences of the binding of xenobiotic to blood constituents. As binding is increased (fu decreased), the curve is displaced to the right. Consequently when binding becomes more avid a corresponding higher CLu_{INT} is required to maintain the same hepatic clearance. Within the linear region of each curve (at extraction ratios below 0.2) the following relationship describes hepatic clearance (CL_H)

$$CL_H = fu \cdot CLu_{INT} \tag{22}$$

Although blood binding restricts hepatic extraction when clearance is low this is not the case when clearance is high. Enzyme activity may be so efficient that the drug is stripped

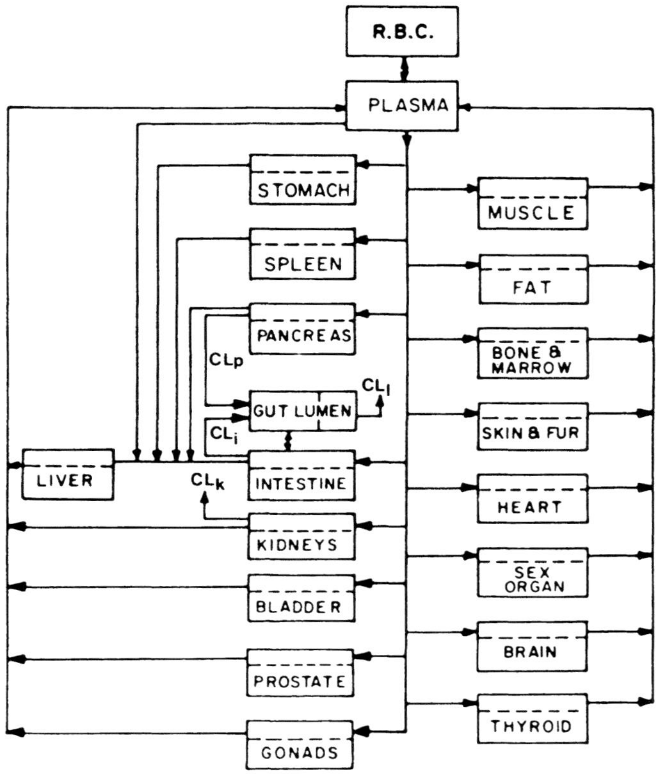

FIGURE 7. Example of a comprehensive physiologically based pharmacokinetic model illustrating the flow network operating between various tissues. (From Jain, R. K., et al., *Ann. Biomed. Eng.*, 9, 347, 1982. Pergamon Press. With permission.)

from its binding sites during transit through the liver. In this respect, binding within the blood provides a delivery system for the compound to the liver and may augment the clearance process. A flow limitation will apply to a highly bound drug when CLu_{INT} exceeds the hepatic clearance by a factor equal to five divided by fu.

III. FACTORS GOVERNING XENOBIOTIC CONCENTRATIONS IN CIRCULATING FLUIDS

In the preceeding section, the various factors influencing xenobiotic uptake into a specific tissue were discussed. By linking the individual tissues in a flow network with a series or parallel arrangement similar to that in the body, the kinetics of distribution of the xenobiotic throughout the entire system may be formulated. Figure 7 shows an example of a physiologically based pharmacokinetic model. This model is comprehensive but in practice models should provide an as simple as possible description of the data. A tissue should only be included in the model if it is a substantial depot for the distribution, metabolism, or excretion of the xenobiotic, or if it is a compartment of special interest such as a tumor. Each tissue contributes its own mass balance equation (of the type shown as Equations 3 and 19) and the resulting set of simultaneous differential mass balance equations solved numerically to generate time curves for all the tissues of interest. Reviews are available which illustrate the fairly extensive application of such approaches.[4,25]

Due to tissue uptake, effluent xenobiotic concentrations differ from arterial concentrations until steady state is achieved. The magnitude of this difference will vary widely between tissues, in accord with their perfusion rates and K_P, and will be greatest in the earlier moments of exposure. Consequently systemic events are best defined with respect to arterial blood concentrations.

Under the conditions of the constant exposure model, there is a rise in the concentration of xenobiotic in the circulating fluids until steady-state is achieved, as exemplified in Figure 1. The actual concentrations achieved depend not only on the exposure rate to the xenobiotic, but also the volume of distribution and systemic clearance of the xenobiotic. These two primary pharmacokinetic parameters describe the extent of distribution and rate of elimination from the circulating fluids. They may be determined with respect to total plasma or unbound concentrations. Conversion between the various reference fluids used to express volume of distribution and clearance may be readily achieved using the blood to plasma concentration ratio and the fu. For example,

$$CL_B = \frac{CL_P}{C_B/C_P} = fu \cdot CLu \qquad (23)$$

where the subscripts B and P relate to blood and plasma entities.

In the remainder of this section, the three commonly used pharmacokinetic parameters, volume of distribution, systemic clearance, and half-life, are discussed within the context of the constant exposure model. In addition, consideration is given to the influence of route of entry of xenobiotic on the systemic availability and hence toxicological response.

A. Volume of Distribution

The concentration of xenobiotic in the circulating fluids is dependent upon the amount of xenobiotic in the body and the extent of tissue distribution. The volume of distribution may therefore be defined as the apparent dilution space occupied by the xenobiotic at a concentration equal to that in the circulating fluids. There have been several methods proposed to calculate volume of distribution, but the only unequivocal method provides an estimate of the volume of distribution at steady-state (Vss). Vss relates to the dilution space occupied when rates of transport in and out of the tissues are equal and hence complications from distribution dynamics and concomitant elimination are minimized.

Vss may be conceived of as the sum of all the effective tissue spaces into which the xenobiotic distributes, and can thus be expressed in terms of the K_P and V_T of each tissue

$$Vss = V_B + \sum_{i=1}^{n} V_{T_i} \cdot K_{P_i} \qquad (24)$$

on the assumption that unbound concentration in tissue and blood are equal, substitutions of Equation 12 yields[26]

$$Vss = V_B + \sum_{i=1}^{n} V_{T_i} \cdot \frac{fu_B}{fu_T} \qquad (25)$$

The first term is much smaller than the second term, therefore Equation 25 may be approximated by

$$Vss = 42 \cdot \frac{fu_B}{fu_T} \qquad (26)$$

where 42 ℓ is total tissue water (in man) and fu represents an average fraction unbound for all tissues. This approach may be extended further when information on which macromolecules are involved in binding and their particular tissue distribution is known.[27]

As a result of tissue binding, Vss is usually a hypothetical volume with no correspondence to a physiological space. The minimum volume of distribution for a compound bound to albumin is 7 ℓ (in man) due to the extravascular distribution of this protein.[24] However, there is no maximum value for volume of distribution; volumes of tens of thousands of liters have been calculated for certain drugs (see Figure 9).

B. Systemic Clearance

Systemic clearance (CL_S) is a measure of the overall rate of elimination of xenobiotic from the body. Together with the exposure rate to xenobiotic (R) it governs the steady-state concentration of xenobiotic

$$Css = \frac{R}{CL_S} \tag{27}$$

Clearance terms are additive and thus total systemic clearance can be regarded as the net effect of the component tissue clearances. When the tissues are arranged in parallel within the body (for example, the liver and kidneys), the total systemic clearance is merely the sum of the individual organ clearances.

$$CL_S = \sum_{i=1}^{n} CL_i \tag{28}$$

When the anatomical arrangement for two eliminating tissues is in series, the total systemic clearance is calculated by

$$CL_S = Q[1 - \sum_{i=1}^{n} (1 - E_i)] \tag{29}$$

where Q is the common blood flow to the tissues under consideration. Although the intestine and liver are arranged in series, there is the added complication of two sources of hepatic blood supply — hepatic artery and hepatic portal vein. Only the latter perfuses the intestine in addition to the liver. Hence, in the calculations of total clearance by these two organs account of the fractional source (f) of hepatic blood supply is necessary

$$CL = Q_H[1 - f_{HPV}(1 - E_I)(1 - E_H) - f_{HA} \cdot E_H] \tag{30}$$

where the subscripts H, I, HPV, and HA relate to hepatic, intestinal, hepatic portal vein, and hepatic artery.

In order to assess the consequences of alterations in tissue blood flows, metabolizing capacity, or blood binding, it is necessary to consider the clearance of each eliminating tissue in isolation. Therefore, the term systemic clearance must be partitioned into individual components of clearance prior to any interpretation of this nature.

C. Half-Life

It is well known that the half-life of a xenobiotic defines the loss of the compound from the body following a bolus dose. The half-life also governs the rate of increase of xenobiotic in the body during a constant input.[10] In Figure 8, the rise in xenobiotic concentrations following a zero order infusion together with the wash out of xenobiotic, when the infusion

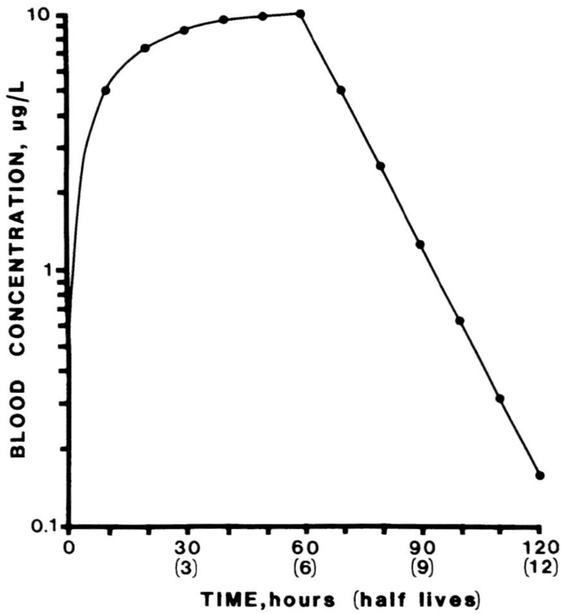

FIGURE 8. Blood concentration-time profile for a xenobiotic during and after a zero order intravenous infusion. The total systemic clearance and infusion rates are 10 ℓ/hr and 100 µg/hr (hence Css = 10 µg/ℓ). The volume of distribution is 140 ℓ and hence the half-life is 10 hr (k = 0.07/hr). Time scale is shown in units of half-life in addition to units of hours. The concentration-time profile is defined by $C = Css\ (1 - e^{-k.t})$ during the infusion and by $C = Css.e^{-k.t}$ when the infusion is stopped.

is ceased, is shown. The time scale is expressed in units of half-life to illustrate the respective increases and decreases of 50% for each unit. The equations given in the legend to Figure 8 have the exponent, k, the total elimination rate constant. The relationship between half-life and k being

$$t^1/_2 = \frac{\ln 2}{k} \tag{31}$$

It is therefore the half-life which governs the time to steady-state in Figures 1 and 8. Fifty percent of steady-state is reached in one half-life and 90% in 3.3 half-lives. It is also the half-life which controls the fluctuations in concentration when there are multiple inputs of xenobiotic in the body.

The dependence of half-life upon both systemic clearance and volume of distribution is a key relationship in pharmacokinetics.[24]

$$t^1/_2 = \frac{\ln 2 \cdot Vss}{CL_S} \tag{32}$$

The hybrid nature of half-life is well illustrated in Figure 9 which summarizes clearance and volume terms for a variety of drugs. The diagonals on this figure show the combination of clearance and volume which result in the same half-life. It can be seen that drugs with a high clearance and large volume of distribution (for example, desmethylimipramine) exhibit similar half-lives to drugs with low clearance and small volumes of distribution (for example, warfarin). Although there are no drugs shown with low clearance and large volumes of

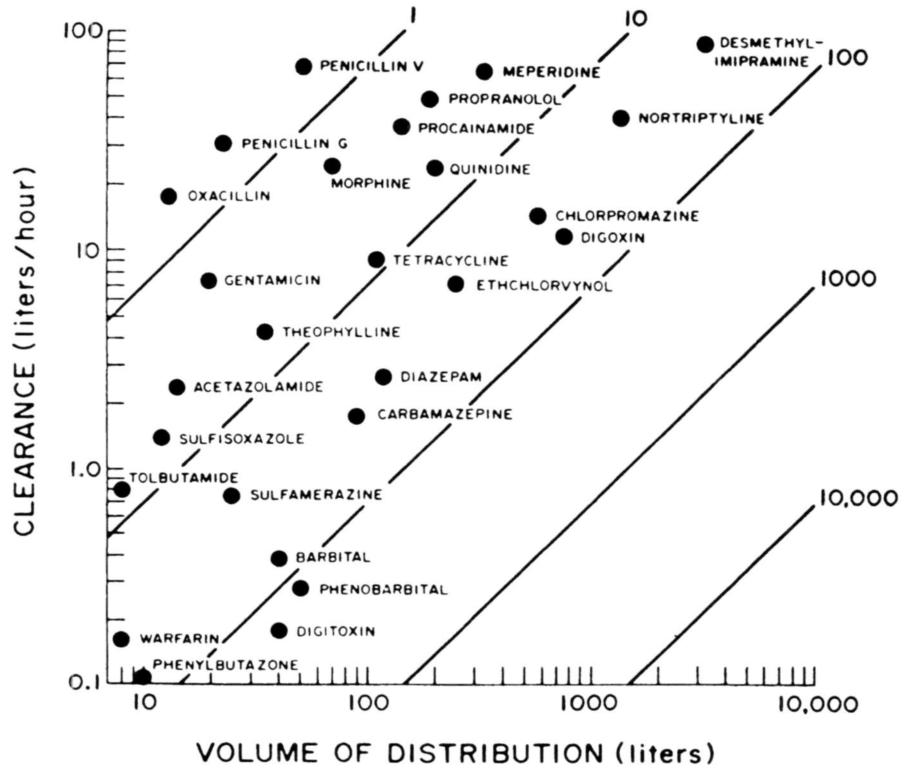

FIGURE 9. Relationship between systemic clearance and volume of distribution for a variety of different drugs. The diagonal lines show the combination of clearance and volume giving the same half-life (in units of hours). (From Tozer, T. N., *Pharmacol. Ther.*, 12, 109, 1981. Pergamon Press. With permission.)

distribution, since such compounds would display half-lives in weeks or more, xenobiotics such as polychlorinated biphenyls possess such characteristics.

D. Systemic Availability

An additional parameter which influences the circulating concentrations of xenobiotic is the systemic availability. The rate and extent of absorption from the site of contact has obvious effects. In addition, the presence of metabolizing enzyme systems within the tissue, which is traversed during the absorption process, may also influence circulating concentrations of xenobiotic. The systemic availability is the fraction of dose, to which the body is exposed, that reaches the circulating fluids. When a xenobiotic is both incompletely absorbed and metabolized during absorption then the systemic availability (F) may be calculated by

$$F = f_A \cdot f_B \tag{33}$$

where f_A is the fraction leaving the site of absorption and f_B is the fraction traversing the absorption barrier intact. F therefore relates exposure rate to input rate in the constant exposure model.

The common portals of entry for xenobiotics are the lungs, skin, and GI tract. All three tissues have been demonstrated to contain various enzymes capable of metabolizing xenobiotics, in particular esterases, glucuronyl transferase, and sulfotransferase, are present in substantial amounts. In addition, xenobiotics which are ingested and absorbed via the hepatic portal system pass through the liver prior to reaching the general circulation (Figure 7). Hence there are two potential sources of enzymatic loss in availability for this route of exposure. The systemic availability following oral ingestion being

$$F = f_A \cdot f_H \cdot f_I \tag{34}$$

where f_H and f_I denote the fractions of absorbed xenobiotic which escape metabolism in the single pass across the liver and intestinal mucosa, respectively.

Mass balance relationships allow calculation of individual tissue f values from their respective extraction ratios

$$f = 1 - E \tag{35}$$

and hence F may be predicted.

IV. CONCLUSION

The potential of pharmacokinetics in rationalizing the tissue distribution of xenobiotics has been discussed. Using the constant exposure model, xenobiotic concentrations within specific tissues and circulating fluids can be related to various physiological parameters and to the physiochemical properties of the xenobiotic. The value of the physiological approach to modeling the disposition of many therapeutic agents is now in evidence.[3,4,25] Much of this experience will be of benefit to toxicology; particularly pertinent is the modeling of uptake of anticancer agents into tumor tissue.[4,17] In the future it is likely that the principles of physiological pharmacokinetics will receive more general application in the assessment of target organ toxicity.

REFERENCES

1. **Holford, N. H. G. and Sheiner, L. B.,** Kinetics of pharmacologic response, *Pharmacol. Ther.*, 16, 143, 1982.
2. **Gehring, P. G., Watanabe, P. G., and Blau, G. E.,** Pharmacokinetic studies in evaluation of the toxicological and environmental hazard of chemicals, in *New Concepts in Safety Evaluation*, Vol. 1, Part 1, Melhman, M. A., Shapiro, R. E., and Blumenthal, H., Eds., Hemisphere Publications, Washington, D.C., 1976, 195.
3. **Rowland, M.,** Physiologic pharmacokinetic models: relevance, experience and future trends, *Drug Metab. Rev.*, 15, 55, 1984.
4. **Gerlowski, L. E. and Jain, R. K.,** Physiologically based pharmacokinetic modelling: principles and applications, *J. Pharm. Sci.*, 72, 1103, 1983.
5. **Tucker, G. T.,** Empirical vs. compartmental vs. physiological models in *Topics in Pharmaceutical Sciences*, Breimer, D. D. and Speiser, P., Eds., Elsevier/North-Holland, Amsterdam, 1981, 33.
6. **Adolph, E. F.,** Quantitative relations in the physiological constituents of mammals, *Science*, 109, 579, 1949.
7. **Boxenbaum, H.,** Interspecies scaling, allometry, physiological time and the ground plan of pharmacokinetics, *J. Pharmacokinet. Biopharm.*, 10, 201, 1982.
8. **Dedrick, R. L.,** Animal scale-up, *J. Pharmacokinet. Biopharm.*, 1, 435, 1973.
9. **Segre, G.,** Pharmacokinetics — compartmental representation, *Pharmacol. Ther.*, 17, 111, 1982.
10. **Rowland, M. and Tozer, T. N.,** *Clinical Pharmacokinetics: Concepts and Applications*, Lea & Febiger, Philadelphia, 1980.

11. **Bischoff, K. B., Dedrick, R. L., and Zaharko, D. S.,** Preliminary model for methotrexate pharmacokinetics, *J. Pharm. Sci.*, 59, 149, 1970.
12. **Jain, R. K., Gerlowski, L. E., Weissbrod, J. M., Wang, J., and Pierson, R. N.,** Kinetics of uptake, distribution and excretion of zinc in rats, *Ann. Biomed. Eng.*, 9, 347, 1982.
13. **Lutz, R. J., Galbraith, W. M., Dedrick, R. L., Shrager, R., and Mellett, L. B.,** A model for the kinetics of distribution of actinomycin-D in the beagle dog, *J. Pharmacol. Exp. Ther.*, 200, 469, 1977.
14. **Bischoff, K. B.,** Some fundamental considerations of the applications of pharmacokinetics to cancer chemotherapy, *Cancer Chemother. Reg.*, 59, 777, 1975.
15. **Kety, S. S.,** The theory and application of inert gases at the lungs and tissues, *Pharm. Rev.*, 3, 1, 1951.
16. **Rall, D. P., Stabenau, J. R., and Zubrod, C. G.,** Distribution of drugs between blood and cerebrospinal fluid: general methodology and effect of pH gradients, *J. Pharmacol.*, 125, 185, 1959.
17. **Lutz, R. J., Dedrick, R. L., and Zaharko, D. S.,** Physiological pharmacokinetics: an in vivo approach to membrane transport, *Pharmacol. Ther.*, 11, 559, 1980.
18. **Wilkinson, G. R.,** Plasma and tissue binding considerations in drug disposition, *Drug Metab. Rev.*, 14, 427, 1983.
19. **Levy, R. and Shand, D., Eds.,** Clinical implications of drug-protein binding, *Clin. Pharmacokinet.*, 9, 1, 1984.
20. **Lin, J. H., Sugiyama, Y., Awazu, S., and Hanano, M.,** In vitro and in vivo evaluation of the tissue-to-blood partition coefficient for physiological pharmacokinetic models, *J. Pharmacokinet. Biopharm.*, 10, 637, 1982.
21. **Chen, H.-S. G. and Gross, J. F.,** Estimation of tissue-to-plasma partition coefficients used in physiological pharmacokinetic models, *J. Pharmacokinet. Biopharm.*, 7, 117, 1979.
22. **Wilkinson, G. R. and Shand, D. G.,** A physiological approach to hepatic drug clearance, *Clin. Pharmacol. Ther.*, 18, 377, 1976.
23. **Rowland, M., Benet, L. Z., and Graham, G. G.,** Clearance concepts in pharmacokinetics, *J. Pharmacokinet. Biopharm.*, 1, 123, 1973.
24. **Tozer, T. N.,** Concepts basic to pharmacokinetics, *Pharmacol. Ther.*, 12, 109, 1981.
25. **Himmelstein, H. J. and Lutz, R. J.,** A review of the applications of physiologically based pharmacokinetic modeling, *J. Pharmacokinet. Biopharm.*, 7, 127, 1979.
26. **Gillette, J. R.,** Factors affecting drug metabolism, *Ann. N.Y. Acad. Sci.*, 179, 43, 1971.
27. **Oie, S. and Tozer, T. N.,** Effect of altered plasma protein binding on apparent volume of distribution, *J. Pharm. Sci.*, 68, 1203, 1979.

Chapter 3

THE ROLE OF METABOLIC ACTIVATION IN TARGET ORGAN TOXICITY

John Caldwell, Susan A. Sangster, and J. David Sutton

TABLE OF CONTENTS

I. INTRODUCTION

Xenobiotics which enter the animal body will undergo one or more possible fates. They may (1) be eliminated unchanged, (2) undergo enzymic metabolism, (3) undergo spontaneous chemical transformation, which may or may not involve the catalytic activity of nonenzymic body constituents, or (4) remain in the body untransformed. Of these various options, it is enzymic metabolism which is by far the most frequently encountered biological fate of foreign compounds.

The literature contains a bewildering number of examples of the types of metabolic transformations which xenobiotics may undergo. Williams[1] pointed out that these may be conveniently divided into two main types, which generally occur sequentially. In the first phase of metabolism, there occurs a reaction of oxidation, reduction, or hydrolysis, which results in the introduction into the molecule of a functional group which is appropriate to serve as the site for the second phase of metabolism, involving the so-called conjugation reactions. These all result in the combination of the xenobiotic with an endogenous moiety, termed the conjugating agent. Although the biphasic sequence is very commonly encountered, there are many examples of compounds whose phase I metabolites are eliminated without being conjugated, and of compounds possessing functional groups able to undergo conjugation directly without first being metabolized by one of the phase I processes. The elimination products of xenobiotics may thus comprise products of (1) phase I metabolism, (2) phase II metabolism, (3) phase I and phase II reactions, and (4) the unchanged compound.

The biological consequences of the interactions of a foreign compound with the body are potentially legion, but they all exhibit a certain specificity depending upon the structure and chemical reactivity of the compound evoking the response in question. It is thus to be expected that the processes of metabolism, by virtue of their ability to change the structure and reactivity of the parent compound, will have a profound influence upon the types of interactions which a compound will undergo with body constituents. In turn, it may be supposed that alteration of these interactions as a consequence of alterations of structure will in turn change the nature of the evoked activity.

The consequences of metabolism for the biological activity of a compound may thus involve the following possibilities:

1. The formation of chemically stable metabolites, which may show reduced activity, loss of activity, changed activity, or increased activity
2. Formation of chemically reactive species able to interact covalently with body constituents
3. Formation, as by-products, of excess quantities of harmful endogenous compounds, e.g., oxygen radicals
4. Alteration of normal body function by altering levels of endogenous compounds, e.g., redox ratios of coenzymes, depletion of sulfate or amino acids used for conjugation

II. METABOLISM OF XENOBIOTICS

It is the purpose of this brief review to consider the various roles of metabolic transformations in toxicity, particularly in the context of the expression of toxicity in specific target organ(s). Even if we restrict ourselves to those metabolic transformations which result in the enhancement of the activity of their xenobiotic substrates, either by the formation of chemically stable active metabolites or by generating reactive species, a survey of the principle reaction types reveals that they all possess the potential capacity to effect metabolic activation. In each and every case, examples may be cited of increased activity or reactivity occurring as a consequence of metabolism. Table 1 lists the most important reaction types

Table 1
SOME EXAMPLES OF METABOLIC ACTIVATION OCCURRING AS A CONSEQUENCE OF THE PHASE I METABOLIC REACTIONS (OXIDATION, REDUCTION, AND HYDROLYSIS)

Oxidations	Ref.
Carbon centers	
Open chain aliphatic	
Safrole, estragole, methyleugenol	2
Double bonds	
Open chain	
vinyl chloride	3
Aromatic	
Bromobenzene	4
Benzo[a]pyrene	5
Heterocycle	
Furan	6
Nitrogen centers	
Aromatic amines	
2-Naphthylamine	7
Aromatic amides	
2-Acetylaminofluorene	8
Heterocycles	
N,N-Diallylmelamine	9
Sulfur centers	
Thiocarbamides	
α-Naphthylthiourea	10
Thioamides	
Thioacetamide	11
Thiocarbamates	
Diallate	12
(metabolism to the mutagen, chloracrolein)	
Desulfuration	
Carbon disulfide	13
Parathion	14
Dealkylation reactions	
O-Dealkylation	
Phenacetin	15
N-Dealkylation	
Butter yellow	16
Dehalogenation	
Halothane	17
Chloramphenicol	18

Reductions	Ref.
Azo	
Prontosil	19
4-Aminoazobenzene	20
Nitro	
4-Nitroquinoline *N*-oxide	21
Oxidized sulfur centers	
Sulindac	22
Reductive dehalogenation	
Carbon tetrachloride	23
Hydrolyses	
Epoxides	
Benzo[*a*]pyrene	24
Amide	
Phenacetin	25
Ester	
Esters of allyl alcohol	26
Aspirin	27

in xenobiotic metabolism, together with examples of compounds activated as a consequence of metabolism by the reaction in question.

A. Oxidative Transformations

Virtually all of the atoms within organic compounds have the potential to undergo oxidation in the animal body. Thus, carbon atoms with protons attached, linked to other carbon atoms, or to heteroatoms, may be oxidized. This may occur either by the direct insertion of an oxygen atom, (as in the case of the oxidation of the 1′-C of the allylic side chain of safrole) giving an alcohol[28] or (as in the cases of isolated or conjugated double bonds) result in the formation of a three-membered oxirane ring (termed an arene oxide in the case of an aromatic ring or an epoxide from an alkene).[29] The reactivity of an alcohol depends very much upon the configuration in question. In general, such compounds are extremely stable, but the 1′-hydroxyallylbenzenes are chemically reactive due to their ability to give rise to carbonium ions at the 1′-center;[30] this property is enhanced by subsequent esterification of the 1′-hydroxyl group (see below).

Oxiranes are often reactive electrophiles, since the three-membered ring is highly strained with unequal charge distribution. In the case of the arene oxides, this generally leads to their facile rearrangement to phenols, exhibiting the "NIH shift" with migration and retention of substituents ortho- to the site of hydroxylation,[29] or hydration to dihydrodiols.[31] It is this reactivity which leads to the ability of oxiranes to interact irreversibly (covalently) with nucleophilic centers in various tissue macromolecules. In many cases, this covalent binding is associated with toxicity. It is important to stress that not all epoxides are harmful. Some are so stable that they are excreted in the urine, e.g., the 10,11-oxide of the anticonvulsant drug carbamazepine,[32] while others, although formed from toxic or carcinogenic compounds, are apparently not involved in the expression of toxicity, e.g., estragole and safrole.[33]

1. Oxidation of Heteroatoms

Next to carbon, the most commonly encountered centers in xenobiotics for oxidative metabolism are nitrogen and sulfur atoms. Nitrogen atoms, in a variety of oxidation states, may be present in rings, exo- to the ring systems of cyclic compounds and in open-chain compounds, and may be oxidized provided there is a proton attached (e.g., $-NH_2$, $-NH.R$, $-NHOH$) or the lone pair is available to form a dative covalent bond with oxygen. Oxidation of nitrogen atoms exo- to ring systems is well documented to result in their metabolic activation, both in terms of stable metabolites with biological activity (e.g., the *N*-hydroxy metabolite of *N*,*N*-diallylmelamine, which is responsible for its hypotensive action) and chemically reactive metabolites which may give rise to electrophilic species (carbonium or nitrenium ions) apparently responsible for the carcinogenicity of compounds such as 2-naphthylamine and 2-acetylaminofluorene. Nitrogen oxidation of aliphatic amines apparently does not result in their metabolic activation,[34,35] although the *N*-oxidation products of certain aliphatic amines do possess CNS activity.[36]

Sulfur occurs in xenobiotics in a variety of oxidation states, from -2 to $+6$, in open-chain compounds and both endo- and exo- to a variety of ring systems. All of these, except those already at the highest oxidation state, may undergo biological oxidation. It is a complicating factor that oxidized sulfur compounds are readily reduced in the body, and it is sometimes difficult to assess the extent of S-oxidation of a particular compound due to this facile reversibility (see below). As listed in Table 1, sulfoxidation reactions may give rise to active metabolites, and the oxidative removal of elemental sulfur in a reactive form from thio compounds can also be responsible for metabolic activation.

2. Cleavage of Carbon-Heteroatom Bonds (Dealkylation Reactions)

Xenobiotics frequently contain within their structures alkyl or arylakyl groups attached to a heteroatom, generally oxygen, nitrogen, or sulfur. The cleavage of such bonds resulting in the appearance of variously an hydroxyl group, a primary (from a secondary) or secondary (from a tertiary) amine, or a thiol is well established.[37] The mechanisms of dealkylation are especially well studied in the case of the *N*-dealkylations[38] and initially involve oxidation of either the nitrogen center or the adjacent alpha-carbon atoms of its substituents, the products of which undergo rearrangement leading to C-N bond cleavage.

One of the classical examples of the metabolic activation of a drug is the *O*-deethylation of the minor analgesic phenacetin (now withdrawn in many countries) to give the widely used compound paracetamol (acetaminophen). Other examples of *N*- and S-dealkylations resulting in metabolic activation are presented in Table 1.

3. Cleavage of Carbon-Halogen Bonds

The removal of halogen atoms from xenobiotics by the cleavage of carbon-halogen bonds is a reaction of considerable toxicological significance. It occurs most frequently with haloalkanes, but dehalogenation of haloaromatics has been reported.[23,39] In mechanistic terms, dehalogenation may proceed by both reductive and oxidative processes, which may be either 1- (radical) or 2-electron reactions.[40] Reactive intermediates are frequently generated during such dehalogenation reactions, and these may include radicals, carbonium ions, and as in the case of CCl_4, radical species formed by the interaction of the metabolically produced CCl_3-radical and oxygen.

B. Reductions

A variety of groupings encountered in xenobiotics undergo metabolic reduction which involves enzymes located in the tissues of the body and/or possessed by the GI microflora. The reductive cleavage of carbon-halogen bonds has been referred to above, and reduction may also be a feature of the metabolism of alkene, carbonyl, azo, nitro, sulfoxide, and sulfone functionalities. With the exception of alkene and carbonyl groups, instances of metabolic activation arising from the reduction of all of these groupings have been reported.

Azo reduction is a reaction of historical importance, since the recognition of the reductive cleavage of the azo dye Prontosil to its active anti-bacterial metabolite, sulfanilamide, by Trefouel and colleagues in 1935,[19] led to the development of the sulfonamide group of chemotherapeutic agents. The reductive cleavage of azo bonds may also result in metabolic activation subsequent to the liberation of free aromatic amines.

The significance of nitrogen oxidation in the metabolic activation of xenobiotics has been discussed earlier, and it appears that the metabolites responsible for the toxicity and carcinogenicity of many aromatic amines are of intermediate oxidation states, i.e., hydroxylamines and nitroso compounds. The metabolic reduction of nitro groups in the body essentially reverses this sequence, thus:

$$NH_2 \leftrightarrows NHOH \leftrightarrows N{=}O \leftrightarrows NO_2$$

It is therefore not surprising that such reductions may result in metabolic activation, as the same reactive metabolites may be obtained by oxidation or reduction. The contribution of the nitro reduction capacity of the gut flora toward the toxicity and carcinogenicity of various nitro aromatic compounds is well documented.[41]

The reduction of various oxidized sulfur compounds is frequently encountered (see above). The toxicological sequelae of such reductions have been little studied, but the anti-inflammatory agent sulindac ((Z)5-fluoro-2-methyl-1-[*p*-(methylsulfinyl)phenyl]-methylene-1H-indene-3-acetic acid) affords a good example of its potential pharmacological consequences.[22] The parent drug is a sulfoxide of an open chain methylthioether, and careful study of the kinetics of the drug and its oxidized (sulfone) and reduced (sulfide) metabolites has shown that the parent drug actually functions as a prodrug for the active pharmacophore, the sulfide. There occurs a redox equilibrium between the sulfoxide and the sulfide, but oxidation to the inactive sulfone is unidirectional, thus:

$$R\text{–}S\text{–}R \leftrightarrows R\text{–}(S\text{—}O)\text{–}R \rightarrow R\text{–}(SO_2)\text{–}R$$

C. Hydrolysis

Hydrolytic cleavage of ester and amide bonds is a common reaction of xenobiotic metabolism,[26] while the hydration of oxirane rings is important in inactivating this group of potentially reactive metabolites.[31,42]

Esters are commonly used as prodrugs, and thus examples of their hydrolysis leading to the liberation of pharmacologically active species are legion.[43] Esters also occur widely in plants, and are thus commonly encountered as food flavor and fragrance chemicals. In cases

Table 2
SOME EXAMPLES OF METABOLIC ACTIVATION OCCURRING AS A CONSEQUENCE OF THE PHASE II (CONJUGATION) REACTIONS

	Ref.
Glucuronic acid conjugation	
Morphine	49
Sulfation	
2-Acetylaminofluorene	50
Safrole	28
Methylation	
Isoprenaline	51
Pyridine	52
***N*-Acetylation**	
Sulfonamides	53
Procainamide	54
2-Naphthylamine	55
Amino acid conjugation	
Hypoglycin A (via the acyl CoA intermediate)	56
Glutathione conjugation	
1,2-Dibromoethane	57

where one or both parts of the ester are toxic, their hydrolysis will be associated with metabolic activation, e.g., esters of allyl alcohol. In addition to these truly xenobiotic esters, many compounds are transformed to esters during their metabolism, and the hydrolysis of these may be associated with a re-exposure of the body to potentially toxic species. This is particularly the case with glucuronides which, if excreted in the bile, are frequently broken down by the enzymes of the gut secretions and microflora. This topic will be discussed in more detail later.

The amide linkage is also used in prodrugs[26,43] and its cleavage will result in the liberation of the active pharmacophore. Amide hydrolysis will give rise to an amine and an acid, and either or both of these may be further metabolized resulting in their activation. This is more commonly encountered with amines liberated from amides, and a good example is seen with phenacetin. Here, amide cleavage gives rise to phenetidine, the metabolite supposedly responsible for the methemoglobinemia and renal damage produced by phenacetin.[25]

Oxirane hydration, which is a hydrolytic reaction, serves to inactivate these potentially reactive metabolites. However, the dihydrodiols so produced may undergo further oxidation to give intermediates far more reactive than the original oxiranes. It is well known that the activation of the polycyclic aromatic hydrocarbons (PAH) involves epoxidation, hydration, and a second epoxidation of the resultant dihydrodiol to give the ultimate carcinogenic electrophile. Thus, in the case of benzo[*a*]pyrene, the metabolite apparently responsible for its carcinogenicity is its 7,8-dihydrodiol-9,10-oxide. For more detailed considerations of these matters, see Jerina et al.[24] and Chapter 10 in Volume II.

D. Conjugation Reactions

As defined in the introduction, the conjugation reactions involve the linkage of a xenobiotic or one of its metabolites to an endogenous moiety, the conjugating agent. An enormous number of endogenous conjugating agents have now been identified,[44,45] but of these only a few are commonly encountered, and these are listed in Table 2. It will be seen that these reactions in general add a bulky, polar moiety to the xenobiotic, a consequence of which is to give a product of markedly greater water solubility than the parent compound, which is more readily excreted. The great changes to the structure of the xenobiotic brought about by conjugation thus reduce its ability to interact with the tissues of the body, and thus conjugation reactions are generally responsible for the inactivation of xenobiotics.[44,46] However, there are examples of compounds (1) whose conjugates are active per se, (2) where conjugation contributes to the formation of reactive intermediates, or (3) where further metabolism of a conjugate results in metabolic activation. These various examples will be discussed for each of the major conjugation reactions.

1. Glucuronidation

The glucuronic acid conjugation mechanism is the most versatile of the conjugation reactions in terms of its zoological distribution and range of substrates.[47] In the majority of cases, glucuronidation results in the detoxication and facile elimination of the compound in question,[44,46-48] with the consequence that glucuronidation capacity, which is determined by factors such as dose size, drug interactions, and species, strain and genetic differences in enzyme activity, is an important determinant of toxicity.[44,46,48] In contrast to this detoxication role, in one case glucuronic acid conjugation has been shown to result in metabolic activation. Thus, morphine-6-glucuronide (but not the isomeric 3-glucuronide), a minor metabolite of morphine, contributes to the analgesic and other effects of morphine. It is important to appreciate that this action is *not* a consequence of the hydrolysis of the conjugate.

In other cases, glucuronides of xenobiotics may act as delivery forms of the parent compound or metabolite to particular sites in the body, where they are hydrolyzed and exert a localized effect, e.g., intestinal ulceration from indomethacin excreted as its glucuronide in the bile, and the occurrence of bladder tumors following the administration of 2-naphthylamine. These matters will be described in more detail later.

2. Sulfation

The formation of sulfate esters of xenobiotics gives products with high polarity and water solubility, which in the great majority of cases are inactive and are rapidly excreted. However, the sulfation of certain metabolically introduced hydroxyl groups can give rise to reactive intermediates which contribute significantly to the toxicity of the parent compound. This phenomenon was first recognized in the metabolism of 2-acetylaminofluorene and related carcinogenic aromatic amines and amides,[50] and has more recently been shown with safrole and related allylbenzenes.[28] These compounds undergo an initial hydroxylation (at the nitrogen atom in the cases of the amines and amides, and at C-1 of the allyl side chain in the cases of the safrole analogues) and are then sulfated at the new hydroxyl group. The *O*-sulfate is a facile leaving group, and its formation thus promotes the production of an electrophilic intermediate, an arylnitrenium ion from the *N*-*O*-sulfates or a carbonium ion from the C-*O*-sulfates. These electrophilic reactive intermediates may then interact covalently with cellular macromolecules including DNA, resulting in the appearance of various toxic manifestations.[58]

3. Methylation

Methylation is a reaction of catechol hydroxyl, thiol, and various nitrogen centers in xenobiotics.[48,59-61] Unlike the preceding reactions, methylation often serves to increase the lipid solubility and reduce the polarity of the substrate; an exception to this is offered by the metabolic *N*-methylation (quaternization) of azaheterocycles. Two instances of metabolic activation by methylation may be quoted: 3-*O*-methylisoprenaline, a metabolite of the β-adrenergic agonist isoprenaline, has β-adrenergic blocking activity and thus acts as an antagonist to the parent drug.[51] The simple azaheterocycle pyridine is methylated to the *N*-methylpyridinium ion, which has markedly greater acute toxicity than pyridine itself.[52]

4. Acetylation

Acetylation is a commonly occurring route of metabolism of the primary amino group ($-NH_2$) in a wide variety of xenobiotics.[62] Like methylation, acetylation reduces polarity and may reduce water solubility of the parent compound, and its function must thus be sought in terms other than facilitation of elimination. Indeed, the crystalluria and consequent nephrotoxicity of various sulfonamides is due to the precipitation of their *N*-acetylated metabolites in the lumen of the kidney tubule.[53]

The *N*-acetyl metabolites of a number of drugs have been shown to retain much of the activity of the parent compound, and in the cases of procainamide[54] and sulfanilamide,[48] the *N*-acetyl compounds are used as drugs in their own right.

The metabolic activation of aromatic amines and amides by *N*-hydroxylation and sulfation has already been mentioned. Many aromatic amines are *N*-acetylated prior to such activation, and in some cases, there can occur transfer of the acetyl group from *N*- to *O*- (the so-called *N,O*-acetyl group transfer), with the result that the *O*-acetyl moiety can act as a leaving group to yield the arylnitrenium ion, analogous with the process described for sulfation.[55]

In only a very few cases are conjugates further metabolized, but a good example with toxic consequences is seen with the anti-tubercular drug isoniazid (isonicotinic acid hydrazide). The free NH_2 group of the hydrazide moiety is acetylated, and the product, *N*-acetylisoniazid, is in part hydrolyzed yielding isonicotinic acid and *N*-acetylhydrazine. The cytochrome P-450-dependent oxidation of *N*-acetylhydrazine yields an intermediate responsible for the hepatitis-like toxicity of isoniazid.[61]

5. Amino Acid Conjugations

The amino acid conjugations are a group of reactions in which the carboxyl group of a xenobiotic acid is linked with the amino group of one of a range of amino acids, most commonly glycine, glutamine, and taurine. The particular amino acid used is a function of both the structure of the xenobiotic acid and animal species in question.[62,63] Although the formation of amino acid conjugates serves to decrease the toxicity and enhance the elimination of the parent acid, the changes to its properties are not dramatic.[62] It is likely that the key to the function of amino acid conjugations lies in an appreciation of their enzymic mechanisms. These reactions have three distinct steps in which the acid is first activated by the formation of an acyl adenylate that is rapidly converted to an acyl CoA thioester. It is the latter high energy intermediate to which the amino acid is transferred, with the liberation of free CoA. Although the final product of this sequence, the amino acid conjugate, is a detoxication product, the intermediate acyl CoA is of considerable biological significance. Many xenobiotic acyl CoAs have inhibitory actions against a number of enzymes of lipid and carbohydrate metabolism, and the formation of these intermediates during the process of amino acid conjugation may assume considerable toxicological significance.[64] An excellent example of this is seen in the metabolism of hypoglycin A, the toxic principle of the unripe Ackee fruit. This fruit is the agent responsible for Jamaican Vomiting Sickness, a frequently fatal syndrome involving hypothermia, hypoglycemia, and abnormal metabolism of amino acids.[65] Hypoglycin A (methylenecyclopropylalanine) is converted to methylenecyclopropylacetyl CoA (MCPA-CoA) by transamination and β-oxidation, the latter involving the formation of the CoA ester. MCPA-CoA is an irreversible inhibitor of a number of fatty acyl-CoA dehydrogenases, and this is directly or indirectly responsible for the metabolic disturbances.[66] It is very relevant to note the ability of exogenous glycine to reverse the various toxic effects of hypoglycin A.[56] This suggests that the toxicity of this agent is a consequence of the failure of MCPA-CoA to undergo amino acid conjugation, perhaps due to limitations upon the availability of the endogenous co-substrate glycine.

6. Glutathione Conjugation

The nucleophilic tripeptide glutathione is able to undergo conjugation with electrophilic centers in a wide variety of xenobiotics, which may be present per se, e.g., nitro- and haloalkanes, or introduced by metabolism, e.g., oxiranes (see above). In the great majority of cases, conjugation with glutathione serves to detoxify these electrophiles, preventing their interaction with nucleophilic centers in macromolecules such as proteins and nucleic acids, and in general leads to the elimination of a mercapturic acid (an S-substituted *N*-acetylcysteine) following metabolism of the S-substituted glutathione.[67] With some compounds, however, the reaction with glutathione does not produce conjugates but gives products resembling those of the phase I reactions, e.g., the glutathione-dependent denitration of glyceryl trinitrate.[68]

In contrast to the above, the glutathione conjugation of a number of dihaloalkanes results in their metabolic activation, as exemplified by 1,2-dibromoethane:[57]

$$BrCH_2CH_2Br \xrightarrow{GSH} G{\cdot}S{\cdot}CH_2CH_2Br$$
$$\downarrow$$
$$G{\cdot}S{\cdot}CH_2CH_2^+ + Br^-$$

Nucleophilic attack by glutathione on a C-Br bond results in the formation of a thioether with the expulsion of Br^-. The thioether is then analogous to a sulfur mustard, which dissociates to yield the carbonium ion shown above.

III. REACTION SEQUENCES IN THE METABOLIC ACTIVATION OF XENOBIOTICS

In the first part of this review, attention was focused on the ability of individual metabolic pathways in isolation to produce metabolites possessing either biological activity or chemical reactivity of toxicological significance. It is important to appreciate that the formation of active metabolites, in particular reactive intermediates, frequently involves the concerted operation of more than one metabolic reaction. This may involve multiple phase I reactions, the sequential operation of phase I and II reactions, or on occasion, the further transformation of conjugates, which are normally considered as end products of metabolism.

PAHs, typified by benzo[*a*]pyrene, exert their carcinogenicity as a consequence of multiple phase I metabolic reactions.[24] Benzo[*a*]pyrene is first converted to its 7,8-oxide by the cytochrome P-450 system, and this oxide is then converted to the 7,8-dihydrodiol by epoxide hydrolase. This diol is then further oxidized by cytochrome(s) P-450 to yield the ultimate carcinogen, the 7,8-dihydrodiol-9,10-oxide.

Activation by the sequential operation of phase I and II reactions is exemplified in the case of carbon centers by safrole and related allylbenzenes, and in the case of nitrogen centers, by 2-acetylaminofluorene and other aromatic amines and acetamides.[30] Both of these series of compounds undergo an initial oxidation at C-1 of the side chain of the allylbenzenes or at the nitrogen atom, to give hydroxylated metabolites, which then undergo sulfation. In both cases, it is these sulfate esters which are the ultimate carcinogens. The *O*-sulfate moiety is a favored leaving group, leaving positively charged ions able to interact covalently with nucleophilic centers in proteins, nucleic acids and other cellular macromolecules. It is the formation of these macromolecular adducts which is presumed to be responsible for many types of toxicity.

The further metabolism of conjugates may take two forms. In the first, metabolic activation may ensue simply as a consequence of the hydrolysis of a conjugate, thereby liberating the parent compound or a metabolite. In the second case, the conjugate is cleaved to yield new products, which then undergo metabolic activation, as is seen in the case of isoniazid. Thus, isonicotinic acid hydrazide undergoes *N*-acetylation, and the *N*-acetylisoniazid so produced is hydrolyzed, not at the acetamido group but rather at the isonicotinic acid hydrazide bond. The products of this are isonicotinic acid and *N*-acetylhydrazine, thus:

$$\mathrm{R{\cdot}CONH{\cdot}NH_2 \rightarrow R{\cdot}CONH{\cdot}NH{\cdot}COCH_3}$$
$$\downarrow$$
$$\mathrm{R{\cdot}COOH + H_2N{\cdot}NHCOCH_3}$$

N-Acetylhydrazine is then converted to electrophilic species by cytochrome(s) P-450, giving metabolites apparently responsible for the hepatotoxicity of isoniazid.[61] In this context, it is interesting to note that the *N*-acetylation of isoniazid exhibits a genetic polymorphism in man, and that the hepatotoxicity is more frequent in fast acetylators, who produce more of the essential intermediate, *N*-acetylisoniazid.[69]

IV. ACTIVE OXYGEN SPECIES IN XENOBIOTIC TOXICITY

While the bulk of this review has dealt with the role of metabolically produced electrophiles in xenobiotic toxicity, it must not be forgotten that a variety of active oxygen species may be of great relevance in toxic mechanisms.[70] In the last 15 years, attention has focused on the role of such species in tissue damage arising from a number of causes, such as ionizing radiation, the inflammatory process, oxygen toxicity and aging, as well as chemical injury. The term "active oxygen species" encompasses the superoxide anion radical $O_2^{\bar{\cdot}}$, singlet oxygen $^1O_2^{\bar{\cdot}}$, the hydroxyl radical OH·, lipid alkoxy and peroxy radicals, and hydrogen peroxide. Although the properties of these various moieties lie outside the scope of this review, they are noteworthy insofar as their formation may be a consequence of the metabolism of xenobiotics in the cell.

A. Superoxide Anion Radical and Hydrogen Peroxide

The superoxide anion radical is produced by the one-electron reduction of molecular oxygen by a variety of processes.

NADPH-cytochrome P-450 reductase is able to donate one electron to quinones to produce semiquinone radicals. Substrates for this reaction include xenobiotic quinones such as menadione,[71] and the anthracycline cytotoxic agents, e.g., adriamycin and daunorubicin,[70], and those produced by oxidative metabolism, such as the 3,6- and 6,12-quinones of benzo[*a*]pyrene.[72] These semiquinone radicals reduce molecular oxygen, giving the superoxide anion radical and regenerating the quinone, which is then able to recycle through this series of reactions (redox cycling). This redox cycling may be initiated by other flavoprotein enzymes in the cell, such as the mitochondrial NADH-ubiquinone oxidoreductase. A number of other xenobiotics are able to undergo such redox cycling leading to the formation of the superoxide anion radical, including aromatic hydroxylamines and nitro compounds and the so-called "redox dyes" typified by paraquat.[73]

Hydrogen peroxide is formed by the dismutation of the superoxide radical by the superoxide dismutases, metalloenzymes of two types. The mitochondrial enzyme contains Mn, while that in the cytosol contains Cu and Zn. In addition, hydrogen peroxide can be formed directly by the two-electron reduction of molecular oxygen by enzymes such as monoamine oxidase.

B. Hydroxyl Radical

The hydroxyl radical is a very potent oxidant generated from a number of sources, including xanthine oxidase, during the oxidation of ethanol and by the microsomal electron transport chain. Due to its high reactivity, it is generally considered that the hydroxyl radical does not diffuse away from its site of formation in the cell.

C. Singlet Oxygen

Singlet oxygen is generally present at very low levels, but is of interest due its postulated role in lipid peroxidation. In addition to its formation during lipid peroxidation, singlet oxygen may also be formed by the reaction of the superoxide anion radical with hydrogen peroxide, and this may assume importance in the cases of agents which promote redox cycling (quinones, paraquat, etc.) and thus lead to elevated levels of both these reactants.

D. Alkoxy and Peroxy Radicals

Alkoxy ($RO^{\cdot}$) and peroxy ($ROO^{\cdot}$) radicals are formed during the decomposition of lipid hydroperoxides, and are of importance in the radical chain reactions of lipids. The polyunsaturated fatty acids which undergo peroxidation in the cell, initiated by active oxygen species, are membrane bound, so that many of the injurious effects of active oxygen species arise from damage to the essential function of membranes. The formation and fate of such radicals within membranes is thus of great significance in such processes. The principal mechanisms giving rise to such radicals are (1) reaction of lipid hydroperoxides with superoxide anion radical:

$$ROOH + O_2^{\bar{\cdot}} \rightarrow RO^{\cdot} + OH^- + O_2$$

and (2) autoxidation (chain) reactions of unsaturated fatty acids.

V. FACTORS CONTROLLING THE LEVELS OF REACTIVE SPECIES WITHIN THE CELL

The potential of a compound to exert a toxic effect in the cell is often determined by the types of metabolites which it is converted to, and this is especially true in the case of chemical carcinogens. These compounds exert their carcinogenic action as a consequence of covalent interaction with cellular macromolecules, notably the nucleic acids. However, the great majority of chemical carcinogens are chemically inert, and require metabolic activation prior to covalent interaction with tissue macromolecules.

The magnitude of any toxic effect, which a compound might exert, will depend upon the time course of the concentration of the active species (unchanged compound or metabolite(s)) at sites of interest. It is thus important to appreciate that there exist enzyme systems able to metabolize and generally detoxify reactive metabolites, and that the effects of such metabolites will depend upon the balance of the reactions leading to their formation and those leading to their inactivation. It is this balance which determines whether exposure to a potentially toxic compound will or will not lead to the expression of toxicity.

Many situations could be cited to illustrate this concept, and the two examples to be discussed further are chosen to illustrate it from the viewpoint of "classical" reactive intermediates on the one hand and from that of active oxygen species on the other.

The metabolic activation of the PAH benzo[*a*]pyrene to metabolites able to interact covalently with nucleic acids has already been described. The levels of the 7,8-dihydrodiol-9,10-oxide will be determined by the activities of a variety of enzyme systems.[72-74] Thus, the initial formation of the 7,8-oxide will be determined by the regioselectivity of form(s) of cytochrome P-450 relative to other forms oxidizing the molecule at other sites. The 7,8-oxide may be metabolized by two distinct reactions, viz., glutathione conjugation and hydration to the dihydrodiol, which is the essential intermediate in the metabolic activation. Once formed the diol may be dehydrated to a catechol, conjugated with either glucuronic acid or sulfate,[75] or undergo further metabolic oxidation to give the ultimate carcinogenic metabolite, the 7,8-dihydrodiol-9,10-oxide. Once formed, the diolepoxide is metabolized either by hydration to the 7,8,9,10-tetrol or by conjugation with glutathione. The levels of the ultimate carcinogen are thus dependent upon the relative activities of a considerable number of enzyme systems under independent regulation. The modulation of the metabolite-mediated mutagenicity by factors affecting the activities of various of these enzyme systems serves to illustrate these principles clearly.[73]

In the case of compounds whose effects are mediated by oxygen radicals produced during their metabolism, it is possible to see a similar balance between reactions favoring the formation of radicals and those serving to inactivate them. The ability of menadione to give rise to cytotoxic oxygen radicals has already been described. As well as undergoing one-electron reduction, ultimately resulting in the formation of superoxide anion radical, menadione, like other quinones, is a substrate for DT-diaphorase, which effects their two-electron reduction, yielding hydroquinones. These hydroquinones may then be sulfated or glucuronidated, leading to their excretion. The levels of superoxide anion radical and related moieties in the cell will be determined in part by the balance of one- to two-electron reduction, and in part by the activities of enzymes such as superoxide dismutase, catalase, and glutathione peroxidase, which inactivate in various ways active oxygen species. The modification of the cytotoxicity of menadione by factors altering the activities of these various enzymes has been well illustrated by the group of Orrenius.[71,76]

VI. THE IMPORTANCE OF METABOLIC ACTIVATION IN TARGET ORGAN TOXICITY

Although the basic events leading to toxicity are the same in all the cells of the body, and it is the case that in principle all cells are susceptible to the harmful effects of chemicals, it is generally observed that there occur marked differences in susceptibility between particular cell types, either within an organ or between organs. The organs affected are generally referred to as the target organs of toxicity, and the differences between these and the other organs of the body may be quantitative or qualitative. Thus, a compound may have specific effects on particular organ(s) at a low dose, but exert a more generalized toxicity at higher doses. In other cases, only particular target organs may be affected regardless of the dose administered.

It is a matter of great practical importance that the mechanisms of such target organ toxicity be understood. In some cases the origins of selectivity are simple: a compound has a high affinity for a particular organ, reaches high concentrations there, and thus causes a localized toxicity. Such tissue-specific uptake is especially frequent in the lung and kidney, and compounds whose target organ toxicity is explained in this way include paraquat in the lung[73] and the cephalosporin antibiotics in the kidney.[77]

The importance of metabolic activation in toxicity has been indicated above, and it is important to understand how metabolism relates to target organ toxicity. Two contradictory problems may be discerned: (1) the liver is the principal site of xenobiotic metabolism in the body, yet it is only rarely a target organ of toxicity, and (2) the extreme reactivity of the intermediates thought to be responsible for toxicity is such that they are unlikely to be stable enough to allow their transport around the body, but are more likely to act directly at the sites at which they are formed. However, it must be pointed out that it is now well understood that virtually all tissues of the body possess the ability to metabolize xenobiotics, and that there occur considerable differences between different organs,[78,79] and between different cell types within the same organ[80,81] in the distribution of the various multiple forms of the principal drug metabolizing enzymes.

In addition, it must also be remembered that while metabolic activation can occur in more than one organ, there may occur organ specificity in the types of reactive metabolites formed. Thus, the minor analgesic phenacetin may exert hepatoxicity as a consequence of its conversion to paracetamol and the metabolic activation of this compound. The renal toxicity of phenacetin is apparently due to the formation of free radical reactive intermediates from *N*-deacetylated metabolites by co-oxygenation reactions (mediated by prostaglandin synthetase) which occur predominantly in extrahepatic organs and are particularly well-developed in the kidney.[82-84]

Three possibilities for the origin of metabolite-mediated tissue-specific toxicity exist: (1) that metabolic activation is a feature of tissue specific metabolism, (2) that reactive metabolites are formed in the liver and subsequently transported to other, more susceptible sites at which toxicity is produced, or (3) that activation involves more than one metabolic step, with the initial step occurring in the liver and further reaction(s) taking place in the target organ itself.

Numerous examples of the tissue-specific metabolism of a toxic chemical have been reported. A classic example is seen with the furan 4-ipomeanol, found in moldy sweet potatoes. This compound is selectively toxic to the Clara cells of lung, resulting in severe lung damage. The Clara cells are rich in form(s) of cytochrome P-450 able to activate the toxin via oxidation of the 2,3-double bond of the furan ring system. The presumed epoxide intermediate is extremely reactive, and its nature may only be discerned indirectly. If the activities of the epoxide forming enzymes in the liver are enhanced artificially by enzyme inducers, then 4-ipomeanol also becomes hepatotoxic. This topic has been thoroughly reviewed by Boyd.[10]

A more recent example of tissue-specific metabolism underlying target organ toxicity is seen with hexachlorobutadiene. This halogenated alkene is nephrotoxic, and this is mediated by its conjugation with glutathione.[85] It appears that the nephrotoxic metabolite is the mercapturic acid, *N*-acetyl-*S*-(pentachlorobutadienyl)-L-cysteine, which accumulates in the renal cortex and causes proximal tubular necrosis.[86] Both the liver and the kidney have the ability to form glutathione conjugates of halogenated alkenes, but it is noteworthy that the final steps of the conversion of S-substituted glutathiones to mercapturic acids, the cleavage of the cysteinylglycine bond, and the acetylation of S-substituted cysteines have long been known to occur principally in the kidney.[67] There is thus a close association between the site of formation of the toxic metabolite and the organ in which the toxicity is expressed.

Reactive intermediates exert their toxicity by virtue of their ability to interact covalently with tissue constituents. It thus seems reasonable to think of these metabolites as essentially "short range toxins", i.e., that they will initiate toxic reactions very close to their sites of formation. However, recent findings suggest that this is not always the case and that there can occur transport of reactive intermediates from their sites of formation to other organs to which they are toxic. The pyrrolidizine alkaloids are a group of compounds occurring in plants of *Senecio, Heliotropium,* and *Crotolaria* species, which can be present in both animal and human foods arising from contamination of food grains with the seeds of these weeds, or by the use of preparations of these plants in herbal teas and medicines, as food supplements, or as salad vegetables.[87] These alkaloids, typified by monocrotaline, are highly toxic to the liver and the lung, and this toxicity is mediated by reactive pyrrole metabolites produced by dehydration. Although the liver is able to convert monocrotaline to the toxic metabolite, dehydromonocrotaline, pulmonary xenobiotic metabolizing enzymes apparently lack this facility. The lung damage produced by these compounds is thus due to the toxic metabolite which is formed in the liver and transported in the circulation to the lung[88] (see Chapter 3 in Volume II).

One of the earliest examples where the hepatoxicity of a xenobiotic was shown to be due to the formation of a reactive intermediate able to bind covalently to hepatic macromolecules was bromobenzene.[89] The toxic intermediate is bromobenzene-3,4-oxide. In addition to its effects on the liver, bromobenzene is covalently bound and produces toxic changes in extrahepatic tissues, and the mechanism of this toxicity has been the subject of conjecture for many years.[90] Monks et al.[91] have shown that despite its presumed high reactivity, bromobenzene-3,4-oxide is able to escape from hepatocytes, the site of its formation. This provides a possible mechanism for the extrahepatic toxicity of bromobenzene in which the compound undergoes metabolic activation in the liver and is then transported to other tissues in which toxicity is expressed. In confirmation of this hypothesis, Lau et al.[92] have shown (1) the presence of bromobenzene-3,4-oxide in the systemic venous blood of rats given bromobenzene, and (2) that the half-life of this reactive intermediate, although short (approximately 13.5 sec), is longer than the mean circulation time in rats (5 to 10 sec). This indicates the availability of hepatically generated bromobenzene-3,4-oxide to extrahepatic tissues, and provides an explanation of the toxicity of bromobenzene to the lung and other tissues.

Finally, it is appropriate to mention the proposed mechanism for the carcinogenicity of 2-naphthylamine and related aromatic amines to the urinary bladder. These agents undergo a two-step metabolic activation in the liver and other tissues, as described earlier. In addition to the formation of *N-O*-conjugates, the *N*-hydroxy metabolites are also converted to *N*-glucuronides, with the *N*-hydroxy group free. These *N*-glucuronides are stable excretion products, and as such are transported to the urinary bladder. Under mildly acidic conditions (pH <6) and in the presence of β-glucuronidase, which is often present in the urine, the *N*-glucuronides break down to liberate the free *N*-hydroxy metabolites. Although not very reactive to intracellular macromolecules in the liver and other organs, at the pH of the bladder, these metabolites are reactive toward nucleic acids. It is proposed that the initiation of tumor formation is due to the liberation in the urinary bladder of these proximate carcinogens by the sequence outlined.[93]

VII. CONCLUDING REMARKS

This brief review has served to describe the major pathways of biotransformation of xenobiotics in the organism, particularly from the viewpoint of their ability to cause metabolic activation. It is seen that all of these pathways have the potential to cause such activation, although this is much more common with oxidative and reductive reactions than with the conjugations. In addition to this, reactive oxygen species may be produced during the metabolism of xenobiotics, and these have important toxicological and pathological consequences. Metabolic activation is often the result of more than one reaction acting in concert to produce a reactive intermediate.

Toxic responses are frequently the result of the formation of active or reactive metabolites, and their tissue specificity arises either from tissue specificity in their formation or tissue specificity in response to circulating reactive metabolites formed in other organs. Although examples of both situations can be found in the literature, the first case is by far the most commonly encountered. An understanding of the metabolic pathways of xenobiotics, of the consequences of these for the activity of the compound in question, and the tissue distribution of such metabolic activity are essential to understanding the phenomenon of organ-directed toxicity.

REFERENCES

1. **Williams, R. T.,** *Detoxication Mechanisms,* 2nd ed., Chapman & Hall, London, 1959.
2. **Miller, J. A., Miller, E. C., and Phillips, D. H.,** The metabolic activation and carcinogenicity of alkenylbenzenes that occur naturally in many spices, in *Carcinogens and Mutagens in the Environment,* Vol. 1, *Food Products,* Stich, H. F., Ed., CRC Press, Boca Raton, Fla., 1982, 83.

3. **Bonse, G., Urban, T., Reichert, D., and Henschler, D.,** Chemical reactivity, metabolic oxirane formation and biological reactivity of chlorinated ethylenes in the isolated perfused rat liver preparation, *Biochem. Pharmacol.*, 24, 1829, 1975.
4. **Jollow, D. J. and Smith, C. A.,** Biochemical aspects of toxic metabolites: formation, detoxication and covalent binding, in *Biological Reactive Intermediates*, Jollow, D. J., Kocsis, J. J., Snyder, R., and Vainio, H., Eds., Plenum Press, New York, 1977, 42.
5. **Sims, P. and Grover, P. L.,** Epoxides in polycyclic aromatic hydrocarbon metabolism and carcinogenesis, *Adv. Cancer Res.*, 20, 165, 1974.
6. **Swenson, D. H., Miller, E. C., and Miller, J. A.,** 2,3-Dihydro-2,3-dihydroxyaflatoxin B1: an acid hydrolysis product of an RNA-aflatoxin B1 adduct formed by hamster and rat liver microsomes in vitro, *Biochem. Biophys. Res. Commun.*, 60, 1036, 1974.
7. **Damani, L. A.,** Oxidation at nitrogen centers, in *Metabolic Basis of Detoxication*, Jakoby, W. B., Bend, J. R., and Caldwell, J., Eds., Academic Press, New York, 1982, 127.
8. **Cramer, J. W., Miller, J. A., and Miller, E. C.,** N-Hydroxylation: a new metabolic reaction observed in the rat with the carcinogen 2-acetylaminofluorene, *J. Biol. Chem.*, 235, 885, 1960.
9. **Zins, G. R., Emment, D. E., and Walk, R. A.,** Hypotensive metabolite production and sequential metabolism of N,N-diallylmelamine in rats, dogs and man, *J. Pharmacol. Exp. Ther.*, 159, 146, 1968.
10. **Boyd, M. R.,** Biochemical mechanisms in chemical-induced lung injury: roles of metabolic activation, *CRC Crit. Rev. Toxicol.*, 7, 103, 1980.
11. **Porter, W. R. and Neal, R. A.,** Metabolism of thioacetamide and thioacetamide S-oxide by rat liver microsomes, *Drug Metab. Disp.*, 6, 379, 1978.
12. **Schuphan, I., Rosen, J. D., and Casida, J. E.,** Novel activation mechanisms for the promutagenic herbicide Diallate, *Science*, 205, 1013, 1979.
13. **Neal, R. A. and Halpert, J.,** Toxicity of thiono-sulfur compounds, *Ann. Rev. Pharmacol. Toxicol.*, 22, 321, 1982.
14. **Kamitaki, T., Lee Lin, M. C., Belcher, D. H., and Neal, R. A.,** Studies of the metabolism of parathion with an apparently homogenous preparation of rabbit liver cytochrome P-450, *Drug Metab. Disp.*, 4, 180, 1976.
15. **Smith, J. N. and Williams, R. T.,** The metabolism of phenacetin (p-ethoxyacetanilide) in the rabbit and a further observation on acetanilide metabolism, *Biochem. J.*, 44, 239, 1949.
16. **Huang, M. T., Miwa, G. T., and Lu, A. Y. H.,** Rat liver cytosolic azoreductase, *J. Biol. Chem.*, 254, 3930, 1978.
17. **Sipes, I. G., Gandolfi, A. J., Pohl, L. R., Krishna, G., and Brown, B. R.,** Comparisons of the biotransformation and hepatotoxicity of halothane and deuterated halothane, *J. Pharmacol. Exp. Ther.*, 214, 716, 1980.
18. **Halpert, J., Naslund, B., and Betner, I.,** Suicide inactivation of rat liver cytochrome P-450 by chloramphenicol in vivo and in vitro, *Mol. Pharmacol.*, 23, 445, 1983.
19. **Trefouel, J., Trefouel, J., Nitti, F., and Bovet, D.,** Activite du p-aminophenylsulfonamide sur les infections streptococciques: experimentales de la souris et du lapin, *C.R. Seances Soc. Biol. Ses. Fil.*, 120, 756, 1935.
20. **Roberts, J. J. and Warwick, G. P.,** Reactions of tritium-labelled butter yellow in vivo and N-hydroxymethylaminoazobenzene in vitro, *Biochem. J.*, 93, 18P, 1964.
21. **Endo, H., Ono, T., and Sugimura, T.,** *Chemistry and Biological Actions of 4-Nitroquinoline N-Oxide*, Springer-Verlag, Berlin, 1971.
22. **Duggan, D. E., Hooke, K. F., Noll, R. M., Hucker, H. B., and Van Arman, C. G.,** Comparative disposition of sulindac and metabolites in five species, *Biochem. Pharmacol.*, 27, 2311, 1978.
23. **Wiersma, D., Schnellmann, R., and Sipes, I. G.,** Pathways of halogenated hydrocarbon metabolism, in *Foreign Compound Metabolism*, Caldwell, J. and Paulson, G. D., Eds., Taylor & Francis, London, 1984, 53.
24. **Jerina, D. M., Yagi, H., Thakker, D. R., Sayer, J. M., Van Bladeren, P., Lehr, R. E., Whalen, D. L., Levin, W., Chang, R. L., Wood, A. W., and Conney, A. H.,** Identification of the ultimate carcinogenic metabolites of the polycyclic aromatic hydrocarbons: bay region (R,S)-diol-(S,R)-epoxides, in *Foreign Compound Metabolism*, Caldwell, J. and Paulson, G. D., Eds., Taylor & Francis, London, 1984, 257.
25. **Devonshire, H. W., Kong, I., Cooper, M., Sloan, T. P., Idle, J. R., and Smith, R. L.,** The contribution of genetically determined oxidation status to inter-individual variations in phenacetin disposition, *Br. J. Clin. Pharmacol.*, 16, 157, 1983.
26. **Heymann, E.,** Hydrolysis of carboxylic esters and amides, in *Metabolic Basis of Detoxication*, Jakoby, W. B., Bend, J. R., and Caldwell, J., Eds., Academic Press, New York, 1982, 229.
27. **Putter, J.,** Extramicrosomal drug metabolism, *Eur. J. Drug Metab. Pharmacokinet.*, 4, 1, 1979.

28. **Wislocki, P. G., Bortchert, P., Miller, J. A., and Miller, E. C.,** The metabolic activation of the carcinogen 1′-hydroxysafrole in vivo and in vitro and the electrophilic reactivities of possible ultimate carcinogens, *Cancer Res.*, 36, 1686, 1976.
29. **Daly, J. W., Jerina, D. M., and Witkop, B.,** Arene oxides and the NIH shift: the metabolism, toxicity and carcinogenicity of aromatic compounds, *Experientia*, 28, 1129, 1972.
30. **Miller, J. A. and Miller, E. C.,** The concept of reactive electrophilic metabolites in chemical carcinogenesis: recent results with aromatic amines, safrole and aflatoxin B1, in *Biological Reactive Intermediates*, Jollow, D. J., Kocsis, J. J., Snyder, R., and Vainio, H., Eds., Plenum Press, New York, 1977, 6.
31. **Oesch, F.,** Mammalian epoxide hydrases: inducible enzymes catalysing the inactivation of carcinogenic and cytotoxic metabolites derived from aromatic and olefinic compounds, *Xenobiotica*, 3, 305, 1973.
32. **Frigerio, A., Fanelli, R., Biandrate, P., and Garattini, S.,** Mass spectrometric characterization of carbamazepine-10,11-epoxide, a carbamazepine metabolite isolated from human urine, *J. Pharm. Sci.*, 61, 1144, 1972.
33. **Phillips, D. H., Miller, J. A., Miller, E. C., and Adams, B.,** Structures of the DNA adducts formed in mouse liver after administration of the proximate hepatocarcinogen 1′-hydroxyestragole, *Cancer Res.*, 41, 176, 1981.
34. **Caldwell, J.,** The metabolism of amphetamines in mammals, *Drug Metab. Rev.*, 4, 219, 1976.
35. **Caldwell, J.,** Comparative aspects of the N-oxidation of aliphatic amines: studies with chlorphentermine, in *Biological Oxidation of Nitrogen*, Gorrod, J. W., Ed., Elsevier/North-Holland, Amsterdam, 1977, 57.
36. **Jenner, P. G.,** Synthetic and metabolic N-oxidation products as centrally active pharmacological agents, in *Biological Oxidation of Nitrogen*, Gorrod, J. W., Ed., Elsevier/North-Holland, Amsterdam, 1977, 383.
37. **Jakoby, W. B., Bend, J. R., and Caldwell, J., Eds.,** *Metabolic Basis of Detoxication*, Academic Press, New York, 1982.
38. **Lindeke, B. and Cho, A. K.,** N-Dealkylation and deamination, in *Metabolic Basis of Detoxication*, Jakoby, W. B., Bend, J. R., and Caldwell, J., Eds., Academic Press, New York, 1982, 105.
39. **Anders, M. W.,** Aliphatic halogenated hydrocarbons, in *Metabolic Basis of Detoxication*, Jakoby, W. B., Bend, J. R., and Caldwell, J., Eds., Academic Press, New York, 1982, 29.
40. **McDonald, T. L.,** Chemical mechanisms of halocarbon metabolism, *CRC Crit. Rev. Toxicol.*, 11, 85, 1983.
41. **Rickert, D. E.,** Reductive reactions, in *Foreign Compound Metabolism*, Caldwell, J. and Paulson, G. D., Eds., Taylor & Francis, London, 1984, 227.
42. **Oesch, F.,** Epoxide hydrolase, in *Enzymatic Basis of Detoxication*, Vol. 2, Jakoby, W. B., Ed., Academic Press, New York, 1980, 277.
43. **Sinkula, A. A. and Yalkowsky, S. H.,** Rationale for design of biologically reversible drug derivatives: prodrugs, *J. Pharm. Sci.*, 64, 181, 1975.
44. **Caldwell, J.,** Conjugation reactions in foreign compound metabolism: definition, consequences and species variations, *Drug Metab. Rev.*, 13, 745, 1982.
45. **Eadsforth, C. V. and Hutson, D. H.,** Formation of carbohydrate, sulphate and other xenobiotic conjugates, in *Foreign Compound Metabolism*, Caldwell, J. and Paulson, G. D., Eds., Taylor & Francis, London, 1984, 171.
46. **Caldwell, J.,** Conjugation reactions in the metabolism of xenobiotics, in *The Liver: Biology and Pathobiology*, Arias, I. M., Popper, H., Schachter, D., and Shafritz, D. A., Eds., Plenum Press, New York, 1982b, 281.
47. **Caldwell, J.,** Glucuronic acid conjugation in the context of the metabolic conjugation of xenobiotics, in *Advances in Glucuronide Conjugation*, Matern, S., Ed., MTP Press, Lancaster, England, 1985, 7.
48. **Caldwell, J.,** Conjugation reactions, in *Concepts in Drug Metabolism*, Part A, Jenner, P. and Testa, B., Eds., Marcel Dekker, New York, 1980, 211.
49. **Mori, M., Oguri, K., Yoshimura, H., Shimomura, K., Kamata, O., and Ueki, S.,** Chemical synthesis ana analgesic effect of morphine ethereal sulphates, *Life Sci.*, 11, 525, 1972.
50. **Irving, C. C.,** Conjugates of N-hydroxy compounds, in *Metabolic Conjugation and Metabolic Hydrolysis*, Vol. 1, Fishman, W. H., Ed., Academic Press, New York, 1970, 53.
51. **Dollery, C. T., Davies, D. S., and Connolly, M. E.,** Differences in the metabolism of drugs, depending upon the routes of administration, *Ann. N.Y. Acad. Sci.*, 179, 108, 1971.
52. **D'Souza, J., Caldwell, J., and Smith, R. L.,** Species differences in the N-methylation and quaternization of pyridine, *Xenobiotica*, 10, 151, 1980.
53. **Dutton, G. J.,** Phase II metabolic reactions in man, in *Drug Metabolism in Man*, Gorrod, J. W. and Beckett, A. H., Eds., Taylor & Francis, London, 1978, 81.
54. **Dreyer, D. E., Kluger, J., Reidenberg, M. M., and Lahita, R.,** Theoretical basis for interest in acetylprocainamide and clinical experience with this new antiarrhythmic agent, *Drug Metab. Rev.*, 10, 239, 1979.

55. **Flammang, T. J. and Kadlubar, F. F.,** Activation of N-hydroxy metabolites of aromatic amines by acetyltransferase(s), in *Microsomes and Drug Oxidations,* Boobis, A. R., Caldwell, J., de Matteis, F., and Elcombe, C. R., Taylor & Francis, London, 1985, 190.
56. **Al-Bassam, S. S. and Sherratt, H. S. A.,** The antagonism of the toxicity of hypoglycin by glycine, *Biochem. Pharmacol.,* 30, 2817, 1981.
57. **Van Bladeren, P. J., Breimer, D. D., Rotteveel-Smijs, G. M. T., de Jong, R., Buijs, W., van der Gen, A., and Mohr, G. R.,** The role of glutathione conjugation in the mutagenicity of 1,2-dibromoethane, *Biochem. Pharmacol.,* 29, 2975, 1980.
58. **Mulder, G.,** The role of sulfation and glucuronidation in toxification of xenobiotics, in *Foreign Compound Metabolism,* Caldwell, J. and Paulson, G. D., Eds., Taylor & Francis, London, 1984, 235.
59. **Bortchardt, R. T.,** N- and O-methylation, in *Enzymatic Basis of Detoxication,* Vol. 2, Jakoby, W. B., Ed., Academic Press, New York, 1980, 43.
60. **Caldwell, J.,** Conjugation reactions of nitrogen centers, in *Metabolic Basis of Detoxication,* Jakoby, W. B., Bend, J. R., and Caldwell, J., Eds., Academic Press, New York, 1982c, 292.
61. **Timbrell, J. A.,** The role of metabolism in the hepatotoxicity of isoniazid and iproniazid, *Drug Metab. Rev.,* 10, 125, 1979.
62. **Caldwell, J., Idle, J. R., and Smith, R. L.,** The amino acid conjugations, in *Extrahepatic Metabolism of Drugs and Other Foreign Compounds,* Gram, T. E., Ed., SP Medical & Scientific, Jamaica, N.Y., 1980, 435.
63. **Caldwell, J.,** Conjugation of xenobiotic acids, in *Metabolic Basis of Detoxication,* Jakoby, W. B., Bend, J. R., and Caldwell, J., Eds., Academic Press, New York, 1982, 271.
64. **Caldwell, J.,** Xenobiotic acyl-coenzymes A: critical intermediates in the biochemical pharmacology and toxicology of carboxylic acids, *Biochem. Soc. Trans.,* 12, 9, 1984.
65. **Sherratt, H. S. A.,** Hypoglycin and related hypoglycaemic compounds, *Br. Med. Bull.,* 25, 250, 1969.
66. **Billington, D., Osmundsen, H., and Sherratt, H. S. A.,** Mechanisms of the metabolic disturbances caused by hypoglycin and by pent-4-enoic acid. In vitro studies, *Biochem. Pharmacol.,* 27, 2879, 1978.
67. **Tate, S. S.,** Enzymes of mercapturic acid formation, in *Enzymatic Basis of Detoxication,* Vol. 2, Jakoby, W. B., Ed., Academic Press, New York, 1980, 95.
68. **Jerina, D. M. and Bend, J. R.,** Glutathione S-transferases, in *Biological Reactive Intermediates,* Jollow, D. J., Kocsis, J. J., Snyder, R., and Vainio, H., Eds., Plenum Press, New York, 1977, 207.
69. **Mitchell, J. R.,** Isoniazid liver injury: clinical spectrum, pathology and probable pathogenesis, *Ann. Intern. Med.,* 84, 181, 1976.
70. **Sies, H. and Cadenas, E.,** Biological basis of detoxication of oxygen free radicals, in *Biological Basis of Detoxication,* Caldwell, J. and Jakoby, W. B., Eds., Academic Press, New York, 1983, 181.
71. **Thor, H., Smith, M. T., Harzell, P., Bellomo, G., Jewell, S. A., and Orrenius, S.,** The metabolism of menadione (2-methyl-1,4-naphthoquinone) by isolated hepatocytes, *J. Biol. Chem.,* 257, 12419, 1982.
72. **Jefcoate, C. R.,** Integration of xenobiotic metabolism in carcinogen activation and detoxication, in *Biological Basis of Detoxication,* Caldwell, J. and Jakoby, W. B., Eds., Academic Press, New York, 1983, 31.
73. **Smith, L. L., Rose, M. S., and Wyatt, I.,** The pathology and biochemistry of paraquat, *Ciba Found. Symp.,* (New Series), 65, 321, 1979.
74. **Oesch, F.,** Role of various enzymes in the control of oxidative reactive metabolites, in *Foreign Compound Metabolism,* Caldwell, J. and Paulson, G. D., Eds., Taylor & Francis, London, 1984, 213.
75. **Nemoto, N. and Gelboin, H. V.,** Enzymatic conjugation of benzo(a)pyrene oxides, phenols and dihydrodiols with UDP-glucuronic acid, *Biochem. Pharmacol.,* 25, 1221, 1976.
76. **Orrenius, S., Thor, H., and Dimonte, D.,** Metabolic activation and inactivation — a critical balance in toxicity, in *Foreign Compound Metabolism,* Caldwell, J. and Paulson, G. D., Eds., Taylor & Francis, London, 1984, 245.
77. **Wold, J. S.,** Cephalosporin nephrotoxicity, in *Toxicology of the Kidney,* Hook, J. B., Ed., Raven Press, New York, 1981.
78. **Gram, T. E., Ed.,** *Extrahepatic Metabolism of Drugs and Other Foreign Compounds,* SP Medical & Scientific, Jamaica, N.Y., 1980.
79. **Wolf, C. R.,** Distribution and regulation of drug-metabolizing enzymes in mammalian tissues, in *Foreign Compound Metabolism,* Caldwell, J. and Paulson, G. D., Eds., Taylor & Francis, London, 1984, 37.
80. **Baron, J., Kawabata, T. T., Redick, J. A., Knapp, S. A., Wick, D. G., Wallace, R. B., Jakoby, W. B., and Guengerich, F. P.,** Localization of carcinogen-metabolizing enzymes in human and animal tissues, in *Extrahepatic Drug Metabolism and Chemical Carcinogenesis,* Rydstrom, J., Montelius, J., and Bengtsson, M., Eds., Elsevier, Amsterdam, 1983, 73.
81. **Baron, J., Kawabata, T. T., Knapp, S. A., Voigt, J. M., Redick, J. A., Jakoby, W. B., and Guengerich, F. P.,** Intrahepatic distribution of xenobiotic-metabolizing enzymes, in *Foreign Compound Metabolism,* Caldwell, J. and Paulson, G. D., Eds., Taylor & Francis, London, 1984, 17.

82. **Moldeus, P., Larsson, R., Ross, D., Andersson, B., Nordenskjold, M., Rahimtula, A., and Lindeke, B.,** Prostaglandin synthase catalyzed metabolism of p-phenetidine to reactive products, in *Extrahepatic Drug Metabolism and Chemical Carcinogenesis*, Rydstrom, J., Montelius, J., and Bengtsson, M., Eds., Elsevier, Amsterdam, 1983, 113.
83. **Moldeus, P., Larsson, R., Ross, D., and Andersson, B.,** Peroxidase-catalyzed metabolism of drugs and carcinogens, *Biochem. Soc. Trans.*, 12, 20, 1984.
84. **Moldeus, P., Ross, D., and Larsson, R.,** Co-oxidation of xenobiotics, *Biochem. Soc. Trans.*, 13, 849 (1985).
85. **Green, T., Nash, J. A., Odum, J., and Howard, E. F.,** The renal metabolism of a glutathione conjugate of the carcinogen hexachloro-1,3-butadiene: evidence for the formation of a mutagenic metabolite in the rat kidney, in *Extrahepatic Drug Metabolism and Chemical Carcinogenesis*, Rydstrom, J., Montelius, J., and Bengtsson, M., Eds., Elsevier, Amsterdam, 1983, 623.
86. **Lock, E. A.,** Effect of probenecid on N-acetyl-S-(pentachlorobutadienyl)-L-cysteine-induced nephrotoxicity, *Br. J. Pharmacol.*, in press.
87. **Culvenor, C. C. J.,** Toxicology of pyrrolizidine alkaloids, in *Toxicology in the Tropics*, Smith, R. L. and Bababunmi, E. A., Eds., Taylor & Francis, London, 1980, 124.
88. **Mattocks, A. R.,** Toxicity and metabolism of Senecio alkaloids, in *Phytochemical Ecology*, Harborne, J., Ed., Academic Press, New York, 1972, 179.
89. **Brodie, B. B., Reid, W. D., Cho, A. K., Sipes, I. G., Krishna, G., and Gillette, J. R.,** Possible mechanism of liver necrosis caused by aromatic organic compounds, *Proc. Natl. Acad. Sci. U.S.A.*, 68, 160, 1971.
90. **Reid, W. D., Christie, B., Eichelbaum, M., and Krishna, G.,** 3-Methylcholanthrene blocks hepatic necrosis induced by administration of bromobenzene or carbon tetrachloride, *Exp. Mol. Pathol.*, 15, 363, 1971.
91. **Monks, T. J., Lau, S. S., and Gillette, J. R.,** Diffusion of reactive metabolites out of hepatocytes: studies with bromobenzene, *J. Pharmacol. Exp. Ther.*, 228, 393, 1984.
92. **Lau, S. S., Monks, T. J., and Gillette, J. R.,** Detection and half-life of bromobenzene-3,4-oxide in blood, *Xenobiotica*, 14, 539, 1984.
93. **Kadlubar, F. F., Miller, J. A., and Miller, E. C.,** Hepatic microsomal N-glucuronidation and nucleic acid binding of N-hydroxy arylamines in relation to urinary bladder carcinogenesis, *Cancer Res.*, 37, 805, 1977.

Chapter 4

ROLE OF EXCRETION

Walter G. Levine

TABLE OF CONTENTS

I. INTRODUCTION

Through the processes of metabolism and excretion, the body is provided with a means of ridding itself of the many foreign chemicals (xenobiotics) to which it is exposed. Interruption of any of the major routes of excretion permits accumulation of the chemicals and/or their metabolites and appreciably increases their toxicity. Quantitatively, the kidney is the most important excretory organ, and in most species the vast majority of xenobiotics and their metabolites are excreted in the urine to some extent. The bile ranks next in quantitative importance, and in addition to an excretory pathway, provides a means of retaining drugs and other xenobiotics within the body by way of the enterohepatic circulation. Other routes of excretion include the expired air, sweat, saliva, feces, and milk. Although these are generally of less quantitative importance than are urine and bile, they become consequential for specific types of compounds and circumstances. The finding that a number of drugs appear in the expired air and in saliva has given rise to a means of assessing plasma levels of drugs in humans by simple noninvasive procedures. The secretion of potentially toxic chemicals into milk is of great concern at a time when breast feeding of infants is again finding increasing acceptance and when intentional and nonintentional exposure of the mother to a wide variety of xenobiotics is becoming commonplace. Limitation of space precludes extensive coverage of all routes and mechanisms of excretion. Biliary excretion will be a major focus of this chapter in view of its relation to hepatic metabolism and toxicity of numerous xenobiotics and the extent of physiological and pharmacological interest in this area. Renal mechanisms will be considered to a lesser extent while other routes are of necessity considered only briefly.

II. RENAL EXCRETION

Mechanisms of renal excretion of xenobiotics have been reviewed recently.[1,2] The reader is referred to a pharmacokinetic analysis of renal processes[3] and a review on transport by isolated renal plasma membranes.[4] The process of urine formation and xenobiotic excretion begins with the passage of blood through the glomerulus to yield a tubular fluid which is essentially an ultrafiltrate of plasma. Included in the filtrate are compounds with a mol wt of less than 5000, which would include most xenobiotics and their metabolites. However, it has been established that compounds with a mol wt higher than 500 to 600 are preferentially excreted in the bile.[5] This appears to be a function of hepatic rather than renal physiology. The driving force for filtration is the hydrostatic pressure of the blood and formation of tubular fluid is sensitive to changes in blood flow and pressure. In the average human approximately 180 ℓ of filtrate are produced daily, 99% of which is reabsorbed by the tubules. Glomerular filtration of xenobiotics not bound to plasma proteins, although a passive process, is nearly 100% efficient. Protein binding is a major limitation in glomerular filtration of xenobiotics since it diminishes the free drug concentration gradient across the capillary membrane. Consequently, drugs such as warfarin, which are extensively bound to plasma proteins, have long half-lives. Volume of distribution also influences the amount of xenobiotic which enters the glomerular filtrate since only compounds within the plasma compartment can be filtered. These factors can be appreciably altered in disease states, affecting the elimination of drugs and other chemicals. Glomerular filtration is commonly measured by clearance of creatinine, an endogenous amino acid derived from creatine phosphate of muscles. Although a small amount of tubular secretion of creatinine occurs, the method is suitable for obtaining a first approximation of glomerular filtration rate.

Xenobiotics also enter the tubular fluid by active processes which normally transport physiological organic anions and cations across tubule cells. Transport is relatively unhampered by plasma protein binding since the process is very rapid and the rate of dissociation from plasma protein binding occurs within milliseconds. A case in point is the organic anion, penicillin G, 65% of which is bound to plasma proteins. Its plasma half-life is only 30 min due to active tubular transport. The half-life of penicillin G can be significantly extended by administration of another organic anion, probenecid, which successfully competes for the tubular anion transport mechanism. Excretion of methotrexate, an organic anion, is inhibited by salicylates which compete for tubular transport. This important drug interaction may lead to toxic blood levels of the antitumor agent. Inhibitors of active transport, such as 2,4-dinitrophenol and ouabain, decrease accumulation of organic anions in tubular cells. According to current theory, organic anions are actively transported into tubular cells but then move into the lumen down an electrochemical gradient by either passive or facilitated diffusion. The renal organic anion transport system has been thoroughly reviewed recently;[6] there is documentation for transport of over 200 compounds.

An analogous tubular system is responsible for the transport of organic cations,[7] mainly primary, secondary, tertiary, and quaternary amines carrying a positive charge within the physiological range of pH. Although not as extensively studied as the anionic system, organic cationic transport appears to follow a similar mechanism. Competitive inhibition by transported cations has been demonstrated, although saturation of the putative carrier has been difficult to demonstrate in view of the toxicity of high concentrations of most organic cations.

Although hepatic enzymes catalyze the metabolism of most xenobiotics, the metabolism of some organic cations by tubular cells adds an interesting aspect to their excretion. For example, morphine is taken up by the cells in cationic form, but conjugated to sulfate and excreted as the anion. The anion inhibitor, probenecid, does not block renal excretion of basic drugs that are sulfated by the kidney, but does block excretion of the preformed sulfate, supporting the concept of active transport of organic anions only in the uptake phase of tubular transport. The cationic and anionic systems are generally considered to be independent of each other, although exceptions may exist. Substances transported by the organic cation system include acetylcholine, creatinine, dopamine, epinephrine, 5-hydroxytryptamine, histamine, atropine, hexamethonium, morphine, neostigmine, paraquat, quinine, and tetraethylammonium.

Both anionic and cationic systems are deficient in the neonate. This may be associated with the efficiency of the mother's kidney and, consequently, the lack of exposure of the fetal kidney to organic anions and cations. However, in view of the transplacental transport of many xenobiotics, their presence in the maternal circulation is considered a potential risk to the fetus. Maturation of these systems begins shortly after birth as the kidney in the neonate assumes responsibility for dealing with xenobiotics.

Xenobiotics within the tubular fluid may be reabsorbed by passive transport through tubular membranes back into the blood, mainly in proximal tubules. The degree of reabsorption is dependent on urine flow, is proportional to the lipid-solubility of the drug, and is influenced by the pH difference between tubular and peritubular fluid and by binding of reabsorbed drug to plasma protein. Highly lipid-soluble compounds are readily reabsorbed and there is little or no urinary excretion despite nearly complete glomerular filtration. For drugs with a suitable pK, the net rate of excretion can be increased or decreased by manipulation of urinary pH. Phenobarbital, a weak organic anion, is slowly excreted unchanged in the urine. Alkalinizing the urine increases the charge on the molecule, resulting in decreased tubular reabsorption and increased total excretion. Diuretics can alter passive reabsorption of other drugs by increasing urinary flow, thus decreasing exposure time to luminal membranes, and by changing urinary pH.

Most xenobiotic metabolism is considered to be conducive to urinary excretion in view of the conversion of relatively lipophilic compounds to polar, charged metabolites. This reduces passive tubular reabsorption and enhances the possibility of tubular active transport into the urine. The relation between metabolism and urinary excretion has been reviewed

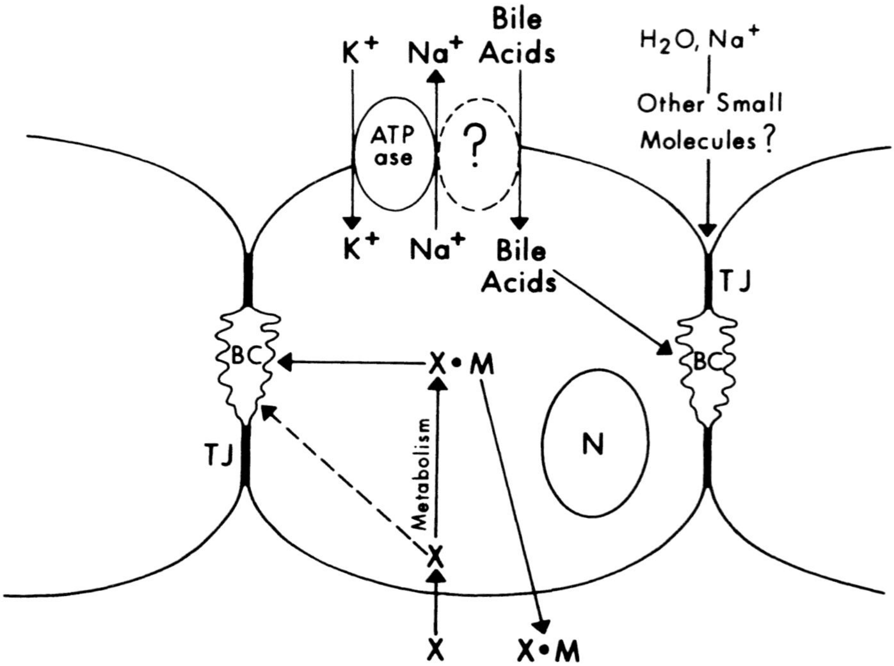

FIGURE 1. Representation of hepatobiliary function. The scheme reflects the dependence of bile acid uptake on Na^+ transport although the actual mechanism of coupling is not as well established as that for Na^+K^+ATPase control of Na^+ and K^+ transport. Movement of H_2O, Na^+, and possibly other small molecules through tight junctions (TJ) is currently accepted. Xenobiotics (X) taken up by the liver are in some cases transported directly to the bile canaliculus (BC), although more commonly, metabolism intervenes and the metabolite (X·M) is then transported to the bile canaliculus. Alternatively, the metabolite may return to the plasma and eventually be excreted in the urine.

by Pritchard and James.[1] The initial metabolism of most xenobiotics is catalyzed by the cytochromes P-450 (mixed function oxidase) system within the hepatic endoplasmic reticulum. Although the resulting metabolites are generally more polar than the parent compounds, conjugation, the second metabolic step, makes the major contribution to increased polarity and charge. In particular, glucuronide and sulfate conjugates are strong anions and many drugs appear in the urine in this form.

III. BILIARY EXCRETION OF XENOBIOTICS

Although humans excrete fewer drugs in the bile than in the urine, it is an important route for many compounds and its function in the enterohepatic circulation is responsible for the persistence of many xenobiotics within the body. Lower animals excrete drugs in the bile to a greater extent, and compounds with mol wt as low as 200 are found in rat bile. Our understanding of biliary excretion is somewhat less than that of urinary excretion since our knowledge of biliary physiology is relatively sketchy. Several recent articles[8-14] review current concepts of formation and flow of bile which will be considered briefly here.

Some of the aspects of hepatobiliary function discussed in this chapter are represented schematically in Figure 1.

A. Physiology of Bile Formation

1. Structural Considerations

The lobule is the functional unit of the liver and is defined by its microcirculation. Blood flows from the portal vein and hepatic artery centrally through sinusoids to the terminal hepatic vein. (Alternatively, the liver acinis may be considered the functional unit with blood flow proceeding from zone 1 through sinusoids toward the terminal hepatic vein in zone 3.) Consequently, the microenvironment of the peripheral (periportal) hepatocytes contains higher concentrations of blood solutes than that of centrilobular hepatocytes. This is thought to contribute to the morphologic and functional heterogeneity of hepatocytes within the lobule. However, anastomotic networks of sinusoids and the branches of the hepatic artery emptying intermittantly along sinusoids suggest that the rigid concept of portal to central gradient within the lobule should be more flexible. Intralobular heterogeneity has been dealt with in a number of recent publications.[15-17] The canaliculus, which is morphologically and functionally distinct from the basolateral area of the hepatocyte membrane, is characterized by microvilli, indicative of an extensive transport system. By virtue of the tight junction between hepatocytes, it had been assumed that essentially all blood to bile transport is transcellular. However, there is evidence that paracellular movement of water[18] and Na[19] may occur. In addition, inulin and sucrose, which equilibrate slowly with cellular water, equilibrate quite rapidly with bile, suggesting paracellular transport. It is likely that uncharged solutes may undergo paracellular transport more readily than do anionic solutes.[20]

Accumulated evidence is consistent with a role for the Golgi complex in biliary function.[13] This includes observations of enlarged Golgi during stimulation of bile secretion, its pericanalicular location and possible localization of solutes within the Golgi prior to appearance in the bile. Chronic treatment with phenobarbital leads to choleresis and a profound proliferation of the smooth endoplasmic reticulum. However, these membranes do not appear to participate in biliary excretion other than through metabolism of xenobiotics. Accumulation within the lysosomes is seen for some basic drugs such as d-tubocurarine, but the bound drug appears not to be readily available for biliary excretion.[21] Acid hydrolases appear in the bile,[22] probably derived from lysosomal vesicles, although the significance of this organelle in biliary excretion is not clear. Electron microscopy reveals a system of microfilaments in the pericanalicular area of the hepatocyte[23] closely associated with microvilli membranes. These may influence biliary transport by contributing to the contraction and relaxation of the canalicular microvilli and to intracytoplasmic migration of vesicles in the pericanalicular region. This conclusion is supported by the observation that phalloidin, which inhibits depolymerization of actin into microfilaments, and norethandrolone, which induces alterations in pericanalicular microfilaments, will each suppress bile flow.[24]

Biochemical mechanisms in bile formation have received relatively little attention. Cellular levels of cAMP may influence bile formation since theophylline, an inhibitor of phosphodiesterase and glucagon, which stimulates adenyl cyclase, in turn, stimulate bile salt-independent bile flow in dogs.[25] On the other hand, there is no parallel between bile flow and cellular levels of cAMP in dogs and rats, casting doubt on a causal relationship.[26] Similarly, no increased biliary cAMP is seen during taurocholate-induced choleresis in the dog.[27] Although uptake of bile acids is associated with the activity of Na^+K^+ATPase in the plasma membrane (see below), the total oxygen requirement for this function, even at maximal rates of uptake, is insignificant relative to that for total hepatic Na^+K^+ATPase activity.[28] This contrasts with other tissues where epithelial transport may account for nearly all of the Na^+K^+ATPase activity.

The contribution of the ductule system to bile formation is poorly defined. In species which eat intermittently (dog, man), ductular mechanisms for modifying bile flow appear to be of greater importance relative to those in continuously eating species (rats, rabbits, guinea pigs).[29] In species other than rabbit and rat, secretin is a potent chloretic and increases bicarbonate and chloride excretion, probably at the ductule level.[30] However, recent work suggests a canalicular origin as well for bicarbonate.[31] Reabsorption of water probably occurs in the ductule as demonstrated directly by retrograde instillation of ^3H_2O into the rat biliary tree.[32] Several xenobiotics are also absorbed by the ductules, indicating bidirectional exchange of solutes within the biliary tree. Ductular reabsorption of electrolytes and glucose has been observed. Thus, the canaliculus-ductule system appears to be functionally analogous to the nephron in that primary bile is formed at the proximal end while compositional change occurs as the bile flows through the ductular system. Molecules as large as inulin and albumin are apparently reabsorbed through the biliary tree,[33] giving rise to speculation that the protein normally found in the bile[34]. may also undergo bidirectional flow within the biliary tree; however, direct evidence is lacking. Even a highly charged compound such as phenolphthalein glucuronide appears in the plasma after retrograde injection into the bile duct.[35] Thus, permeability to macromolecules and to highly charged metabolites of xenobiotics has been demonstrated, although it has yet to be proven whether the canaliculi or the ductules, or both, are involved.

2. Bile Salt-Dependent and -Independent Bile Formation

Bile secretion, unlike that of the urine, is relatively independent of hydrostatic pressure but is dependent upon facilitated and active transport mechanisms. A primary step in bile formation is the transport of bile acids across the canalicular membrane resulting in an osmotic movement of water into the canalicular lumen.[36] The relationship between bile flow and the secretion of bile salts is referred to as *bile salt-dependent flow*. Numerous other organic anions such as sulfobromophthalein (BSP) and iodipamide exert a similar osmotic effect. Bile salts are normally present as micelles within the bile, and depending upon their concentration, their osmotic activity is consequently somewhat lower than would be predicted on a molar basis. Bile salts which do not form micelles cause a proportionally greater increase in bile flow than do those which form micelles. Counter ion transport with bile salts may make a major contribution to the osmotic gradient. Nonosmotic factors may also play a role in bile-salt dependent flow.

From analysis of bile flow at varying levels of bile salt infusion, it has been calculated that bile flow can occur in the absence of bile salts, i.e., the *bile salt-independent flow*. This has been recently reviewed.[12] Quantitation of this flow is difficult since it assumes a linear relationship between bile salt secretion and bile flow, even at low concentration of bile salts, i.e., below that critical for micelle formation. However, this relationship may be curvilinear at low concentrations of bile salt, casting doubt on the conventional procedure for determination of bile salt-independent flow.[37] Furthermore, nearly normal bile flow is maintained in the rat in the virtual absence of bile salts after prolonged bile drainage and in the isolated perfused rat liver. The theory of bile salt-independent flow assumes that the major driving force is the active transport of sodium associated with Na^+K^+ATPase activity. Induction by phenobarbital of synthesis of Na^+K^+ATPase and of bile salt-independent flow are both blocked by inhibitors of protein synthesis.[38] However, ouabain, and ethacrynic acid, each of which inhibit Na^+K^+ATPase, and therefore should suppress bile salt-independent flow, actually increase flow under certain circumstances,[39] due possibly to the osmotic effect of their own excretion. Recently, it has been shown that Na^+K^+ATPase exists exclusively in the basalateral membrane and not in the canalicular membranes,[9] casting doubt on this mechanism in canalicular bile formation. However, the membrane potential across the canaliculus consequent to the sodium gradient may facilitate transport processes. Present evidence indicates that the canalicular lumen is electrically negative with respect to the sinusoids[8] and may account for some of the cation concentration gradients observed. These mechanisms, however, are insufficient to account for all of the bile salt-independent flow. Paracellular transport of sodium and probably other electrolytes and water has been accepted.[8,12] Consideration of the specific dependence on sodium in bile formation is further complicated by the observation that bile secretion in the perfused rat liver continues unabated when lithium is substituted for sodium in the perfusion medium.[12,40] It has even been proposed that secretion of endogenous organic anions is the major contributor to bile salt-independent flow, whereas inorganic ion transport results mainly from solvent drag.[12] Such organic anions have not been identified but may warrant investigation in view of the uncertainty of the role of sodium and other inorganic ions on bile salt-independent flow.

The question has been raised as to whether there actually is a meaningful distinction between bile salt-independent and -dependent flow.[11] Bile salt-dependent flow relies on Na^+K^+ATPase creating a Na gradient across the cell membrane upon which bile salt uptake depends. The negative intracellular potential produced by this gradient favors the passive transport of bile salts across the canalicular membrane. In this manner, there is a coupling between ATP energy and bile salt-dependent flow. Bile acid-independent flow may have a similar dependence on Na^+K^+ATPase, although as indicated above, the mechanism is unclear. Thus, there is a continuum between the two concepts of bile formation and rigid adherence to separate mechanisms is no longer tenable. In addition, all future theories on bile formation must be consistent with Na^+K^+ATPase localization entirely on the sinusoidal basolateral aspect of the hepatocyte cell membrane.

B. Characteristics of Compounds Excreted in the Bile

A previous review listed over 200 xenobiotics or their metabolites which had been detected in the bile.[41] Table 1 lists compounds reported in the bile only during the past 3 years. This list will certainly continue to grow in view of the importance of this route of excretion and the ever-increasing number of xenobiotics tested for toxicity and developed for therapeutic application. These findings do not necessarily imply that the bile is the sole, or even the main route of excretion for these compounds, only that they are excreted in the bile in detectable quantities. Biliary compounds include nearly every class of pharmacologic agent as well as numerous environmental, carcinogenic, and otherwise toxic chemicals. Many metals, toxic as well as physiological, are excreted in the bile. Excretion has been studied in rodents and other common laboratory animals, fish, several domesticated and wild animals, and humans, but there is considerable variation among species.

The original classification of biliary solutes by Brauer[95] according to their bile/plasma concentration ratios is not as significant for xenobiotics as it once was for physiological substances. The finding of a low ratio for a particular substance reflects a low rate of hepatic transport but does not imply a lack of significance for the biliary route. For most excreted xenobiotics, distinction is made between organic anions, cations, and noncharged compounds. Each group is considered to be transported into bile by an independent mechanism although a number of subgroups and exceptions to the theory are seen. Bile salts are transported by mechanisms independent of that for other organic anions. However, even two bile acids, dehydrocholate and taurocholate, may not entirely share the same pathway.[96] Biliary excretion of organic cations, although proceeding by an apparently independent mechanism, also exhibits multiplicity of pathways, as exemplified by those for the mono- and diquaternary amines, procainamide ethobromide and d-tubocurarine.[97,98] Another intragroup distinction is seen in the biliary excretion of the anions, morphine sulfate, and morphine glucuronide; phenobarbital treatment enhances excretion of the former but not the latter.[99] In the rat, BSP may be transported by two pathways, only one of which is shared with bilirubin.

After treatment with nafenopin, the isolated perfused rat liver takes up BSP rather poorly, whereas bilirubin transport is unaffected.[100] The transport mechanism for uncharged compounds is typically studied using ouabain, a cardiac glycoside. It has a steroid-like structure and a number of steroid hormones inhibit its uptake into liver slices as well as its secretion in the bile. A distinct pathway for noncharged compounds has been questioned in view of findings that the organic anions, BSP and phenolphthalein, inhibit the uptake of ouabain into liver slices and into the bile.[101] It is apparent that the question of independent hepatic

Table 1
XENOBIOTICS EXCRETED IN BILE

	Ref.
Antibiotics and other antibacterial compounds	
Piperacillin	42
Moxalactam	43
Cefoxitin	44
Cefmetazole	45
Azlocillin	46
Amoxycillin	47
Ceftriaxone	47
Ureidopenicillin	48
Kanamycin	49
Amikacin	49
Cefoperazone	50
Cefuroxime	51
Cephtizoxime	52
Cardiovascular drugs	
Gitoxin	53
Bretylium tosylate	54
Mecinarone	55
Butaprozine	56
Heptaminol	57
Anti-inflammatory drugs	
2-Acetamido-4-(chloromethyl)thiazole	58
Sulindac	59
Etodolac	60
Fentiazac	61
Drugs affecting nervous system	
Bromocryptine	62
Tripamide	63
Temazepam	64
Methamphetamine	65
Atracurium	66
Vecuronium	67
Pancuronium	67
Haloperidol	68
Buprenorphine	69
(−)-3-Phenoxy-*N*-methylmorphinan	70
Naltrexone	71
Valproic acid	72
Ketamine	73
Antineoplastic agents	
Adriamycin	74,75
Cisplatin	76
1-(2-Chloroethyl)-3-(2′,3′,4′-tri-*o*-acetyl. ribopyranosyl)-1-nitrosourea	77
Ethidium bromide	78
5-Methyltetrahydrohomofolate	79
Vincamycin	80
Doxorubicin	81
4-Epi-doxorubicin	81
9-Hydroxyellipticine	82
2-Methyl-9-hydroxyellipticine	82

Table 1 (continued)
XENOBIOTICS EXCRETED IN BILE

	Ref.
Melphalan	83
Miscellaneous compounds	
8-Methoxypsoralen	84
Aminoglutethimide	85
Sulfinpyrazone	86
Levamisole	87
Epomediol	88
Captopril	89
Tiflorex	90
Azelastine	91
Clofibride	92
Cimetidine	93
Thiazinamium	94

mechanisms for transport of anions, cations, and noncharged compounds has yet to be resolved.

A mol wt threshold exists for biliary organic anions which varies considerably among species. The minimum mol wt is 325, 400, and 475, and 500 to 700, respectively, for rats, guinea pigs, rabbits, and humans.[102] These figures are only approximate and there is considerable latitude within each species. In general, compounds below these thresholds are excreted in the urine. For organic cations, the threshold is 200 in the rat and appears to exhibit little or no species variation.[103] The mechanism for such distinction based on molecular weight remains unknown.

C. Hepatic Uptake vs. Biliary Transport

Biliary excretion has frequently been studied from the viewpoint of movement from plasma to bile, overlooking the distinction between hepatic uptake and canalicular transport. Plasma disappearance rates are often equated with hepatic uptake, neglecting extrahepatic distribution. The latter problem can be overcome using the isolated perfused liver. Recent studies on xenobiotic uptake involve the use of isolated hepatocytes.[104,105] In many cases, the data obtained compare well with those from perfused liver. However, the structural and functional polarity of the hepatocyte in vivo is lost during isolation. In suspension, both sinusoidal and canalicular faces are exposed to the medium. Therefore, it is impossible to determine if release of a compound into the medium reflects efflux into the plasma or secretion into the canaliculus in vivo. Consequently, uptake studies must be carried out over a short period to minimize release. Cell to medium ratios of more than one are often obtained, implying active transport. Uptake of bile acids (taurocholate) occurs against an apparent concentration gradient. However, the true gradient, i.e., for unbound compound, is a matter for conjecture since binding to plasma and intracellular proteins is not accurately known. Bile acid uptake depends on the activity of Na^+K^+ ATPase and a transmembranal Na^+ gradient while transport of BSP, indocyanine green, and ouabain does not. This suggests different mechanisms and confirms findings in vivo and in the perfused liver which indicate that bile salts are transported up by mechanisms differing from those for most xenobiotics. Some investigators have found primary cultures of rat hepatocytes to be of considerable value in studying biliary function. For example, bile acid secretion by cultured hepatocytes was shown to be regulated by cholesterol availability.[106] There are report of reconstitution of functional canaliculi during the culture period.[107] This system, if judiciously used, may overcome some of the stated disadvantages of hepatocyte suspensions and may provide a means of distinguishing sinusoidal uptake from canalicular secretion.

To distinguish between hepatic uptake and canalicular transport, use was made of probenecid, an inhibitor of renal transport of organic anions. It was found to depress hepatic uptake of ouabain in vivo and in liver slices,[108] but not biliary excretion. Nafenopin, on the other hand, inhibits the biliary excretion of bilirubin in vivo,[109] but not its hepatic uptake in the isolated perfused liver.[100] Bucolome, an anti-inflammatory drug, enhances the biliary excretion of ouabain but inhibits its plasma disappearance,[110] although clearance of BSP and bilirubin is enhanced. The selective response to these drugs undoubtedly reflects multiple transport mechanisms within the hepatocyte plasma membrane, suggesting a system far more complex than just a simple distinction between cations, anions, and noncharged molecules. The independence or interdependence of these systems requires additional study.

D. Intracellular Binding

For bile salts, bilirubin and most xenobiotics, uptake into the liver is followed by intracellular binding prior to metabolism and excretion. A major cytosolic binding protein is ligandin.[111,112] It normally comprises about 5% of hepatic cytosolic protein and may increase to 10 to 15% after treatment with phenobarbital. It binds bilirubin, glutathione, and a number of xenobiotics and is identical with an azo dye binding protein and a cortisol metabolite binding protein. Rat liver ligandin is identical to glutathione S-transferase B, one of several such transferases found within the cytosol. Indeed, all of the glutathione transferases appear to have similar binding activity. Treatment of rats with phenobarbital, DDT, dieldrin, or pregnenolone-16-α-carbonitrile, inducers of the hepatic mixed-function oxidase system, increases hepatic but not renal levels of ligandin. Treatment with TCDD (tetrachlorodibenzo-*p*-dioxin) induces both hepatic and renal ligandin.[113] Phenobarbital also enhances uptake of bilirubin and BSP, ostensibly through decreased efflux.[114] However, such direct evidence for a role of ligandin in hepatic uptake is not available for most other xenobiotics. Elasmobranchs (sharks, skates) are virtually devoid of hepatic ligandin, although these species are able to concentrate and excrete organic anions.[115] This suggests that alternative mechanisms for hepatic transport are operative in elasmobranchs.

A second cytosolic binding protein, Z (amino azo dye binding protein A), also binds a diverse group of xenobiotics.[116] There is little evidence as to its role in hepatic transport. Treatment with nafenopin markedly increases hepatic concentration of Z[117] and leads to decreased uptake of BSP in the perfused liver while bilirubin uptake is unaffected.[100] Z protein also binds fatty acids,[118] possibly reflecting one aspect of the hypolipidemic response to nafenopin. Other cytosolic proteins bind covalently a large number of highly reactive electrophilic compounds nonenzymically. This group may simply represent a general pool of proteins with accessible nucleophilic sites and no function in hepatobiliary transport has been ascribed them.

The bile contains over 12 proteins,[34,119] but their role, if any, in the binding and transport of xenobiotics has received little attention. Indocyanine green is bound approximately 50% to biliary proteins in guinea pigs,[120] but the contribution to dye excretion is unknown. Differential binding of metabolites of the carcinogenic azo dye, *N,N*-dimethyl-4-aminoazobenzene, to fractions of biliary proteins has been demonstrated, as well as the covalent binding of one or more DAB metabolites to a specific biliary protein.[121] The specificity of binding suggests a possible function for certain biliary proteins in the biliary excretion of azo dye metabolites.

E. Metabolism

Most xenobiotics are converted to metabolites prior to excretion in bile or urine. The enzymatic and chemical aspects of biotransformation have been reviewed elsewhere.[122,123] Briefly, metabolism generally may be divided into two major steps. In the first, the mixed-function oxidase (cytochrome P-450) system catalyzes hydroxylation, dealkylation, sulfoxidation, aromatization, or nitro and azo reduction reactions. Following this are several synthetic reactions which form a number of conjugates, glucuronides, sulfates, or glutathione-S ethers among others. The appropriate enzymes are found in the microsomal and cytosolic fractions of the liver. The second group of reactions greatly increase the polarity and molecular size of the compound, thus facilitating biliary excretion (Chapter 3).

The importance of biotransformation in biliary excretion is well illustrated by the Gunn rat which remains jaundiced throughout its lifetime due to the absence of bilirubin-UDP-glucuronyl transferase. Induction of drug metabolizing enzymes frequently leads to increased biliary excretion of xenobiotics. This has been demonstrated for a number of inducing agents and xenobiotics,[41] although the significance of such studies is sometimes difficult to assess. Phenobarbital induces many drug metabolizing enzymes,[124] but also increases liver size, hepatic blood flow,[125,126] and bile flow[127,128] unrelated to effects on cytochrome P-450. Since clearance of compounds with high hepatic extraction is dependent on blood flow, enhanced biliary excretion after phenobarbital might in some cases be attributable mainly to increased hepatic blood flow. A study with six barbiturates with varying properties showed that most stimulated biliary excretion of organic anions independent of their effect on microsomal enzyme induction.[129] Paradoxically, phenobarbital and *trans*-stilbene oxide induce glucuronidation of morphine in vitro but depress excretion of the conjugate in vivo,[130,131] possibly by inhibition of biliary transport. Other inducing agents, spironolactone and pregnenolone-16-α-carbonitrile, also increase bile flow.[132] On the other hand, polycyclic hydrocarbons do not increase liver weight or bile flow in the short period of time required for significant enzyme induction, usually less than 24 hr.

In the absence of enhanced bile flow or increased liver size, an increase in the rate of excretion of a metabolized xenobiotic after pretreatment with these agents leads to the conclusion that metabolism is the rate-limiting step in the hepatobiliary fate of that particular xenobiotic. This is seen for methadone[133] and a number of chemical carcinogens.[134-138] For example, benzo[*a*]pyrene, which appears in the bile entirely in the form of its metabolites, is excreted more rapidly after induction with 3-methylcholanthrene and with phenobarbital even when increased liver size and bile flow are taken into account.[134] Conversely, inhibitors of the oxidase system depress the rate of biliary excretion of benzo[*a*]pyrene metabolites. Sex differences for benzo[*a*]pyrene metabolism in rats are also reflected in rates of biliary excretion. Similar observations have been made with the carcinogens, 3-methylcholanthrene, 7,12-dimethylbenzanthracene, 2-acetylaminofluorene, and *N,N*-dimethyl-4-aminoazobenzene.[134-138] On the other hand, the rate of biliary excretion of phenolphthalein, 3-methylumbelliferone, and 8-hydroxyquinoline, which appear in the bile as glucuronides, is not limited by metabolism.[139] In general, conjugation reactions are carried out far more rapidly than are the initial oxidative steps. Therefore, it is less likely that compounds requiring only conjugation would be limited by metabolism in their hepatobiliary excretion. Generalizations concerning rate-limiting steps for xenobiotic excretion must take into account whether initial or conjugation phases (or both) are involved as well as the nonmetabolic effects of inducing agents. A case in point is the effect of nafenopin, which inhibits biliary transport but not metabolism of xenobiotics. Oxidative metabolism is rate-limiting in the excretion of benzo[*a*]pyrene; nafenopin, predictably, does not alter the rate of biliary excretion. However, if metabolism is markedly induced with 3-methylcholanthrene or is entirely bypassed through injection of preformed metabolites, canalicular transport becomes rate-limiting and inhibition by nafenopin is seen.[140]

It follows from the above that glutathione conjugation, for example with BSP, would not be rate limiting in excretion of the dye. However, depletion of hepatic glutathione by diethylmaleate or methyliodide depresses dye excretion[141,142] attributable to loss of substrate (glutathione) for the glutathione S-transferase. *N,N*-Dimethyl-4-aminoazobenzene undergoes *N*-demethylation and ring hydroxylation and is excreted in the bile as a mixture of metabolites, including conjugates of glutathione, glucuronic acid, and sulfate.[143,144] Depletion of hepatic glutathione specifically depresses excretion of *N*-demethylated metabolites while ring hydroxylated metabolites are relatively unaffected.[145] Some of the inhibition of excretion is probably attributable to depressed formation of glutathione conjugate although the total mechanism of this interaction has yet to be elucidated. A direct toxic effect of depleting agents has been ruled out since the effect is readily reversed upon restoration of hepatic glutathione and very high concentrations of these agents are required to inhibit *N*-demethylation in vitro. A complicating factor is the fact that diethylmaleate, frequently used to deplete hepatic glutathione, causes morphological changes in the Golgi complex and an increase in bile flow shortly after administration.[146] At present, it is not known whether the Golgi participates directly in the biliary excretion of xenobiotics even though indirect evidence suggests a role in bile salt excretion.[13] Specific inhibition of cytochrome P-450 may be a consequence of glutathione depletion, possibly mediated through heme oxygenase and lipid peroxidation. Both are increased after glutathione depletion[147-149] and each could lead to loss of cytochrome P-450 activity. In particular, the cytochrome is sensitive to effects of endogenously formed, as well as to exogenous lipid peroxides. A dependence upon glutathione has been recently demonstrated for both the *N*-demethylation and 4′-hydroxylation of *N,N*-dimethyl-4-aminoazobenzene in vitro.[150] The effect is not associated with inhibition of lipid peroxidation. Clearly, the relationship of glutathione to the metabolism and biliary excretion of this azo dye is multifaceted and will only be resolved by future investigation.

F. Effects of Other Xenobiotics

Many xenobiotics directly or indirectly influence hepatobiliary excretion. Some of these compete for transport based on similarity of charge. Others induce or inhibit drug-metabolizing enzymes. Still others are hepatotrophic and increase the functional size of the liver, and some are choleretic. A xenobiotic may affect its own excretion. BSP is commonly used to assess hepatobiliary function but is toxic in high doses. During infusion for measurement of transport maximum, decreased bile flow and dye excretion have been seen.[151,152] Liver ATP falls precipitously,[153] possibly through an action of the dye on the mitochondrial membrane.[154] The glutathione conjugate of BSP does not show similar toxicity. In view of these effects, the use of BSP to determine transport capacity of the liver is open to question.[152]

Administration of nafenopin increases liver size and evokes a marked choleresis accompanied by inhibition of biliary excretion of organic anions and uncharged xenobiotics, but not organic cations.[109,155,156] The effect is seen in vivo and in the isolated perfused liver. Unlike most hepatotrophic chemicals, nafenopin does not induce hepatic drug-metabolizing enzymes[109] and there is no effect on excretion where metabolism is rate limiting. In contrast to other organic anions, chlorothiazide[157] and harmol sulfate[158] are excreted more rapidly after nafenopin pretreatment, probably in association with increased bile flow. This provides further evidence for the multiplicity of biliary anion transport mechanisms discussed above. Recent studies show that nafenopin blocks both the uptake[100] and the biliary excretion[109] of BSP; bilirubin uptake is not affected although its excretion is depressed. Since nafenopin inhibits transport of only highly charged compounds, there is no influence on the movement of unchanged bilirubin, whereas excretion of its highly charged glucuronide is suppressed. The choleretic effect of nafenopin is apparently bile salt-independent but otherwise not entirely explained. The increased flow is of canalicular origin and requires pretreatment with the drug as well as the presence of the drug and/or its metabolites in the bile.[156] Thus, a change in the functional state of the liver as well as a possible osmotic effect may contribute to the choleretic response.

Many other xenobiotic and physiological compounds influence hepatobiliary function[41] mainly, although not entirely, in a detrimental manner. These include carbon tetrachloride, the pesticide, kepone, the toxic chemicals, 2,3,7,8-tetrachlorodibenzo-*p*-dioxin (TCDD) and PCBs, organic forms of certain metals, and others. Commonly used drugs, such as diazepam and acetaminophen, and hormones, such as aldosterone, hydrocortisone, and estrone, also affect biliary function in experimental animals. Extrapolation of these results to humans is difficult, since relatively few biliary studies have been carried out in humans. Pregnancy and chronic administration of estrogens are accompanied by decreased bile flow and depressed biliary excretion of a number of substances.[159] Sensitive women may become jaundiced with oral contraceptives or during the last trimester of pregnancy. Some of this response may relate to the known inhibitory effect of estrogens on drug-metabolizing enzymes. However, estradiol-17β-glucuronide, ordinarily considered a detoxication product, produces toxic effects in the hepatobiliary system, including cholestasis.[160,161] Other estrogens and their conjugates are not inhibitory. The question as to whether this is responsible for estrogen-induced depression of biliary function in women remains an open one.

Ethinyl estradiol suppresses bile salt-independent bile flow,[162] biliary excretion of bile acids,[163] bilirubin[163] and BSP,[164] and hepatic Na^+K^+ATPase activity.[165] Administration of Triton® WR-1339 restores bile flow and bile acid excretion whereas pretreatment with phenobarbital corrects bile flow only.[166] The Triton effect is independent of protein synthesis. It has been inferred that ethinyl estradiol primarily alters lipid structure and, consequently, fluidity of the cell membrane which leads to changes in Na^+K^+ATPase and transport activities. Similarly, reversal of ethinyl estradiol-induced suppression of Na^+K^+ATPase activity and bile flow is brought about by administration of *S*-adenosyl-1-methionine. The reversal has been attributed to increased Na^+K^+ATPase activity and membrane fluidity[167] as well as to enhanced formation of methoxy derivatives of the estrogen.[168] The synthetic estrogen, diethylstilbestrol, selectively inhibits biliary excretion of the glucuronide conjugate of hydroxyphenytoin but not of the hydroxymetabolite itself.[169] However, the response was secondary to the cholestatic effect of the drug.

In addition to estrogens, many xenobiotics induce cholestasis.[162] Of considerable clinical importance are the propionate and estolate forms of erythromycin, and phenothiazines such as chlorpromazine and tricyclic antidepressants. Less significant are the responses to novobiocin, rifampicin, and nitrofurantoin, each of which may elicit mild jaundice and retention of BSP. An extensively studied cholestatic agent is α-naphthylisothiocyanate (ANIT) which affects rats and mice more than other species. With inhibition of biliary function, there is dilation of bile ducts with a loss of microvilli. Hyperplasia of bile ducts is seen after chronic administration. Inhibition of the mixed-function oxidase system is also seen, although the relationship to cholestasis is unknown. Phenobarbital pretreatment potentiates and SKF 525-A reduces hepatotoxicity, implying metabolic activation of ANIT. Inhibitors of protein synthesis block the effect of ANIT as well as excretion of its metabolites. The active metabolite(s) requires an intact enterohepatic circulation for effect. The cholestatic effect of chlorpromazine was shown to be associated with the parent compound and the 7,8-dihydroxy metabolite, but not the sulfoxide,[170] illustrating a highly selective relationship between metabolism and toxicity of this compound. Ticrynafen, used for a brief period as a uricosuric diuretic, exhibits considerable hepatic toxicity. The depression in bile flow and biliary dye excretion is exacerbated by pretreatment with phenobarbital, suggesting that here too a metabolite is the toxic agent.[171]

Chelating agents, as might be predicted, often influence the fate of metals which ordinarily are excreted in the bile to a minor extent. Trisodium calcium diethylenetriaminepentaacetic acid (DTPA) increases the excretion of plutonium in bile[172] and desferoxamine promotes the excretion of iron.[173] On the other hand, neither citrate, cysteine, nor penicillamine alters biliary levels of cadmium, although each readily forms a complex with the metal.[174] Excretion of methyl mercury is altered only slightly by penicillamine, dimercaprol, and diethyldithiocarbamate.[175] Spironolactone, although not a chelating agent, enhances biliary excretion of mercury,[176] decreases excretion of silver, and does not affect excretion of lead, manganese, and arsenic.[177]

Pretreatment of homozygous and heterozygous Gunn rats with PCBs (polychlorinated biphenyls) accelerates biliary excretion and depresses plasma levels of thyroxine,[178] which is excreted in great part as its glucuronide. However, homozygous Gunn rats are incapable of conjugating thyroxine with glucuronic acid, and it was found that PCBs act directly on the thyroid gland to lower plasma thyroxine levels.

The vehicle in which a xenobiotic is administered may affect disposition. When benzo[*a*]pyrene and methylcholanthrene are injected as solutions in dimethylsulfoxide, biliary excretion of metabolites is more rapid than when an albumin suspension of the carcinogens is injected.[179] This is attributable to facilitation of binding of the carcinogen to microsomes in the presence of solvent, leading to greater metabolism and, consequently, more rapid biliary excretion.

Of considerable practical importance to studies of biliary excretion is the effect of anesthetics. After pentobarbital or ether, biliary excretion of xenobiotics is somewhat diminished relative to that seen after urethane.[180-182] Ether and halothane each diminish bile flow and bile salt excretion but also produce a sharp rise in biliary β-glucuronidase and acid phosphatase. These enzymes probably are of lysozomal origin and would be of particular importance in assessing the metabolic form in which xenobiotics appeared in the bile. TCDD, an extremely toxic industrial by-product, inhibits bile flow and biliary excretion of a number of xenobiotics. TCDD-suppressed excretion of ouabain can be reversed with spironalactone and pregnenolone-16α-carbonitrile, which do not return bile flow to normal levels.[183] Although both of these agents induce hepatic cytochrome P-450 levels, phenobarbital and 3-methylcholanthrene do not exhibit this effect. Furthermore, ouabain is excreted metabolically unchanged.

G. Effects of Bile Salts

Uptake, conjugation, and secretion of bile salts is a primary function of the hepatobiliary system. Effects of bile salts on xenobiotic disposition have been studied extensively. Taurocholate infusion facilitates hepatic uptake of BSP and of its nonmetabolized analogue DBSP in the dog,[184] although inhibition has also been reported.[185] The apparent discrepancy may relate to dose. High doses of bile salts inhibit hepatic uptake of DBSP in the rat, whereas low doses stimulate.[186] In the rat, taurocholate infusion increases excretion of BSP and its conjugate[187] and lowers hepatic levels of the dye, implying depressed uptake. Large doses of taurocholate also reduce biliary excretion of succinylsulfathiazole.[188] Enhanced excretion of the cation, acetyl procainamide ethobromide, is seen during taurocholate infusion[189] although inhibited excretion during high rates of bile salt infusion (260 mol/hr) has also been reported.[190] The variation in techniques (perfused liver vs. in vivo), species, and dosage precludes comparison among various laboratories and leaves the total picture of bile salt-xenobiotic interaction essentially unsettled. Taurocholate-stimulated BSP excretion[191,192] is not simply due to a choleretic effect since other choleretics (theophylline,[193] umbelliferone,[194] and SC2614[195]) do not increase dye excretion. A direct interaction between the dye and the bile salts has been suggested, the latter acting as a carrier. The micellar sink theory[196] proposed that certain xenobiotics could be incorporated into mixed micelles in which form bile salts and lipid appear in the bile. Several studies to test this theory[197,198] have yielded conflicting information. Among the problems is the dependence of micellar formation on a critical concentration of bile salts, which varies from one bile salt to another. Dilution or concentration of the micellar constituents during analysis is difficult to avoid, which impairs interpretation of results. Cerulein increases the Tm of BSP and rose bengal in chickens, but only if bile salts are present,[199] again supporting a specific role for bile salts in the effect despite its unknown mechanism.

An association of quaternary amines with bile salt micelles[200,201] has been demonstrated as a possible means for intestinal absorption of quaternary amines; the mechanism may be applicable to biliary excretion as well.

H. Miscellaneous Factors

Sex differences in biliary function, not directly related to drug metabolism, are rarely seen. The azo dye, tartrazine, is excreted unchanged in the bile far more rapidly in female than in male rats,[202] but the sex difference is not seen in guinea pigs or rabbits. The difference is hormonally controlled; estradiol enhances excretion in males and testosterone depresses excretion in females. The effect is highly specific since excretion of the structural analogue, lissamine fast, exhibits no sex difference. During liver regeneration, tartrazine excretion is depressed in both male and female rats. The sex difference disappears on the second and third day[203] but reappears with time and is equal to that of control animals upon restoration of original liver size. No sex differences are seen in the biliary excretion of DBSP, ouabain, amaranth, or indocyanin green in rats although the glucuronide of 2,4-dinitrotoluene is transported into bile more rapidly by male than female rats.[204]

Sex-related variations in drug metabolism have been thoroughly reviewed;[205] they are mainly restricted to studies with rats since mice, guinea pigs, hamsters, rabbits, dogs, monkeys, and humans show no clear cut sex differences. The differences probably are related to sex differences in cytochrome P-450 since specific, hormone-dependent isozymes of the cytochrome have been isolated from male and female rats.[206] Metabolism of drugs exhibiting substantial sex differences in rats is less susceptible to induction with 3-methylcholanthrene than are those with smaller sex differences. This supports the idea of a cytochrome P-450-dependent sex difference since 3-methylcholanthrene induces a highly selective isozyme of cytochrome P-450. Sex differences in xenobiotic excretion are also seen where xenobiotic excretion is limited by metabolism. Metabolites of benzo[*a*]pyrene are excreted more rapidly in male rats than in females in parallel to their relative rates of metabolism in vitro.[134] In contrast, aldosterone, which appears in the bile mainly in the form of its metabolites, is excreted more rapidly in female rats than in males.[207] Since metabolites were not identified, it is not known whether this observation is in accord with the generalizations concerning sex differences in metabolism. Male rats metabolize and excrete pentobarbital more rapidly than do females.[208] Livers from female rats metabolically reduce corticosteroids more rapidly than do those from male rats,[209] while oxidative metabolism of steroids and drugs proceeds more rapidly in males than in females.[205] These differences directly relate to levels of estradiol and testosterone which depress and stimulate, respectively, the microsomal mixed-function oxidase system.

The fate of xenobiotics in the body varies considerably with age. Many aspects of drug disposition are poorly developed in the newborn and, after attainment of adult levels, begin to wane with the onset of old age. The toxicity of a number of xenobiotics is higher in immature compared to adult animals although similar comparisons have not been made for older animals. Developmental aspects of oxidative[210] and conjugative metabolism[211] have been reviewed, as have the consequences of aging for pharmacokinetics.[212] Eosine, which is excreted in the bile unmetabolized, is excreted more efficiently in 30- to 60-day-old rats than in animals somewhat younger or older.[213] Older, but not younger, rats excrete the dye more rapidly after phenobarbital pretreatment, indicating an ability to respond to the inducing agent in older rats which is absent in younger rats. However, bile flow does increase in younger animals in the face of unchanged dye excretion. Decreased ouabain excretion is seen in older rats, but this returns to adult levels after pretreatment with spironolactone.[214] Development of the enterohepatic circulation of diethylstilbestrol in the rat proceeds with age parallel to rates of glucuronidation, conjugate transport, and conjugate hydrolysis by gut flora.[215] Carbon tetrachloride inhibits plasma to liver transport of ouabain in young rats ($<$35 days) and liver to bile transport in older rats ($>$35 days).[216] The inability of the neonatal rat to excrete methylmercury has been attributed to a lack of a suitable transport system for glutathione in the neonate[217] since methylmercury is excreted as the glutathione conjugate.

I. Enterohepatic Circulation

The recycling of biliary constituents by reabsorption from the intestine, passage into the portal circulation, and transport again into bile is referred to as the enterohepatic circulation

(EHC).[218,219] Methods for obtaining unequivocal evidence for EHC include (1) infusion of bile of a donor animal into the duodenum of a recipient animal, (2) establishment of a portal/systemic circulation concentration gradient after parenteral administration, and (3) more rapid plasma clearance following total diversion of bile flow. Bile salts undergo extensive EHC such that over 90% of the biliary output is conserved. The system is very efficient and the half-life of bile salt transfer from intestine to bile or plasma half-life after intravenous injection is only a few minutes. Any compound absorbed from the intestine and present in the bile might be expected to undergo EHC, although convincing evidence is available for only a few xenobiotics. Metabolism is a major consideration; hepatic pathways typically decrease lipophilicity of a xenobiotic which should impede intestinal absorption by passive diffusion. However, further metabolism by bacteria of the large intestine[220,221] often reverses this process and is a key step in determining which compounds actually enter the EHC. During metabolism within the gut, the original compound may be regenerated or oxidative metabolites may be liberated through conjugate hydrolysis and further metabolized. The reactions catalyzed by intestinal flora may be considered complementary and often opposite to those of the liver. For example, hepatic reactions include hydroxylation, oxidation, and conjugation, whereas the intestinal flora carry out dehydroxylation, reduction, and conjugate hydrolysis. Hydrolysis of glucuronides is probably the most common reaction which bears on the EHC of the compounds studied.

Evidence for EHC is obtained in a number of ways.[221] Where hydrolysis of glucuronides is believed to be crucial, glucuronidase from intestinal bacteria is the focus of study. Administration to rats of D-glucaro-1,4-lactone, an inhibitor of β-glucuronidase, leads to a shortened duration of action of phenobarbital and progesterone, each of which are partially excreted in the bile as glucuronides.[222] Response to zoxazolamine is unchanged by this inhibitor since it is excreted in the bile as the glucuronide of the pharmacologically inactive 6-hydroxy metabolite. Suppression of the flora by antibiotics or use of germ-free animals is also used to assess EHC. Under such conditions, there are changes in the disposition of morphine,[223] diethylstilbestrol,[224] mestranol,[225] estradiol,[225] estrone,[226] and etorphine.[227] Cautious interpretation of such findings is required since the antibiotics may cause diarrhea, altering transit time in the gut. Blood levels and urinary excretion are also measured with and without bile drainage. Also used is the technique of channeling bile directly into the intestine of a second animal and again measuring biliary excretion. Differences between biliary and fecal excretion are also interpreted as a reflection of EHC. Considerable urinary excretion of many xenobiotics which readily appear in the bile is due to extensive EHC. Each recycling will result in transport of some of the compound to the kidney, provided the first pass effect of the liver is less than 100%.

A previous review[41] listed 33 xenobiotics known to undergo EHC. Since then there have been additional reports for buprenorphine,[69] sulindac,[59] methotrexate,[79,228] gitoxin,[53] clofibride,[92] cimetidine,[93] aminosalicylic acid,[229] 2,4-dinitrotoluene,[230] epomedial,[88] and haloperidol.[68]

For the most part, investigations have been performed in a single species and extrapolation to other species is not advisable since species variation is seen. The intestinal lesions observed after feeding indomethacin are attributed to its extensive EHC[231] and show a parallel species variation. On the other hand, although there is extensive EHC of acetylmethadol,[232] a drug which requires metabolic activation, the EHC makes little contribution to maintenance of blood levels of active metabolites. One study showed that the prodrug, sulindac, and its active metabolite, sulindac sulfide, undergo EHC.[59] Since conversion to the sulfide is reversible, the amount of active metabolite in the circulation depends on both metabolism and EHC. Kinetic evaluation of the disposition of the drug in relation to therapeutic efficacy becomes quite complex. However, use of the prodrug is advantageous since gastric irritation is associated mainly with the sulfide metabolite. Thus, metabolism and EHC provide sufficient therapeutically active metabolite while avoiding gastric exposure.

In an interesting clinical application, the ion exchange resin, cholestyramine, is used to bind bile salts within the intestine, interrupting their EHC and consequently reducing plasma levels of cholesterol.[233] This has led to the use of cholestyramine in overdose of digitalis and poisoning with the pesticide, kepone.[234]

J. Toxic Xenobiotics in Bile

Biliary metabolites, especially conjugates, of xenobiotics are traditionally regarded as detoxication products passing into their final route of excretion. A major exception is seen in the enterohepatic circulation which serves to retain many compounds within the body. In recent years, it has become evident that bile contains substances of considerable toxic potential. This has recently been reviewed.[235] Bile acids have been implicated in carcinogenic and other toxic effects. Qualitative changes in the normal spectrum of bile acids have been associated with experimental biliary conditions. Dietary lithocholic acid induces ductule cell reactions in several species[236] and high levels of specific bile acids have been seen in various hepatic diseases.[237] Feeding of the azo dye carcinogen, 3′-methyl-*N*,*N*-dimethyl-4-aminoazobenzene, is associated with a great increase in fecal and urinary bile acids which correlates with cell proliferation.[238] Lithocholic and deoxycholic acids are found in high concentrations in the human bowel and experimentally, are promotors of colon carcinogenesis.[239,240] It has been proposed that gut bacteria metabolize bile acids to products which are carcinogenic or co-carcinogenic.[241] Large population studies indicate a correlation between colorectal cancer and fecal concentrations of acid steroids and deoxycholic acid.[241] Lithocholic acid binds to tissue and high levels of such binding may be associated with tumors of such tissues.[242]

A number of potentially toxic aromatic amines are excreted in the bile as glucuronide conjugates which are susceptible to attack by β-glucuronidase of intestinal bacteria. Methylazoxymethanol, a metabolite of the colon carcinogen, 1,2-dimethylhydrazine, is transported as its glucuronide in the bile to the intestine. Within the colon, bacterial β-glucuronidase hydrolyzes the conjugate, providing a high concentration of the putative proximal carcinogen.[243] 2,3-Dimethyl-4-aminobiphenyl is carcinogenic to normal but not germ-free animals. Biliary metabolites of several aromatic amines[244] and polycyclic hydrocarbons[245] are mutagenic. Certain simple alkyl halides are activated through conjugation with glutathione and bile from animals given these compounds is highly mutagenic.[246] Apparently, the bile is effective as a delivery system for these toxic xenobiotics, but for some, intervention of gut bacterial enzymes is obviously required. Indomethacin causes intestinal lesions in humans and experimental animals. This has been attributed to biliary excretion of the drug leading to high concentrations within the gut.[231] On the other hand, *N*-methylnitrosourea-induced colon carcinogenesis in rats is blocked by indomethacin.[247] The mechanism is unknown but does not involve metabolism since *N*-methylnitrosourea is a direct acting carcinogen not requiring metabolic activation.

IV. EXCRETION INTO MILK

The concept of target organs for drugs and other xenobiotics is unique when considering excretion in the milk since the entire nursing infant is the target. In another sense, milk should be considered a link in the nutritional chain rather than a route of excretion. The reader is referred to several recent reviews on this subject.[248-254] Analogous to the liver, each of the numerous glands within the breast is divided into lobules which contain the functional units consisting of secretory cells, lumen, ducts, and blood supply. A role for microtubules in milk secretion was demonstrated in rats using colchicine.[255] In addition to loss of microtubules, the drug caused cessation of secretion and accumulation of milk constituents within the cytoplasm. Although not specifically examined, some xenobiotics may be transported into milk by microtubule-dependent means. The mammary duct is permeable to water but milk is isosmotic with plasma. Estrogens and progesterone facilitate development of mammary glands during pregnancy and the high levels of prolactin following delivery and suckling promote secretion. Prolactin secretion is suppressed by dopaminergic drugs and increased by certain neuroleptic drugs. Multiple interactions of other hormones on milk secretion have been described.[252] These include growth hormone, ACTH, insulin, cortisol, thyroid, and parathyroid. The quantity of milk secreted also depends on blood flow which is very high to the breast during lactation in response to prolactin.

Very few preformed endogenous systemic substances are found in milk, suggesting in effect a blood-milk barrier. Xenobiotics diffuse across the capillary wall into the alveolar breast cells and then into the milk mainly by passive diffusion or reverse pinocytosis. Concentrations within the milk are usually independent of milk volume. If these were the sole considerations for plasma to milk transport, then concentrations of unbound xenobiotic in milk and plasma would be equal. However, the milk/plasma ratio is somewhat less than one for a number of drugs.[254] Notable exceptions are erythromycin, ethanol, ether, isoniazid, meprobamate, sulfapyridine, thiopental, and thiouracil, each of which has a milk/plasma concentration ratio of one or more. Lipid solubility and pK of organic anions or cations influence the extent of passive diffusion. The pH of milk is slightly below that of plasma and influences the secretion of these substances accordingly. The high fat content of milk certainly affects the secretion of highly lipid-soluble xenobiotics. In milk, lipid is contained within globules surrounded by a lipoprotein membrane. Thus a drug may be contained within the globule in the lipid or bound externally to the lipoprotein membrane. Binding of xenobiotics to milk proteins is far less than to plasma proteins and probably has little influence on their diffusion from plasma to milk. Free diffusion of ionized xenobiotics is limited to substances with a mol wt of less than 200. The diffusion of larger molecules is dependent on lipid solubility, which in turn is dependent upon pK. The diurnal variation in composition of milk may influence the concentration of drugs within the milk. Little is known concerning xenobiotic metabolism by the breast, although the alveolar cells contain considerable endoplasmic reticulum.

A considerable number of drugs appear in milk but few have been studied systematically, and the significance to the nursing infant is not always clear. A study of antipyrene, a freely diffusable drug, showed similar milk and plasma concentration over 24 hr after a single oral dose.[251] Sulfasalazine, on the other hand, is undetectable in the milk although it readily passes into the urine.[251] A major metabolite, sulfapyridine, is found in plasma and milk in approximately equal concentrations but does not appear to be hazardous to the infant. Recent publications list over 50 drugs, radioisotopes, and other xenobiotics which appear in milk. The compounds listed in Table 2 are those detected in milk in recent years. Over 30 drugs and environmental pollutants were also reported in earlier reviews.[254,256] The rapid growth of this list reflects the great concern for exposure of a nursing infant to maternal xenobiotics. As with other routes, the spectrum of xenobiotics excreted is very broad, ranging from commonly used therapeutic agents to several potentially dangerous environmental pollutants. These findings raise questions as to the advisability of nursing an infant while taking drugs. For example, the concentrations of codeine and morphine in human milk are more than twice those in plasma.[276] Several environmental contaminants including DDT and PCB are also found in breast milk.[277] The high but variable fat content of milk probably facilitates their secretion. Nevertheless, reports on overt toxicity from compounds transmitted in the milk are rare. Obstructive jaundice due to chlorinated hydrocarbon in milk has been seen.[278] Methemoglobin was found in an infant whose mother was treated with phenytoin.[279] Porphyria and death have been reported in nursing infants whose mothers had eaten wheat seed contaminated with hexachlorobenzene.[280] The toxicity of some xenobiotics for the infant is so great that they are proscribed during nursing; if required by the mother, nursing should cease. Suppression of infants' thyroid glands is possible after administration of radioactive

Table 2
XENOBIOTICS SECRETED IN MILK

	Ref.
Antibiotics and other antibacterial compounds	
Clindamycin	257
Cardiovascular drugs	
Mexiletine	258
Hydralazine	259
Digoxin	260
Drugs affecting nervous system	
Salicylates	261
Caffeine	261
Antipyrene	262
Amitriptyline	263
Nortriptyline	263
Baclofen	264
Cis(Z)-flupentixol	265
Mepindalol	266
Primadone	267
Phencyclidine	268
Ethosuximide	269
Miscellaneous	
Metronidazole	270
Dyphylline	271
Medroxyprogesterone	272
Cyproterone	273
Methimazole	274
Nor-testosterone	275

iodine or propylthiouracil to nursing mothers. Other proscribed drugs include lithium, which is concentrated in milk, methadone, isoniazid, chloramphenicol, and certain antineoplastic agents.[281] There are no reports that tetrahydrocannabinol passes into human milk, but it does appear in the milk of sheep[282] and monkeys.[283] Alcohol passes freely into the milk although there is little evidence that an occasional drink by the mother is harmful to a nursing infant. A woman who smokes 10 to 20 cigarettes per day will have 0.4 to 0.5 mg of nicotine per liter of milk.

Quantitative information on the excretion of xenobiotics in milk, although still sporadic, is gradually increasing. The critical question is whether the amount of drug ingested by a nursing infant is of pharmacologic or toxicologic significance. A recent study of several drugs contained in commonly used analgesic preparations (salicylic acid, phenacetin, acetaminophen, caffeine, codeine, morphine) indicated that less than 1% of the maternal dose was excreted in the milk during a 12 hr period.[276] It was concluded that a nursing infant would not be at risk under such circumstances. They felt that even the limited ability of the neonate to metabolize and excrete xenobiotics would readily dispose of these drugs received through the milk. Clearly, additional quantitative studies are needed to assess infant risk, particularly when larger doses of readily available drugs are taken by the mother. Another consideration is the possibility of induction of the drug-metabolizing enzymes in the neonate consequent to chronic exposure to certain xenobiotics in the maternal milk. This has been seen in experimental animals[284] and possible effects in infants should be considered.

V. SALIVARY EXCRETION

The need to monitor levels of drugs in humans for investigational as well as clinical purposes has led to studies on the salivary route of excretion. Many xenobiotics are found in the saliva[285-291] and concentrations often reflect those in the plasma. The review of Horning et al.[291] describes much of the earlier work in this area. Transport from plasma to saliva is mainly passive and is related to the plasma/saliva concentration gradient, lipid solubility, and pK of the xenobiotic. Only the unbound fraction within the plasma is readily diffusable, and salivary concentrations can be expected to reflect this fraction and thus be a relatively easy means of determining the degree of binding to plasma proteins. Salivary protein concentration is less than 10% that of plasma and should have little influence on passive diffusion of xenobiotics. One study with phenobarbital indicated that 80% of salivary drug was unbound while 60% of plasma drug was unbound.[292] It was also shown that salivary protein concentration varies with the type of stimulus used to induce salivary flow. Rate of salivary flow may influence excretion.[293] Based on a study of six drugs, it was found that salivary concentrations approach those of the unbound plasma fraction as salivary flow increases. This is especially true for drugs relatively excluded from the saliva, while conversely, drugs relatively concentrated in the saliva exhibit lower salivary concentrations when flow is accelerated. The means of promoting salivary flow from drug analysis must be considered. Chewing on a piece of wax or paraffin is commonly used to stimulate sufficient flow. Diminished flow after anticholinergic drugs may create a drawback for this technique. However, 5% citric acid solution in the mouth was found to stimulate salivary flow in spite of the presence of anticholinergic drugs.[294] Experimentally, the use of pilocarpine as a sialogogue results in unchanged salivary excretion of procainamide in rats.[295]

Decreased, increased, and no change in salivary pH have been observed with increased salivary flow in humans.[292,296,297] It is therefore not possible to predict the effect of these parameters on xenobiotic transport into saliva. Although the average salivary pH is thought to be on the acid side, pH has been reported as low as 5.1 and as high as 8.3. This may be expected to influence the saliva/plasma concentration ratios for xenobiotics with pKs in this range. Organic bases in this group tend to be concentrated in the saliva while organic acids diffuse into the saliva less readily. Examples of drugs within this group are the organic acids, phenobarbital and salicylic acid, salivary concentrations of which are less than those predicted based on plasma protein binding alone.[291] On the other hand, lithium ion concentrations in saliva are higher than those in plasma. In a study of six antibiotics, little correlation was obtained between salivary and plasma concentrations.[298] Although secretion was predicted on the basis of pK and partitioning, several antibiotics were not detectable in saliva.

The growing interest in analyzing drugs within the body by noninvasive techniques is exemplified by the list in Table 3 of compounds recently reported in saliva. Salivary xenobiotics should not be considered as excreted in the usual sense, since they are mainly swallowed and probably reabsorbed. The contribution of such an "entero-salivary" recirculation to the retention of xenobiotics in the circulation is unknown. At present, salivary analysis simply provides an accurate, noninvasive means of determining plasma levels of xenobiotics. Individual variations in results cautions against use of this technique for a particular drug until the relationship between salivary and plasma concentrations has been suitably established. Although salivary and plasma concentrations of many drugs are not equal in most cases, the discovery of the correlation between the two has permitted a reasonably accurate assessment of plasma levels of theophylline,[326] lidocaine,[321] amitriptyline,[302] diazepam,[305] amphotericin B,[300] erythromycin,[298] chlorpromazine,[314] cyclophosphamide,[327] indomethacin,[328] and others. A distinct advantage in salivary sampling is timed measurement of cortisol, where the trauma of repeated blood sampling would itself alter plasma cortisol levels.[322]

For a number of drugs, saliva to plasma concentration ratios during the absorption phase vary somewhat more than might be predicted from steady-state measurements. These include acetaminophen,[286] theophylline,[329] amphetamine,[330] and others. It has been hypothesized that there are higher drug concentrations in arterial compared to venous blood and that saliva

Table 3
XENOBIOTICS EXCRETED IN SALIVA

	Ref.
Antibiotics and other antibacterial compounds	
Moxalactam	43
Erythromycin	298
Trimethoprim	299
Sulfomethopyrazine	299
Ampicillin	297
Tetracycline	297
Clindamycin	297
Rifampicin	297
Amphotericin B	300
Cardiovascular drugs	
Quinidine	301
Drugs affecting nervous system	
Amitriptyline	302
Nortriptyline	303
Lithium	304
Diazepam	305—307
Desmethyl diazepam	306,307
Pethidine	308
Caffeine	309,310
Carbamazepine	311
Chlordiazepoxide	312
Antipyrine	313
Chlorpromazine	314
Valproic acid	315
Primidone	316
Phenytoin	317
Phenobarbital	318
Trimethadione	319
Nicotine	320
Lidocaine	321
Hormones	
Cortisol	322
Miscellaneous	
Clofibric acid	323
Fluoride	324
Promethazine	325
Theophylline	326
Cotinine	320
Cyclophosphamide	327
Indomethacin	328

is in equilibrium with arterial blood.[331] With this in mind, it may be possible to determine bioavailability of a drug by calculating the area under the curve based entirely on salivary concentration.

With the expectation that salivary drug concentrations reflect unbound plasma concentrations, it might be predicted that displacement of a drug from plasma protein binding, a common drug interaction, would alter salivary excretion. However, although aspirin displaces phenytoin from plasma protein binding, there is no change in salivary phenytoin concentration which is maintained at 10% that of plasma.[317]

VI. EXCRETION INTO THE EXPIRED AIR

Analysis of drugs in the expired air is an additional noninvasive technique for assessing plasma levels of drugs in humans and, by inference, their rate of metabolism.[332-337] The drug most commonly used is aminopyrine which is *N*-demethylated; the methyl groups are rapidly oxidized to carbon dioxide which appears in the expired air. A tracer dose of (*N*-dimethyl-^{14}C) aminopyrine is administered and the $^{14}CO_2$ in expired air collected over a suitable period of time is trapped in a hyamine solution and counted. It must be appreciated that $^{14}CO_2$ analysis reflects the endpoint of a complex series of steps which begins with *N*-demethylation. Quantitation of the first step by this method is only possible if there is assurance that none of the succeeding reactions is rate-limiting. The technique is also applicable to deacetylation and decarboxylation. CO_2 production is a normal physiological process and varies with physical activity, metabolic state, and body temperature; therefore, these factors must be considered when utilizing this technique, particularly in humans. The "aminopyrine breath test" is now an accepted clinical procedure. It has been suggested that it is suitable for assessment of liver function in cases of cirrhosis, hepatitis, and liver cancer,[334] although patients with cholestasis are not readily distinguished from controls. The technique is also useful in experimental animals, and changes in aminopyrine demethylation in response to phenobarbital, bile duct ligation, partial hepatectomy, and portacaval shunt reflect those predicted from in vitro drug metabolism studies.[335,336] Assessment of other aspects of liver function through breath analysis has been discussed,[334,337] and wider application of the technique can be expected.

One notable application has been in the noninvasive determination of lipid peroxidation during which saturated short-chain hydrocarbons are split from unsaturated fatty acids such as linoleic acid and passed into the expired air.[338] The results do not always agree with in vitro measurement of lipid peroxidation[339] and further work is needed in this area.

VII. SUMMARY

In recent years it has become apparent that metabolism of xenobiotics is no longer synonymous with "detoxication", as implied in the classic work of R. T. Williams[340] nearly 40 years ago. Metabolic activation has been documented or strongly implied for most carcinogenic and toxic compounds. So too is it now becoming clear that passage of a xenobiotic and/or its metabolites into one of the several routes of excretion does not guarantee that the body will be free of such compounds. In many cases it implies delivery to another area of the body or even retention of the compounds within the body. Thus, biliary excretion serves to expose the upper intestine to high concentrations of many xenobiotics and to retain some compounds through the enterohepatic circulation. Even renal excretion permits exposure of the bladder to carcinogenic compounds. Certainly excretion into saliva involves retention within the body. Excretion into milk involves transfer to an individual whose abilities to cope with many xenobiotics is not fully developed, implying a somewhat greater likelihood of toxicity. Investigations into excretion of drugs in breast milk are expanding rapidly. Since over half of the infants in the world are breast fed, and the numbers are increasing, possible exposure to maternal xenobiotics is a major consideration. There is a lack of pharmacokinetic studies in this area and of information on the effects in infants of long-term exposure to maternal xenobiotics through breast feeding. Another area which lacks sufficient quantitative information is the normal excretion of xenobiotics into human bile. Nearly all collection of human bile is associated with surgical procedures and some drug treatment. Since both of these factors could influence biliary excretion, the studies published must be interpreted with caution. Recognition of the need to assess noninvasively plasma drug levels and drug metabolizing capacity has led to consideration of saliva and expired air as routes of excretion. Obviously present success in the area, however limited, will lead to future development. This is particularly stimulated by greater recognition of the necessity to monitor blood levels of numerous drugs and intoxicants as well as hepatic capacity to metabolize these xenobiotics.

REFERENCES

1. **Pritchard, J. B. and James, M. O.,** Metabolism and urinary excretion, in *Metabolic Basis of Detoxication*, Jakoby, W. B., Bend, J. R., and Caldwell, J., Eds., Academic Press, New York, 1982.
2. **Irish, J. M. and Grantham, J. J.,** Renal handling of organic anions and cations, in *The Kidney*, Brenner, B. M. and Rector, F. C., Eds., W.B. Saunders, Philadelphia, 1981, 619.
3. **Garrett, E. R.,** Pharmacokinetics and clearances related to renal processes, *Int. J. Clin. Pharmacol.*, 16, 155, 1978.
4. **Ross, C. R. and Halohan, P. D.,** Transport of organic anions and cations in isolated renal plasma membranes, *Ann. Rev. Pharmacol. Toxicol.*, 23, 65, 1983.
5. **Smith, R. L.,** *The Excretory Function of Bile*, Chapman & Hall, London, 1973.
6. **Moller, J. V. and Sheikh, M. I.,** Renal organic anion transport system: pharmacological, physiological, and biochemical aspects, *Pharmacol. Rev.*, 34, 315, 1983.
7. **Rennick, B. R.,** Renal tubule transport of organic cations, *Am. J. Physiol.*, 240, F83, 1981.
8. **Blitzer, B. L. and Boyer, J. L.,** Cellular mechanisms of bile formation, *Gastroenterology*, 82, 346, 1982.
9. **Boyer, J. L.,** New concepts of mechanisms of hepatocyte bile formation, *Physiol. Rev.*, 60, 303, 1980.
10. **Erlinger, S.,** Hepatocyte bile secretion; current views and controversies, *Hepatology*, 1, 352, 1981.
11. **Erlinger, S.,** Does Na^+-K^+-ATPase have any role in bile secretion, *Am. J. Physiol.*, 243, G243, 1982.
12. **Graf, J.,** Canalicular bile salt-independent bile formation: concepts and clues from electrolyte transport in rat liver, *Am. J. Physiol.*, 244, G233, 1983.
13. **Jones, A. L., Schmucker, D. L., Renston, R. H., and Murakami, T.,** The architecture of bile secretion. A morphological perspective in physiology, *Dig. Dis. Sci.*, 25, 609, 1980.
14. **Reichen, J. and Paumgartner, G.,** Excretory function of the liver, *Int. Rev. Physiol.*, 21, 103, 1980.
15. **Gumucio, J. J., Miller, D. L., Krauss, M. D., and Zanolli, C. C.,** Transport of fluorescent compounds into hepatocytes and the resultant zonal labeling of the hepatic acinus in the rat, *Gastroenterology*, 80, 639, 1981.
16. **Gumucio, J. J. and Miller, D. L.,** Functional implications of liver cell heterogeneity, *Gastroenterology*, 80, 393, 1981.
17. **Groothius, G. M. M., Weitering, J. G., Keulemans, K. P. T., Hardonk, M. J., Mulder, D., and Meijer, D. K. F.,** Heterogeneity of rat hepatocytes in bile acid and DBSP transport studied after induction of selective acinar damage by N-hydroxy-acetylaminofluorene and carbon tetrachloride, *Naunyn-Schmiedeberg's Arch. Pharmacol.*, 322, 310, 1983.
18. **Layden, T. J., Elias, E., and Boyer, J. L.,** Bile formation in the rat. The role of the paracellular shunt pathway, *J. Clin. Invest.*, 62, 1375, 1978.
19. **Graf, J. and Peterlik, M.,** Mechanism of transport of inorganic ions into bile, in *The Hepatobiliary System*, Taylor, W., Ed., Plenum Press, New York, 1976, 43.
20. **Bradley, S. E. and Herz, R.,** Permselectivity of biliary canalicular membrane in rats: clearance probe analysis, *Am. J. Physiol.*, 235, E570, 1978.
21. **Weitering, J. G., Lammers, W., Meijer, D. K. F., and Mulder, G. J.,** Localization of d-tubocurarine in rat liver lysozomes. Lysozomal uptake, biliary excretion and displacement by quinacrine *in vivo*, *Naunyn-Schmiedeberg's Arch. Pharmacol.*, 299, 277, 1977.
22. **La Russo, N. F. and Fowler, S.,** Coordinate secretion of acid hydrolases in rat bile, *J. Clin. Invest.*, 64, 948, 1979.
23. **Oda, M., Price, V. M., Fisher, M. M., and Phillips, N. J.,** Ultrastructure of bile canaliculi with special reference to the surface coat and the pericanalicular web, *Lab. Invest.*, 31, 314, 1974.
24. **Phillips, M. J., Oda, M., and Funatsu, K.,** Evidence for microfilament involvement in norethandrolone-induced intrahepatic cholestasis, *Am. J. Pathol.*, 93, 729, 1978.
25. **Barnhart, J. L. and Combes, B.,** Characteristics common to choleretic increments of bile induced by theophylline, glucagon and SQ-20009 in the dog, *Proc. Soc. Exp. Biol. Med.*, 150, 591, 1975.
26. **Poupon, R. E., Dol, M. L., Dumont, M., and Erlinger, S.,** Evidence against a physiological role of cAMP in choleresis in dogs and rats, *Biochem. Pharmacol.*, 27, 2413, 1978.

27. **Kaminski, D. L., Ruwart, M. J., and Deshpande, Y. G.,** The role of cyclic AMP in canine secretin-stimulated bile flow, *J. Surg. Res.*, 27, 57, 1979.
28. **Van Dyke, R. B. and Scharschmidt, B.,** Oxygen consumption by rats, *Am. J. Physiol.*, 244, G525, 1983.
29. **Forker, E. L.,** Mechanisms of hepatic bile formation, *Ann. Rev. Physiol.*, 39, 323, 1977.
30. **Chenderovitch, J.,** Bile secretion, *Clin. Gastroenterol.*, 2, 31, 1973.
31. **Barnhart, J. L. and Combes, B.,** Characterization of SC2644-induced choleresis in the dog. Evidence for canalicular bicarbonate secretion, *J. Pharmacol. Exp. Ther.*, 206, 190, 1978.
32. **Peterson, R. E. and Fujimoto, J. M.,** Retrograde intrabiliary injection: absorption of water and other compounds from the rat biliary tree, *J. Pharmacol. Exp. Ther.*, 185, 150, 1973.
33. **Olson, J. R. and Fujimoto, J. M.,** Evaluation of hepatobiliary function in the rat by the segmented retrograde intrabiliary injection technique, *Biochem. Pharmacol.*, 29, 205, 1980.
34. **Kakis, G. and Yousef, I. M.,** Protein composition of rat bile, *Can. J. Biochem.*, 56, 287, 1978.
35. **Gustafson, J. H. and Benet, L. Z.,** Biliary excretion kinetics of phenolphthalein glucuronide after intravenous and retrograde biliary administration, *J. Pharm. Pharmacol.*, 26, 937, 1974.
36. **Sperber, I.,** Secretion of organic anions in the formation of urine and bile, *Pharmacol. Rev.*, 11, 109, 1959.
37. **Balabaud, C., Kron, K. A., and Gumucio, J. J.,** The assessment of the bile salt nondependent fraction of canalicular bile water in the rat, *J. Lab. Clin. Med.*, 89, 393, 1977.
38. **Simon, F., Sutherland, E., and Accatino, L.,** Stimulation of hepatic sodium and potassium-activated adenosine triphosphatase activity by phenobarbital — its possible role in regulation of bile flow, *J. Clin. Invest.*, 59, 849, 1977.
39. **Graf, J., Korn, P., and Peterlik, M.,** Choleretic effects of ouabain and ethacrynic acid in the isolated perfused rat liver, *Naunyn-Schmiedeberg's Arch. Pharmacol.*, 272, 230, 1972.
40. **Anwer, M. S. and Hegner, D.,** Role of inorganic electrolytes in bile acid-independent canalicular bile formation, *Am. J. Physiol.*, 244, G116, 1983.
41. **Levine, W. G.,** Biliary excretion of drugs and other xenobiotics, *Prog. Drug. Res.*, 25, 361, 1981.
42. **Taylor, E. W., Poxon, V., Alexander-Williams, J., and Jackson, D.,** Biliary excretion of piperacillin, *J. Int. Med. Res.*, 11, 28, 1983.
43. **Shimada, J. and Ueda, Y.,** Moxalactam — absorption, excretion, distribution and metabolism, *Rev. Infect. Dis.*, 4, S569, 1982.
44. **Hansbrough, J. F. and Clark, J. E.,** Concentration of cefoxitin in gallbladder bile of cholecystectomy patients, *Antimicrob. Agents Chemother.*, 22, 709, 1982.
45. **Shindo, H., Kawai, K., Ikeda, T., Igarashi, I., and Sugawara, S.,** Absorption, distribution, excretion and metabolism of cefmetazole in cynomolgus monkeys, *J. Antibiot.*, 35, 742, 1982.
46. **Gundert-Reiny, U. and Weber, E.,** Elimination of azlocillin in patients with biliary t-tube drainage, *Eur. J. Pharmacol.*, 22, 435, 1982.
47. **Maudgal, D. P., Maxwell, J. D., Lees, L. G., and Wild, R. N.,** Biliary excretion of amoxycillin and ceftriaxone after intravenous administration in man, *Br. J. Clin. Pharmacol.*, 14, 213, 1982.
48. **Gundert-Reiny, U., Frohnapfel, F., Jourdan, W., Weber, E., and Stiehl, A.,** Estimation of biliary excretion of ureidopenicillin in healthy volunteers using marker dilution technique, *Br. J. Clin. Pharmacol.*, 13, 795, 1982.
49. **Hansbrough, J. F., Clark, J. E., and Reimer, L. G.,** Concentration of kanamycin and amikacin in human gallbladder bile and wall, *Antimicrob. Agents Chemother.*, 20, 515, 1981.
50. **Craig, W. A. and Gerber, A. U.,** Pharmacokinetics of cefoperazone: a review, *Drugs*, 22 (Suppl. 1), 35, 1981.
51. **Thomas, M. H., Dash, C. H., Burnard, K. G., and Woodyer, A. B.,** The excretion of cefuroxime in human bile, *Br. J. Surg.*, 68, 290, 1981.
52. **Murakawa, T., Sakamoto, H., Fukada, S., Nakamato, S., Hirose, T., Itah, N., and Nishida, M.,** Pharmacokinetics of ceftizoxime after parenteral dosing, *Antimicrob. Agents Chemother.*, 17, 157, 1980.
53. **Kadima, L. N. and Lesne, M.,** Studies on the pharmacokinetics and the metabolism of gitoxin in the guinea-pig. II. The enteral absorption and the enterohepatic recirculation, *Arch. Int. Pharmacodyn. Ther.*, 258, 4, 1982.
54. **Kamath, B. L., Stampfli, H. F., Lai, C. M., and Yacobi, A.,** Pharmacokinetics of [^{14}C]bretylium tosylate in rats, *J. Pharm. Sci.*, 70, 667, 1970.
55. **Hawkins, D. R., Weston, K. T., Chausseaud, L. F., Darragh, A., and O'Kelly, A.,** The absorption and excretion of the peripheral vasodilator ^{14}C-mecinarone (^{14}C-6809 MD) in rat, dog and man, *Eur. J. Drug Metab. Pharmacokinet.*, 5, 135, 1980.
56. **Drenth, B. F. and deZeeuw, R. A.,** High-performance liquid chromatography in the elucidation of the metabolic fate of butoprozine: screening for unknown metabolites in the rat, *J. Chromatogr.*, 18, 109, 1980.
57. **Chanoine, F., Benedetti, M. S., Ancher, J. F., and Dostert, P.,** Absorption, distribution, elimination and metabolism of ^{14}C-heptaminol hydrochloride in rat, *Arzneimittelforschung*, 31, 1430, 1981.

58. **Rafter, J. J. and Bakke, J. E.,** Biliary metabolites of the anti-inflammatory drug, 2-acetamido-4-(chloromethyl) thiazole, *Drug Metab. Dispos.*, 10, 654, 1982.
59. **Dujovne, C. A., Pitterman, A., Vincek, W. C., Dobrinska, M. R., Davies, R. O., and Duggan, D. E.,** Enterophepatic circulation of sulindac metabolites, *Clin. Pharmacol. Ther.*, 33, 172, 1983.
60. **Cayen, M. N., Kraml, M., Ferdinandi, E. S., Greselin, E., and Dvornik, D.,** The metabolic disposition of etodolac in rats, dogs, and man, *Drug Metab. Rev.*, 12, 339, 1981.
61. **Zanolo, G., Giachetti, C., Mondiano, A., Silvestri, S., Bianchi, E., Segre, G., Gomarasca, P., and DeMarchi, O.,** Pharmacokinetics of fentiazac in rats and monkeys, *Arzneimittelforschung*, 31, 1098, 1981.
62. **Maurer, G., Schreier, E., Delaborde, S., Loosli, H. R., Nufer, R., and Shukla, A. P.,** Fate and disposition of bromocriptine in animals and man. I. Structure elucidation of the metabolites, *Eur. J. Drug. Metab. Pharmacokinet.*, 7, 281, 1982.
63. **Horie, T., Ohno, T., and Kinoshita, K.,** Studies on metabolism of tripamide. III. Metabolic fate of ^{14}C-tripamide in rats and rabbits, *Jpn. J. Pharmacol.*, 32, 1041, 1982.
64. **Tse, F. L., Ballaro, F., and Skinn, J.,** Estimating the fraction reabsorbed in drug undergoing interohepatic circulation, *J. Pharmacokinet. Biopharm.*, 10, 455, 1982.
65. **Sakai, T., Niwaguchi, T., and Murata, T.,** Distribution and excretion of methamphetamine and its metabolits in rats. I. Time course of concentrations in blood and bile after oral administration, *Xenobiotica*, 12, 233, 1982.
66. **Neill, E. A. and Chapple, D. J.,** Metabolic studies in the cat with atracurium, a neuromuscular blocking agent designed for non-enzymic inactivation at physiological pH, *Xenobiotica*, 12, 203, 1982.
67. **Upton, A., Nguyen, T. L., Miller, R. D., and Castognoli, N., Jr.,** Renal and biliary elimination of vecuronium (CRG NC 45) and pancuronium in rats, *Anesth. Analg.*, 61, 313, 1982.
68. **Forswan, A., Larsson, M., Lundborg, H., and Renstrom, P.,** On the distribution and elimination of haloperidol in cholecystectomized patients, *Eur. J. Drug Metab. Pharmacokinet.*, 6, 249, 1981.
69. **Brewster, D., Humphrey, M. J., and McLeavy, M. A.,** Biliary excretion, metabolism and enterohepatic circulation of buprenorphine, *Xenobiotica*, 11, 189, 1981.
70. **Leineuber, F. J., Szuna, A. J., Williams, T. H., Sasso, G. J., and DeBarbieri, B. A.,** The metabolism of radioactive (−)-3-phenoxy-*N*-methyl[2-^{3}H] morphinan in dogs, *Drug Metab. Dispos.*, 9, 284, 1981.
71. **Rodgers, R. M., Taylor, S. M., Dickenson, R. G., Ilias, A. M., Lynn, R. K., and Gerber, N.,** Metabolism and choleretic effect of naltrexone in the isolated perfused rat liver, *Drug Metab. Dispos.*, 8, 390, 1980.
72. **Lawyer, C. H., Gerber, N., Lynn, R. K., and Dickenson, R. G.,** Application of the two-compartment open model with Michaelis-Menten elimination kinetics to valproic acid in the bile-exteriorized rat, *Res. Commun. Chem. Pathol. Pharmacol.*, 27, 469, 1980.
73. **Ireland, S. J. and Livingston, A.,** Effect of biliary excretion on ketamin anaesthesia in the rat, *Br. J. Anaesth.*, 52, 23, 1980.
74. **Tavoloni, N. and Guarino, A. M.,** Bile secretory function: a determinent of adriamycin disposition, *Arch. Int. Pharmacodyn. Ther.*, 245, 180, 1980.
75. **Parker, R. J., Priester, E. R., and Sieber, S. M.,** Effect of route of administration and liposome entrapment on the metabolism and disposition of adriamycin in the rat, *Drug Metab. Dispos.*, 10, 499, 1982.
76. **Kelsen, D. P., Hoffman, J., Alcock, N., Cheng, E., Boiley, E., Young, C., Golbey, R., and Fortner, J.,** Pharmacokinetics of cisplatin regional hepatic infusion, *Am. J. Clin. Oncol.*, 5, 173, 1982.
77. **Godeneche, D., Moreau, M. F., Madelmont, J. C., Duprat, J., and Plagne, R.,** Disposition and metabolism of 1-(2-chloroethyl)-3-(2′,3′,4′-tri-*o*-acetyl, ribopyranosyl)-1-nitrosourea in rats, *Cancer Res.*, 42, 525, 1982.
78. **Fraire, C., Lecointe, P., and Paoletti, C.,** Metabolism of ethidium bromide in rats, *Drug Metab. Dispos.*, 9, 156, 1981.
79. **El-Dareer, S. M., Tillery, K. F., and Hill, D. L.,** Disposition of 5-methyltetrahydrohomofolate and methotrexate in rats, *Cancer Treat. Rep.*, 65, 101, 1981.
80. **Vereczkey, L., Tamas, J., Czera, G., and Szporny, L.,** Metabolism of vincamine in the rat in vivo and in vitro, *Arzneimitte Forschung*, 30, 1860, 1980.
81. **Broggine, M., Colombo, T., Martini, A., and Doneli, M. G.,** Studies in the comparative distribution and biliary excretion of doxorubicin and 4′-epi-doxorubicin in mice and rats, *Cancer Treat. Rep.*, 64, 897, 1980.
82. **Van-Bac, N., Moisand, C., Gouyette, A., Muzard, G., Dat-Xuong, N., LePecq, J. B., and Paoletti, C.,** Metabolism and disposition studies of 9-hydroxyellipticine and 2-methyl-9-hydroxyellepticinium acetate in animals, *Cancer Treat. Rep.*, 64, 879, 1980.
83. **Brown, R. K., Duncan, G., and Hill, D. L.,** Distribution and elimination of melphalan in rats and monkeys and distribution in tumors of mice bearing L1210 or P388 leukemias sensitive and resistant to this agent, *Cancer Treat. Rep.*, 64, 643, 1980.

84. **Malinin, G. I., Glew, W. B., Roberts, W. B., Townsend, L., and Nigra, T. P.**, Kinetics of [^{14}C-5]8-methoxyporalen distribution in rabbits, *Arch. Dermatol. Res.*, 273, 319, 1982.
85. **Egger, H., Bartlett, F., Itterly, W., Rodebaugh, R., and Shimanskas, C.**, Metabolism of aminoglutethimide in the rat, *Drug Metab. Dispos.*, 10, 405, 1982.
86. **Wong, L. T., Buttar, H. S., Solomonraj, G., and Thomas, B. H.**, Effect of salicylate on the fate of [^{14}C]sulfinpyrazone in the rat, *Can. J. Physiol. Pharmacol.*, 60, 763, 1982.
87. **Galtier, P., Coche, Y., and Camguilhem, R.**, Incidence of experimental cirrhosis on hepatic desposition of [^{3}H]levamisole in rats, *J. Pharm. Pharmacol.*, 34, 310, 1982.
88. **Ventura, P. and Selva, A.**, Biotransformation of epomediol. I. Isolation and identification of main metabolites in the rat, *Biomed. Mass Spectrom.*, 9, 18, 1982.
89. **Singhvi, S. M., Peterson, A. E., Ross, J. J., Shaw, J. M., Keim, G. R., and Migdalof, B. H.**, Pharmacokinetics of captopril in dogs and monkeys, *J. Pharm. Sci.*, 70, 1108, 1981.
90. **Mas-Chamberlin, C., Gillet, G., Andre, J., Gomeni, R., Dring, L. G., and Morselli, P. L.**, The metabolism and kinetics of tiflorex in the rat, *Drug Metab. Dispos.*, 9, 150, 1981.
91. **Tatsumi, K., Ou, T., Yamada, H., and Yoshimura, H.**, Studies of metabolic fate of a new antiallergic agent, azelastine (4-(*p*-chlorobenzyl)-2-[*N*-methyl-perhydroazepinyl-4(4)]-1-(2H)-phthalazinone hydrochloride), *Jpn. J. Pharmacol.*, 30, 37, 1980.
92. **Loiseau, G. P., Millischer, R. J., Lohier, P. Y., Mardiguian, J. S., Gilede, A. M., and Ginocchio, A. V.**, On the metabolism of clofibride, a hypolipidaemic drug, *J. Pharm. Pharmacol.*, 32, 483, 1980.
93. **Pederson, P. V. and Miller, R.**, Pharmacokinetics and bioavailability of cimetidine in humans, *J. Pharm. Sci.*, 69, 394, 1980.
94. **Jonkman, J. H., Wijsbeek, J., Dezeeuw, R. A., van Borke, L. E., Orie, N. G., Soeterboek, A. M., and Bender, J.**, Metabolism and excretion of the quaternary ammonium compound thiazinamium methylsulfate (Multergan) in man. I. Parenteral administration, *Pharm. Weekbl.*, 4, 122, 1982.
95. **Brauer, D. S.**, Mechanisms of bile secretion, *JAMA*, 169, 1462, 1959.
96. **Meijer, D. K. F., Vonk, R. J., Scholtens, E. J., and Levine, W. G.**, The influence of dehydrocholate on hepatic uptake and biliary excretion of ^{3}H-taurocholate and ^{3}H-ouabain, *Drug Metab. Dispos.*, 4, 1, 1976.
97. **Cohen, E. N., Winstow, B. H., and Smith, D.**, The metabolism and elimination of d-tubocurarine-H^{3}, *Anesthesiology*, 28, 309, 1967.
98. **Meijer, D. K. F. and Weitering, J. G.**, Curare-like agents: relation between lipid solubility and transport into bile in perfused rat liver, *Eur. J. Pharmacol.*, 10, 283, 1970.
99. **Peterson, R. E. and Fujimoto, J. M.**, Biliary excretion of morphine-3-glucuronide and morphine-3-ethereal sulfate by different pathways in the rat, *J. Pharmacol. Exp. Ther.*, 184, 409, 1973.
100. **Gartner, U., Stockert, R. J., Levine, W. G., and Wolkoff, A.**, Effect of nafenopin on the uptake of bilirubin and sulfobromophthalein by isolated perfused rat liver, *Gastroenterology*, 81, 1163, 1982.
101. **Erttmann, R. R. and Damm, K. H.**, On the problem of a common hepatic transport process for steroids. Uptake of ^{3}H-ouabain, ^{3}H-taurocholic acid and ^{3}H-corticosterone into rat liver slices, *Arch. Int. Pharmacodyn. Ther.*, 214, 232, 1975.
102. **Millburn, P.**, Excretion of xenobiotic compounds in bile, in *The Hepatobiliary System*, Taylor, W., Ed., Plenum Press, New York, 1975, 109.
103. **Hughes, R. D., Millburn, P., and Williams, R. T.**, Molecular weight as a factor in the excretion of monoquaternary ammonium cations in the bile of the rat, rabbit and guinea pig, *Biochem. J.*, 136, 967, 1973.
104. **Vonk, R. J., Jekel, P. A., Meijer, D. K. F., and Hardonk, M. J.**, Transport of drugs in isolated hepatocytes, the influence of bile salts, *Biochem. Pharmacol.*, 27, 397, 1978.
105. **Schwenk, M.**, Transport systems of isolated hepatocytes. Studies on the transport of biliary compounds, *Arch. Toxicol.*, 44, 113, 1980.
106. **Davis, R. A., Hyde, P. M., Kuan, J. W., Malone-McNeal, M., and Archambault-Schexnayder, J.**, Bile acid secretion by cultured rat hepatocytes. Regulation by cholesterol availability, *J. Biol. Chem.*, 258, 3661, 1983.
107. **Gebhardt, R.**, Primary cultures of rat hepatocytes as a model system of canalicular development, biliary secretion, and intrahepatic cholestasis. III. Properties of the biliary transport of immunoglobulin A revealed by immunofluorescence, *Gastroenterology*, 84, 1462, 1983.
108. **Erttmann, R. R. and Damm, K. H.**, Probenecid-induced effects on bile flow and biliary excretion of ^{3}H-ouabain, *Arch. Int. Pharmacodyn. Ther.*, 223, 96, 1976.
109. **Levine, W. G.**, Effect of the hypolipidemic drug nafenopin {2-Methyl-2-[p-(1,2,3,4-tetrahydro-1-naphthyl)phenoxy]propionic acid; TPIA:su-13,437} on the hepatic disposition of foreign compounds in the rat, *Drug Metab. Dispos.*, 2, 178, 1974.
110. **Kitani, K., Kanai, S., and Miura, R.**, Increased biliary excretion of ouabain induced by bucolome in the rat, *Clin. Exp. Pharmacol. Physiol.*, 5, 117, 1978.

111. **Arias, I. M.,** Ligandin: review and uptake of a multifunctional protein, *Med. Biol.*, 57, 328, 1979.
112. **Maruyama, H., Inoue, M., Arias, I. M., and Listowsky, I.,** Ligandins or glutathione S-transferases: a family of multifunctional proteins in the rat, in *Functions of Glutathione: Biochemical, physiological, Toxicological, and Clinical Aspects*, Larsson, A., Orrenius, S., Holmgren, A., and Mannervik, B., Eds., Raven Press, New York, 1983, 89.
113. **Kirsch, R., Kamisaka, K., Fleischner, G., and Arias, I. M.,** Structural and functional studies of ligandin, a major renal organic anion binding protein, *J. Clin. Invest.*, 55, 1009, 1975.
114. **Meijer, D. K. F., Vonk, R. J., Keulemans, K., and Weitering, J. G.,** Hepatic uptake and biliary excretion of dibromosulphthalein. Albumin dependence, influence of phenobarbital and nafenopin pretreatment and the role of Y and Z protein, *J. Pharmacol. Exp. Ther.*, 202, 8, 1977.
115. **Boyer, J. L., Schwartz, J., and Smith, N.,** Biliary secretion in elasmobranchs. II. Hepatic uptake and biliary excretion of organic anions, *Am. J. Physiol.*, 230, 974, 1976.
116. **Ketterer, B., Tipping, E., Hackney, J. F., and Beale, D.,** A low molecular weight protein from rat liver that resembles ligandin in its binding properties, *Biochem. J.*, 155, 511, 1976.
117. **Fleischner, G. M., Meijer, D. K. F., Levine, W. G., Gatmaitan, Z., Gluck, R., and Arias, I. M.,** Effect of hypolipidemic drugs, nafenopin and clofibrate, on the concentration of ligandin and Z protein in rat liver, *Biochem. Biophys. Res. Commun.*, 67, 1401, 1976.
118. **Ockner, R. K., Manning, J. A., Poppenhauser, R. B., and Ho, W. K. L.,** A binding protein for fatty acids in cytosol of intestinal mucosa, liver, myocardium and other tissues, *Science*, 177, 56, 1972.
119. **Mullock, B. M., Dobrota, M., and Hinton, R. H.,** Sources of the protein and rat bile, *Biochim. Biophys. Acta*, 543, 497, 1978.
120. **Hwang, S. W.,** Plasma and hepatic binding of indocyanine green in guinea pigs of different ages, *Am. J. Physiol.*, 228, 718, 1975.
121. **Samuels, A. R., Freedman, J. H., and Bhargava, M. M.,** Purification and characterization of a novel abundant protein in rat bile that binds azo dye metabolites and copper, *Biochim. Biophys. Acta*, 759, 23, 1983.
122. **Jakoby, W. B., Ed.,** *Enzymatic Basis of Detoxication*, Academic Press, New York, 1980.
123. **Jakoby, W. B., Bend, J. R., and Caldwell, J., Eds.,** *Metabolic Basis of Deotxication*, Academic Press, New York, 1982.
124. **Estabrook, R. W. and Lindenlaub, E., Eds.,** *The Induction of Drug Metabolism*, F. K. Schattauer Verlag, Stuttgart, 1979.
125. **Branch, R. A., Shand, D. G., Wilkinson, G. R., and Nies, A. S.,** Increased clearance of antipyrine and d-propranolol after phenobarbital treatment in the monkey. Relative contributions of enzyme induction and increased hepatic blood flow, *J. Clin. Invest.*, 53, 1101, 1974.
126. **Ohnhaus, E. E. and Tilvis, R.,** Liver blood flow, metabolic heat production and body temperature before, during and after phenobarbitone administration, *Acta Hepato-Gastroenterol.*, 23, 404, 1976.
127. **Berthelot, P., Erlinger, S., Dhumeaux, D., and Preaux, A.,** Mechanism of phenobarbital-induced hypercholeresis in the rat, *Am. J. Physiol.*, 219, 809, 1970.
128. **Maxwell, J. D., Hunter, J., Stewart, D. A., Carrella, M., and Williams, R.,** Effect of phenobarbitone on bile flow and bilirubin metabolism in man and the rat, *Digestion*, 9, 138, 1973.
129. **Fischer, E., Gregus, Z., and Varga, F.,** Effects of barbiturates on the hepatic cytochrome P-450 dependent enzyme system and biliary excretion of exogenous organic anions in rats, *Digestion*, 20, 115, 1980.
130. **Roerig, D., Hasegawa, A., Peterson, R., and Wang, R.,** Effect of chloraquine and phenobarbitol on morphine glucuronidation and biliary excretion in the rat, *Biochem. Pharmacol.*, 23, 1331, 1974.
131. **Fuhrman-Lane, C. and Fujimoto, J. M.,** Trans-stilbene oxide administration increased hepatic glucuronidation of morphine but decreased biliary excretion of morphine glucuronide in rats, *J. Pharmacol. Exp. Ther.*, 222, 526, 1982.
132. **Zsgmond, G. and Solymoss, B.,** Effect of spironolactone, pregnenolone-16α-carbonitrile and cortisol on the metabolism and biliary excretion of sulfobromophthalein and phenol-3,6-dibromophthalein disulfonate in rats, *J. Pharmacol. Exp. Ther.*, 183, 499, 1972.
133. **Roerig, D. L., Hasegawa, A. T., and Wang, R. I. H.,** Effect of phenobarbital pretreatment on the metabolism and biliary excretion of methadone, *Biochem. Pharmacol.*, 24, 355, 1975.
134. **Levine, W. G.,** The role of microsomal drug-metabolizing enzymes in the biliary excretion of 3,4-benzpyrene in the rat, *J. Pharmacol. Exp. Ther.*, 175, 301, 1970.
135. **Levine, W. G.,** Metabolism and biliary excretion of *N*-2-fluorenylacetamide and *N*-hydroxy-2-fluorenylacetamide, *Life Sci.*, 10, 727, 1971.
136. **Levine, W. G.,** Hepatic uptake, metabolism, and biliary excretion of 7,12-dimethylbenzanthracene in the rat, *Drug Metab. Dispos.*, 2, 169, 1974.
137. **Levine, W. G.,** Biliary excretion of 3-methylcholanthrene as controlled by its metabolism, *J. Pharmacol. Exp. Ther.*, 183, 420, 1972.
138. **Levine, W. G.,** Induction and inhibition of the metabolism and biliary excretion of the azo dye carcinogen, *N,N*-dimethyl-4-aminoazobenzene (DAB) in the rat, *Drug Metab. Dispos.*, 8, 212, 1980.

139. **Mulder, G. J.**, The rate-limiting step in the biliary elimination of some substrates of uridine diphosphate glucuronyltransferase in the rat, *Biochem. Pharmacol.*, 22, 1751, 1973.
140. **Levine, W. G. and Bognacki, J.**, Biliary excretion of 3,4-benzpyrene in nafenopin treated rats, *J. Pharmacol. Exp. Ther.*, 196, 486, 1976.
141. **Priestly, B. G. and Plaa, G. L.**, Sulfobromophthalein metabolism and excretion in rats with iodomethane-induced depletion of hepatic glutathione, *J. Pharmacol. Exp. Ther.*, 174, 221, 1970.
142. **Varga, F., Fisher, E., and Szily, T. S.**, Biliary excretion of bromsulphthalein and glutathione conjugate of bromsulphthalein in rats pretreated with diethyl maleate, *Biochem. Pharmacol.*, 23, 2617, 1974.
143. **Levine, W. G. and Finkelstein, T. T.**, Biliary excretion of *N,N*-dimethyl-4-aminoazobenzene (DAB) in the rat, *Drug Metab. Dispos.*, 6, 265, 1978.
144. **Ketterer, B., Kadlubar, F., Flammang, T., Carne, T., and Enderby, G.**, Glutathione adducts of *N*-methyl-4-aminoazobenzene formed *in vivo* and by reaction of *N*-benzoyl-*N*-methyl-4-aminoazobenzene with glutathione, *Chem. Biol. Interact.*, 25, 7, 1979.
145. **Levine, W. G. and Finkelstein, T. T.**, A role for liver glutathione in the hepatobiliary fate of *N-N*-dimethyl-4-aminoazobenzene, *J. Pharmacol. Exp. Ther.*, 208, 399, 1979.
146. **Jezequel, A.-M., Bonazzi, P., Amabili, P., Venturini, C., and Orlandi, F.**, Changes of the golgi apparatus induced by diethylmaleate in rat hepatocytes, *Hepatology*, 2, 856, 1982.
147. **Younes, M. and Siegers, C. P.**, Lipid peroxidation as a consequence of glutathione depletion in rat and mouse liver, *Res. Commun. Chem. Pathol. Pharmacol.*, 27, 119, 1980.
148. **Burk, R. F. and Correia, M. A.**, Stimulation of rat hepatic microsomal heme oxygenase by diethyl maleate, *Res. Commun. Chem. Pathol. Pharmacol.*, 24, 205, 1979.
149. **Levine, W. G.**, Glutathione and hepatic mixed-function oxidase activity, *Drug Metab. Rev.*, 14, 909, 1983.
150. **Levine, W. G. and Lee, S.**, Effect of glutathione on the metabolism of *N,N*-dimethyl-4-aminoazobenzene by rat liver microsomes, *Drug Metab. Dispos.*, 11, 239, 1983.
151. **Schulze, P. and Czok, G.**, Reduced bile flow in rats during sulfobromophthalein infusion, *Toxicol. Appl. Pharmacol.*, 32, 213, 1975.
152. **Dhumeaux, D., Berthelot, P., Preaux, A. M., Erlinger, S., and Fauvert, R.**, A critical study of the concept of maximal biliary transport of sulfobromophthalein (BSP) in the Wistar rat, *Rev. Eur. Etud. Clin. Biol.*, 15, 279, 1970.
153. **Laperche, Y. and Oudea, P.**, Inhibition by sulfobromophthalein of mitochondrial translocation of anions and adenine dinucleotides: effects of liver adenosine triphosphate and possible correlation with inhibition of bile flow in the rat, *J. Pharmacol. Exp. Ther.*, 197, 235, 1974.
154. **Schwenk, M., Burr, R., Baur, H., and Pfaff, E.**, Interaction of bromosulfophthalein with mitochondrial membranes: effect on ion movements, *Biochem. Pharmacol.*, 26, 825, 1977.
155. **Meijer, D. K. F., Bognacki, J., and Levine, W. G.**, Effect of nafenopin (Su-13,437) on liver function. Hepatic uptake and biliary excretion of ouabain in the rat, *Drug Metab. Dispos.*, 3, 220, 1975.
156. **Levine, W. G., Braunstein, I. R., and Meijer, D. K. F.**, Effect of nafenopin (Su-13,437) on liver function. Mechanism of the choleretic effect, *Naunyn-Schmiedeberg's Arch. Pharmacol.*, 290, 221, 1975.
157. **Uesugi, T. and Levine, W. G.**, Effect of nafenopin (Su-13,437) on liver function. Influence on the hepatic transport of phenolphthalein glucuronide and chlorothiazide, *Drug Metab. Dispos.*, 4, 107, 1976.
158. **Jorritsma, J., Meerman, J. H. N., Vonk, R. J., and Mulder, G. J.**, Biliary and urinary excretion of drug conjugates: effect of diuresis and choleresis on excretion of harmol sulfate and harmol glucuronide in the rat, *Xenobiotica*, 9, 247, 1979.
159. **Mueller, M. N. and Kappas, A.**, Estrogen pharmacology. I. The influence of estradiol and estriol on hepatic disposal of sulfobromophthalein (BSP) in man, *J. Clin. Invest.*, 43, 1905, 1964.
160. **Meyers, M., Slikker, W., Pascoe, G., and Vore, M.**, Characterization of cholestasis induced by estradiol-17-β-D-glucuronide in the rat, *J. Pharmacol. Exp. Ther.*, 214, 87, 1980.
161. **Slikker, W. and Vore, M.**, Hepatotoxic effects of estradiol glucuronide in the rat and monkey, *J. Pharmacol. Exp. Ther.*, 225, 138, 1983.
162. **Plaa, G. L. and Priestly, B. G.**, Intrahepatic cholestasis induced by drugs and chemicals, *Pharmacol. Rev.*, 28, 207, 1976.
163. **Gumucio, J. J., Accatino, L., Macho, A. M., and Contreras, A.**, Effect of phenobarbital on the etinyl estradiol-induced cholestasis in the rat, *Gastroenterology*, 65, 651, 1973.
164. **Forker, E. L.**, The effect of estrogen on bile formation in the rat, *J. Clin. Invest.*, 48, 654, 1969.
165. **Davis, R. A., Kern, F., Showalter, R., Sutherland, E., Sinensky, M., and Simon, F. R.**, Alterations of hepatic Na^+, K^+-ATPase and bile flow by estrogen: effects on liver surface membrane lipid structure and function, *Proc. Natl. Acad. Sci. U.S.A.*, 75, 4130, 1978.
166. **Simon, F. R., Gonzalez, M., Sutherland, E., Accatino, L., and Davis, R. A.**, Reversal of ethinyl estradiol-induced bile secretory failure with Triton WR-1339, *J. Clin. Invest.*, 65, 851, 1980.

167. **Boelsterli, U. A., Rakhit, G., and Balazs, T.,** Modulation by S-adenosyl-l-methionine of hepatic Na^+,-K^+-ATPase, membrane fluidity, and bile flow in rats with ethinyl estradiol-induced cholestasis, *Hepatology*, 3, 12, 1983.
168. **Stramentronoli, G., Padova, C. D., Gualano, M., Rovagnati, P., and Galli-Kienle, M.,** Ethinylestradiol-induced impairment of bile secretion in the rat: protective effects of S-adenosyl-l-methionine and its implication in estrogen metabolism, *Gastroenterology*, 80, 154, 1981.
169. **Montgomery, G. and Vore, M.,** The effect of diethylstilbestrol treatment on the metabolism and biliary excretion of [^{14}C]phenytoin in the isolated perfused rat liver, *Toxicol. Appl. Pharmacol.*, 58, 510, 1981.
170. **Tavoloni, N. and Boyer, J. L.,** Relationship between hepatic metabolism of chlorpromazine and cholestatic effects in isolated perfused rat liver, *J. Pharmacol. Exp. Ther.*, 214, 269, 1980.
171. **Zimmerman, H. J., Abernathy, C. O., Lukacs, L., and Ezekiel, M.,** Effects of ticrynafen on hepatic excretory function in the isolated perfused rat liver, *Hepatology*, 2, 255, 1982.
172. **Bhattacharyya, M. H. and Peterson, D. P.,** Action of DTPA on hepatic plutonium. III. Evidence for a direct chelation mechanism for DTPA-induced excretion of monomeric plutonium into rat bile, *Radiat. Res.*, 80, 108, 1979.
173. **Figueroa, W. G. and Thompson, J. H.,** Biliary iron excretion in normal and iron loaded rats after desferrioxamine and Ca DTPA, *Am. J. Physiol.*, 215, 807, 1968.
174. **Kiyozuma, M. and Kojima, S.,** Studies on poisonous metals. Excretion of cadmium through bile and gastrointestinal mucosa and effect of chelating agents on its excretion in cadmium pretreated rats, *Chem. Pharm. Bull*, 26, 3410, 1978.
175. **Norseth, T.,** The effect of diethyldithiocarbamate on biliary transport, excretion and organ distribution of mercury in the rat after exposure to methyl mercuric chloride, *Acta Pharmacol. Toxicol.*, 34, 76, 1974.
176. **Kitani, K., Miura, R., Kanai, S., and Morita, Y.,** The effect of spironolactone pretreatment on the biliary excretion and renal accumulation of inorganic mercury in the rat, *Biochem. Pharmacol.*, 26, 1823, 1977.
177. **Klaassen, C. D.,** Effect of spironolactone on the biliary excretion and distribution of metals, *Toxicol. Appl. Pharmacol.*, 50, 41, 1979.
178. **Collins, W. T. and Capen, C. C.,** Biliary excretion of ^{125}I-thyroxine and fine structure alterations in the thyroid glands of Gunn rats fed polychlorinated biphenyls, *Lab. Invest.*, 43, 158, 1980.
179. **Levine, W. G.,** Effect of DMSO on the hepatic disposition of chemical carcinogens, *Ann. N.Y. Acad. Sci.*, 243, 185, 1975.
180. **Bailey, D. G., Pual, H., and Johnson, G. E.,** Pentobarbital inhibits the biliary excretion of organic acids: a study with succinylsulfathiazole in the rat, *Can. J. Physiol. Pharmacol.*, 53, 470, 1975.
181. **Cooper, B., Eakins, M. N., and Slater, T. F.,** The effect of various anesthetic techniques on the flow rate, constituents and enzymic composition of rat bile, *Biochem. Pharmacol.*, 25, 1711, 1976.
182. **Cooke, W. J. and Cooke, L. M.,** Effects of anesthetics on the hepatic metabolism and biliary excretion of iopanoic acid enantiomers in rat, *J. Pharmacol. Exp. Ther.*, 225, 85, 1983.
183. **Peterson, R. E., Hamada, N., Yang, K. H., Madhukar, B. V., and Matsumura, F.,** Reversal of 2,3,7,8-tetrachlorodibenzo-*p*-dioxin induced depression of ouabain biliary excretion by pregnenolone-16α-carbonitrile and spironolactone in isolated perfused rat livers, *Toxicol. Appl. Pharm.*, 50, 407, 1979.
184. **Marinovic, Y., Glasinovic, J., Semelle, B., Boivieux, J., and Erlinger, S.,** Facilitation of hepatic uptake of phenol-3,6-dibromophthalein disulfonate by taurocholate, *Am. J. Physiol.*, 232, E560, 1977.
185. **Delage, Y., Erlinger, S., Duval, M., and Benhamon, J. P.,** Influence of dehydrocholate and taurocholate on bromosulphthalein uptake, storage, and excretion in the dog, *Gut*, 16, 105, 1976.
186. **Vonk, R. J., Danhof, M., Coenraads, T., Van Doorn, A. B. D., Keulemans, K., Scof, A. H. J., and Meijer, D. K. F.,** Influence of bile salts on hepatic transport of dibromosulphthalein, *Am. J. Physiol.*, 237, E524, 1979.
187. **Gregus, Z., Fischer, E., and Varga, F.,** Effect of sodium taurocholate on the hepatic transport of bromosulphthalein in rats, *Arch. Int. Pharmacodyn. Ther.*, 238, 124, 1979.
188. **Abou-El-Makarem, M. M., Millburn, P., and Smith, R. L.,** Biliary excretion of [^{14}C]succinyl-sulphathiazole in the rat and rabbit, *Biochem. J.*, 105, 1295, 1967.
189. **Kuo, S. H. and Johnson, G. E.,** The influence of [^{3}H]taurocholate on the biliary excretion of [^{14}C]acetylprocainamide ethobromide, *Can. J. Physiol. Pharmacol.*, 53, 888, 1975.
190. **Vonk, R. J., Scholtens, E., Kuelemans, G. T. P., and Meijer, D. K. F.,** Choleresis and hepatic transport mechanism. IV. Influence of bile salt chloresis on the hepatic transport of the organic cations, d-tubocurarine and N^4-acetyl procainamide ethobromide, *Naunyn-Schmiedeberg's Arch. Pharmacol.*, 302, 1, 1978.
191. **Boyer, J., Schieg, R. L., and Klatskin, G.,** Effect of sodium taurocholate on the hepatic metabolism of sulfobromophthalein sodium (BSP). The role of bile flow, *J. Clin. Invest.*, 49, 206, 1970.
192. **O'Maille, E. R. L., Richards, T. G., and Short, A. H.,** Factors determining the maximal rate of organic anion secretion by the liver and further evidence on the hepatic site of action of the hormone secretion, *J. Physiol. (London)*, 186, 424, 1966.

193. **Barnhart, J., Ritt, D., Ware, A., and Combes, B.,** A comparison of the effects of taurocholate and theophylline on BSP excretion in dogs, in *The Liver: Quantitative Aspects of Structure and Function*, Paumgartner, G. and Preisig, R., Eds., S. Karger, Basel, 1973, 315.
194. **Erlinger, S. and Dumont, M.,** Influence of canalicular bile flow on sulfobromophthalein transport maximum in bile in the dog, in *The Liver: Quantitative Aspects of Structure and Function*, Paumgartner, G. and Preisig, R., Eds., S. Karger, Basel, 1973, 306.
195. **Gibson, G. E. and Forker, E. L.,** Canalicular bile flow and bromosulfophthalein transport maximum: the effect of a bile salt independent choleretic, SC-2644, *Gastroenterology*, 66, 1046, 1974.
196. **Scharschmidt, B. F. and Schmid, R.,** The micellar sink. A quantitative assessment of the association of organic anions with mixed micelles and other macromolecular aggregates in rat bile, *J. Clin. Invest.*, 62, 1122, 1978.
197. **Vonk, R. J., Van Doorn, A. B. D., Scaf, A. H. J., and Meijer, D. K. F.,** Choleresis and hepatic transport mechanisms. III. Binding of ouabain and K-strophanthoside to biliary micelles and influence of choleresis on their biliary excretion, *Naunyn-Schmiedeberg's Arch. Pharmacol.*, 300, 173, 1977.
198. **Binet, S., Delage, Y., and Erlinger, S.,** Influence of taurocholate, taurochenodeoxycholate and taurodehydrocholate on sulfobromophthalein transport into bile, *Am. J. Physiol.*, 236, E10, 1979.
199. **Angelucci, L., Linari, G., and Baldieri, M.,** Caerulein effect on the biliary excretion of colorants, *Pharmacol. Res. Commun.*, 7, 311, 1975.
200. **Kellaway, I. W. and Marriott, C.,** The interaction of quaternary ammonium compounds with bile salts. I. The interaction of dodecyltrimethylammonium bromide with sodium deoxycholate, *Can. J. Pharm. Sci.*, 12, 70, 1977.
201. **Kellaway, I. and Marriott, C.,** The interaction of quaternary ammonium compound with bile salts. II. The influence of alkyl chain length on the interaction of alkylmethylammonium bromides with sodium deoxycholate, *Can. J. Pharm. Sci.*, 12, 74, 1977.
202. **Bertagni, P., Hirom, P. C., Millburn, P., Osiyemi, F. O., Smith, R. L., Turbert, H. B., and Williams, R. T.,** Sex and species differences in the biliary excretion of tartrazine and lissamine fast yellow in the rat, guinea pig, and rabbit. The influence of sex hormones in tartrazine excretion in the rat, *J. Pharm. Pharmacol.*, 24, 620, 1972.
203. **Uesugi, T., Bognacki, J., and Levine, W. G.,** Biliary excretion of drugs in the rat during liver regeneration, *Biochem. Pharmacol.*, 25, 1187, 1976.
204. **Bond, J. A., Medinsky, M. A., Dent, J. G., and Rickert, D. E.,** Sex-dependent metabolism and biliary excretion of [2,4-^{14}C]dinitrotoluene in isolated perfused rat livers, *J. Pharmacol. Exp. Ther.*, 219, 598, 1981.
205. **Kato, R.,** Sex-related differences in drug metabolism, *Drug Metab. Rev.*, 3, 1, 1974.
206. **Kato, R. and Kamataki, T.,** Cytochrome P-450 as a determinant of sex difference in drug metabolism in the rat, *Xenobiotica*, 12, 787, 1982.
207. **Morris, D. H. and Silverman, J. A.,** Biliary excretion of [^{3}H] aldosterone and its sex dependence in adrenalectomized rats, *Endocrinology*, 96, 1386, 1975.
208. **Buttar, H. S., Coldwell, B. B., and Thomas, B. H.,** Excretion of ^{14}C-pentobarbital and its metabolites into the bile and urine of male and female rats, *Arch. Int. Pharmacodyn. Ther.*, 208, 279, 1974.
209. **Colby, H. D. and Kitay, J. T.,** Sex and substrate effects on hepatic corticosteroid metabolism in the rat, *Endocrinology*, 90, 473, 1972.
210. **Neims, A. H., Warner, M., Loughnan, P. M., and Aranda, J. V.,** Developmental aspects of the hepatic cytochrome P-450 monooxygenase system, *Annu. Rev. Pharmacol. Toxicol.*, 16, 427, 1976.
211. **Dutton, G. J.,** Developmental aspects of drug conjugation with special reference to glucuronidation, *Annu. Rev. Pharmacol. Toxicol.*, 18, 17, 1978.
212. **Richey, D. P. and Bender, A. D.,** Pharmacokinetic consequences of aging, *Annu. Rev. Pharmacol. Toxicol.*, 17, 49, 1977.
213. **Fischer, E., Barth, A., Varga, F., and Klinger, W.,** Age dependence of hepatic transport in control and phenobarbital-pretreated rats, *Life Sci.*, 24, 557, 1979.
214. **Kitani, K., Kanai, S., Miura, R., Morita, Y., and Kasahara, M.,** The effect of aging on the biliary excretion of ouabain in the rat, *Exp. Gerontol.*, 13, 9, 1978.
215. **Fischer, L. J. and Weissinger, J. L.,** Development in the newborn rat of the conjugation and de-conjugation processes involved in the enterohepatic circulation of diethylstilbestrol, *Xenobiotica*, 2, 399, 1972.
216. **Cagen, S. Z. and Gibson, J. E.,** Effect of carbon tetrachloride on hepatic transport of ouabain in developing rats, *Proc. Soc. Exp. Biol. Med.*, 154, 188, 1977.
217. **Ballatori, N. and Clarkson, T. W.,** Developmental changes in the biliary excretion of methylmercury and glutathione, *Science*, 216, 61, 1982.
218. **Dowling, R. H.,** The enterohepatic circulation, *Gastroenterology*, 62, 122, 1972.
219. **Plaa, G. L.,** The enterohepatic circulation, in *Handbook of Experimental Pharmacology*, Part 3, Vol. 28, Gillette, J. R. and Mitchell, J. R., Eds., Springer-Verlag, New York, 1975, 130.

220. **Scheline, R. R.,** Metabolism of foreign compounds by gastrointestinal microorganisms, *Pharmacol. Rev.*, 25, 451, 1973.
221. **Goldman, P.,** The role of intestinal microflora, in *Metabolic Basis of Detoxication*, Jakoby, W. B., Bend, J. R., and Caldwell, J., Eds., Academic Press, New York, 1982, 323.
222. **Marselos, M., Dutton, G., and Hanninen, O.,** Evidence that D-glucaro1,4-lactone shortens the pharmacological action of drugs being disposed via the bile as glucuronides, *Biochem. Pharmacol.*, 24, 1885, 1975.
223. **Walsh, C. T. and Levine, R. R.,** Studies on the enterohepatic circulation of morphine in the rat, *J. Pharmacol. Exp. Ther.*, 195, 303, 1975.
224. **Fischer, L. J., Kent, T. H., and Weissinger, J. L.,** Absorption of diethylstilbestrol and its glucuronide conjugate from the intestine of five- and twenty-five-old rats, *J. Pharmacol. Exp. Ther.*, 185, 163, 1973.
225. **Brewster, D., Jones, R. S., and Symons, A. M.,** Effect of neomycin on the biliary excretion and enterohepatic circulation of mestranol and 17β-estradiol, *Biochem. Pharmacol.*, 26, 943, 1977.
226. **Sim, S. M., Huijghebarert, S., Back, D. V., and Eyssen, H. J.,** Gastrointestinal absorption of estrone sulfate in germfree and conventional rats, *J. Steroid Biochem.*, 18, 499, 1983.
227. **Dobbs, H. E., Hall, J. M., and Steiger, B.,** Enterohepatic circulation of etorphine, a potent analgesic, in the rat, *Proc. Eur. Soc. Study Drug Toxicol.*, 11, 73, 1970.
228. **Steinberg, S. E., Campbell, C. L., Bleyer, W. A., and Hillman, R. S.,** Enterohepatic circulation of methotrexate in rats in vivo, *Cancer Res.*, 42, 1279, 1982.
229. **Shafii, A., Chowdhury, J. R., and Das, K. M.,** Absorption, enterohepatic circulation and excretion of 5-aminosalicylic acid in rats, *Am. J. Gastroenterol.*, 77, 297, 1982.
230. **Medinsky, M. A. and Dent, J. G.,** Biliary excretion and enterohepatic circulation of 2,4-dinitrotoluene metabolites in Fischer-344 rats, *Toxicol. Appl. Pharmacol.*, 68, 359, 1983.
231. **Duggan, D., Hooke, K., Noll, R., and Kwan, K.,** Enterohepatic circulation of indomethacin and its role in intestinal irritation, *Biochem. Pharmacol.*, 25, 1749, 1975.
232. **Roerig, D. L., Hasegawa, A. T., and Wang, R. I. H.,** Enterohepatic circulation of 1-α-acetylmethadol in the rat, *J. Pharmacol. Exp. Ther.*, 213, 284, 1980.
233. **Levy, R. I., Fredrickson, D. S., Stone, N. J., Bilheimer, D. W., Brown, M. V., Glueck, C. J., Gotto, A. M., Herbert, P. N., Kwiterovich, P. O., Langer, T., La Rosa, J., Lux, S. E., Rider, A. K., Shulmon, R. S., and Sloan, H. R.,** Cholestyramine in type II hyperlipoproteinemia — a double-blind trial, *Ann. Int. Med.*, 79, 51, 1973.
234. **Cohn, W. J., Boylan, J. J., Blanke, R. V., Fariss, M. W., Howell, J. R., and Guzelian, P. S.,** Treatment of chlordecone (Kepone) toxicity with cholestyramine. Results of a controlled clinical trial, *N. Engl. J. Med.*, 298, 243, 1978.
235. **Chipman, J. K.,** Bile as a source of potential reactive metabolites, *Toxicology*, 25, 99, 1982.
236. **Hunt, R. D., Leveille, G. A., and Sauerlick, H. E.,** Dietary bile acids and lipid metabolism. III. Effects of lithocholic acid in mammalian species, *Proc. Soc. Exp. Biol. Med.*, 115, 277, 1964.
237. **Carey, J. B. and Williams, G.,** Metabolism of lithocholic acid in bile fistula patients, *J. Clin. Invest.*, 42, 450, 1963.
238. **Holland, J. C. and Spain, J. D.,** Bile acid involvement in azo dye carcinogenesis, *Cancer Res.*, 31, 538, 1971.
239. **Reddy, B. S.,** Role of bile metabolites in colon carcinogenesis, *Cancer*, 36, 2401, 1975.
240. **Narisawa, T., Magadia, N., Weisburger, J., and Wynder, E. L.,** Promoting effect of bile acids on colon carcinogenesis after intrarectal instillation of *N*-methyl-*N'*-nitro-*N*-nitrosoguanidine in rats, *J. Natl. Cancer Inst.*, 53, 1093, 1974.
241. **Hill, M. J.,** The role of colon anaerobes in the metabolism of bile acids and steroids and its relation to colon cancer, *Cancer*, 36, 2387, 1975.
242. **Turjman, N. and Nair, P. P.,** Nature of tissue-bound lithocholic acid and its implications in the role of bile acids in carcinogenesis, *Cancer Res.*, 41, 3761, 1981.
243. **Fiala, E. S.,** Investigations into the metabolism and mode of action of the colon carcinogen, 1,2-dimethylhydrazine, *Cancer*, 36, 2407, 1975.
244. **Conner, T. H., Forti, G. C., Sitra, P., and Legator, M. S.,** Bile as a source of mutagenic metabolites produced in vivo and detected by *Salmonella typhimurium*, *Environ. Mutagen.*, 1, 269, 1979.
245. **Forti, G. C. and Trieff, N. M.,** Kinetics of uptake and biliary excretion of benzo(a)pyrene and mutagenic metabolites in isolated perfused rat liver, *Teratogen. Carcinogen. Mutagen.*, 1, 269, 1980.
246. **Rannug, U. and Beije, B.,** The mutagenic effect of 1,2-dichlorethane on *Salmonella* typhimurium. II. Activation by the isolated perfused liver, *Chem. Biol. Interact.*, 24, 265, 1979.
247. **Narisawa, T., Satoh, M., Sano, M., and Takahashi, T.,** Inhibition of initiation and promotion by *N*-methylnitrosourea-induced colon carcinogenesis in rats by the non-steroid antiinflammatory agent indomethacin, *Carcinogenesis*, 4, 1225, 1983.
248. **O'Brien, T. E.,** Excretion of drugs in human milk, *Am. J. Hosp. Pharm.*, 31, 844, 1974.

249. **Vorherr, H.,** Drug excretion in breast milk, *Postgrad. Med.*, 56, 97, 1974.
250. **Giacoia, G. P. and Catz, C. S.,** Drugs and pollutants in breast milk, *Clin. Perinatol.*, 6, 181, 1979.
251. **Berlin, C. M.,** The excretion of drugs in human milk, *Prog. Clin. Biol. Res.*, 36, 115, 1980.
252. **Wilson, J. T., Brown, R. D., Cherek, D. R., Dailey, J. W., Hilman, B., Jobe, P. C., Manno, B. R., Manno, J. E., Redetzki, H. M., and Stewart, J. J.,** Drug excretion in human breast milk. Principles, pharmacokinetics and projected consequences, *Clin. Pharmacokin.*, 5, 1, 1980.
253. **Kochenour, N. K.,** Drugs in lactating women, *Obstet. Gynecol. Annu.*, 10, 107, 1981.
254. **Welch, R. M. and Findlay, J. W. A.,** Excretion of drugs in human breast milk, *Drug Metab. Rev.*, 12, 261, 1981.
255. **Nickerson, S. C., Smith, J. J., and Keenan, T. W.,** Role of microtubules in milk secretion. — action of colchicine on microtubules and exocytosis of secretory vesicles in rat mammary epithelial cells, *Cell Tissue Res.*, 207, 361, 1980.
256. **Levine, W. G.,** Excretion mechanisms, in *Biological Basis of Detoxication*, Caldwell, J. and Jakoby, W. B., Eds., Academic Press, New York, 1983, 251.
257. **Steen, B. and Rane, A.,** Clindamycin passage into human milk, *Br. J. Clin. Pharmacol.*, 13, 661, 1982.
258. **Lewis, A. M., Patel, L., Johnston, A., and Turner, P.,** Mexiletine in human blood and breast milk, *Postgrad. Med. J.*, 57, 546, 1981.
259. **Liederholm, H., Wahlin-Boll, E., Hanson, A., Ingemarsson, I., and Melander, A.,** Transplacental passage and breast milk concentrations of hydralazine, *Eur. J. Clin. Pharmacol.*, 21, 417, 1982.
260. **Reinhardt, D., Richter, O., Geny, T., and Potthoff, S.,** Kinetics of the translactal passage of digoxin from breast feeding mothers to their infants, *Eur. J. Pediatr.*, 138, 49, 1982.
261. **Bailey, D. N., Weibert, R. T., Naylor, A. J., and Shaw, R. F.,** A study of salicylate and caffeine excretion in the breast milk of two nursing mothers, *J. Anal. Toxicol.*, 6, 64, 1982.
262. **Berlin, C. M. and Vesell, E. S.,** Antipyrine disposition in milk and saliva of lactating women, *Clin. Pharmacol. Ther.*, 31, 38, 1982.
263. **Brixen-Rasmussen, L., Halgrener, J., and Jorgensen, A.,** Amitriptyline and nortriptyline excretion in human breast milk, *Psychopharmacology*, 76, 94, 1982.
264. **Eriksson, G. and Swahn, C. G.,** Concentrations of baclofen in serum and breast milk from a lactating women, *Scand. J. Clin. Lab. Invest.*, 41, 185, 1981.
265. **Kirk, L. and Jorgensen, A.,** Concentrations of cis(Z)-flupentixol in maternal serum, amniotic fluid, umbilical cord serum and milk, *Psychopharmacology*, 72, 107, 1980.
266. **Krause, W., Stoppelli, I., Milia, S., and Rainer, E.,** Transfer of mepindolol to newborns by breast-feeding mothers after single and repeated daily doses, *Eur. J. Clin. Pharmacol.*, 22, 53, 1982.
267. **Nau, H., Kuhnz, W., Egger, H. J., Rating, D., and Helge, H.,** Anticonvulsants during pregnancy and lactation. Transplacental, maternal and neonatal pharmacokinetics, *Clin. Pharmacokinet.*, 7, 508, 1982.
268. **Nicholas, J. M., Lipshitz, J., and Schreiber, E. C.,** Phencyclidine: its transfer across the placenta as well as into breast milk, *Am. J. Obstet. Gynecol.*, 15, 143, 1982.
269. **Rane, A. and Tunell, R.,** Ethosuximide in human milk and in plasma of a mother and her nursed infant, *Br. J. Clin. Pharmacol.*, 12, 855, 1981.
270. **Erickson, S. H., Oppenheim, G. L., and Smith, G. H.,** Metronidazole in breast milk, *Obstet. Gynecol.*, 57, 48, 1981.
271. **Jarboe, C. H., Cook, L. N., Malesic, I., and Fleischaker, J.,** Dyphylline elimination kinetics in lactating women: blood to milk transfer, *J. Clin. Pharmacol.*, 21, 405, 1981.
272. **Schwallie, P. C.,** The effect of depot-medroxyprogesterone acetate on the fetus and nursing infant: a review, *Contraception*, 23, 375, 1981.
273. **Stoppelli, I., Rainer, E., and Humpel, M.,** Transfer of cyproterone acetate to the milk of lactating women, *Contraception*, 22, 485, 1980.
274. **Johansen, K., Anderson, A. N., Kampwann, J. P., Molholm-Hansen, J. M., and Mortensen, H. B.,** Excretion of methimazole in human milk, *Eur. J. Clin. Pharmacol.*, 23, 339, 1982.
275. **Toddywalla, N. S., Mehta, S., Virkar, K. D., and Saxena, B. N.,** Release of 19-nor-testosterone type of contraceptive steroids through different drug delivery systems into serum and breast milk of lactating women, *Contraception*, 21, 217, 1980.
276. **Findlay, J. W. A., De Angelis, R. L., Kearney, M. F., Welch, R. M., and Findlay, J. M.,** Analgesic drugs in breast milk and plasma, *Clin. Pharmacol. Ther.*, 29, 625, 1981.
277. **Rogan, W. J., Bagniewska, M., and Damstra, T.,** Pollutants in breast milk, *N. Engl. J. Med.*, 302, 1450, 1980.
278. **Bagnell, P. C. and Ellenberger, H. A.,** Obstructive jaundice due to a chlorinated hydrocarbon in breast milk, *Can. Med. Assoc. J.*, 117, 1047, 1977.
279. **Mirkin, B. L.,** Diphenylhydantoin: placental transport, fetal localization, neonatal metabolism, and possible teratogenic effects, *J. Pediatr.*, 78, 329, 1971.
280. **Knowles, J. A.,** Breast milk: a source of more than nutrition for the neonate, *Clin. Toxicol.*, 7, 69, 1974.

281. **Amato, D. and Niblett, J. S.,** Neutropenia from cyclophosphosamide in breast milk, *Med. J. Aust.*, 1, 383, 1977.
282. **Jakubovic, A., Tait, R. M., and McGeer, P. L.,** Excretion of THC and its metabolites in ewe's milk, *Toxicol. Appl. Pharmacol.*, 28, 38, 1974.
283. **Chao, F. C., Green, D. E., Forest, I. S., and Kaplan, J. N.,** The passage of [^{14}C]-Δ-9-tetrahydrocannabinol into the milk of lactating squirrel monkeys, *Res. Commun. Chem. Pathol. Pharmacol.*, 15, 303, 1976.
284. **Hart, L. and Fouts, J.,** Hepatic drug metabolism during the perinatal period, *Ann. N.Y. Acad. Sci.*, 123, 245, 1965.
285. **Borzellica, J. F. and Cherrick, H. M.,** The excretion of drugs in saliva, *J. Oral Ther. Pharmacol.*, 2, 180, 1965.
286. **Glynn, J. P. and Bastain, W.,** Salivary excretion of paracetamol in man, *J. Pharm. Pharmacol.*, 25, 420, 1973.
287. **Koysooka, R., Ellis, E. F., and Levy, G.,** Relationship between theophylline concentration in plasma and saliva of man, *Clin. Pharmacol. Ther.*, 15, 454, 1974.
288. **Matin, S. B., Wan, S. H., and Karam, J. H.,** Pharmacokinetics of tolbutamide: prediction by concentration in saliva, *Clin. Pharmacol. Ther.*, 16, 1052, 1974.
289. **Huffman, D. H.,** Relationship between digoxin concentration in serum and saliva, *Clin. Pharmacol. Ther.*, 17, 310, 1975.
290. **Cook, C. E., Amerson, E., and Poole, W. K.,** Phenytoin and phenobarbital concentrations in saliva and plasma measured by radioimmunoassay, *Clin. Pharmacol. Ther.*, 18, 742, 1975.
291. **Horning, M. G., Brown, L., and Nowlin, J., Lertratanangkoon, K., Kellaway, P., and Zion, T. E.,** Use of saliva in therapeutic drug monitoring, *Clin. Chem.*, 23, 157, 1977.
292. **Watanabe, J., Nakase, Y., Urasaki, Y., Hayashi, Y., Iwamoto, K., and Ozeki, S.,** Protein binding effects on salivary excretion of phenobarbital in dogs, *J. Pharmacobiodyn.*, 4, 968, 1981.
293. **Wood, J. H., Flora, K. P., Narasimhachari, N., and Baker, C. A.,** Dependence of salivary drug concentrations on salivary flow rate, *Met. Find. Exp. Clin. Pharmacol.*, 4, 255, 1982.
294. **Selinger, D., Simmons, S., Harler, A. W., Nurnberger, J. I., and Gershon, E. S.,** An effective method for measuring salivary lithium in patients on anticholinergic drugs, *Biol. Psychiatr.*, 17, 1145, 1982.
295. **Basseches, P. J. and DiGregorio, G. J.,** Excretion of procainamide into bile and saliva in rats with chronic renal failure, *Arch. Int. Pharmacodyn. Ther.*, 257, 180, 1982.
296. **Dawes, C. and Jenkins, G. N.,** The effects of different stimuli on the composition of saliva in man, *J. Physiol. (London)*, 170, 86, 1964.
297. **Mucklow, J. C., Bending, M. R., Kahn, G. C., and Dollery, C. T.,** Drug concentration in saliva, *Clin. Pharmacol. Ther.*, 24, 563, 1978.
298. **Stephen, K. W., McCrossan, J., Mackenza, D., MacFarlane, C. B., and Spiers, C. F.,** Factors determining the passage of drugs from blood into saliva, *Br. J. Clin. Pharmacol.*, 9, 51, 1980.
299. **Sardi, A., Vettaro, M. P., Alesina, R., Grasso, S., Meinardi, G., deCarneri, I., and Tamassia, V.,** Serum and saliva levels of trimethoprim-sulfamethopyrazine combination in man, *Eur. J. Clin. Pharmacol.*, 20, 113, 1981.
300. **deVries-Hosper, H. G. and van der Waaij, D.,** Salivary concentrations of amphotericine B following its use as oral lozenge, *Infection*, 8, 63, 1980.
301. **Yosselson-Superstine, S., Yamuka, Y., and Ishai, S.,** Relationship between quinidine concentrations measured in saliva and erythrocytes, and in serum, *Int. J. Clin. Pharmacol.*, 20, 181, 1982.
302. **Baumann, P., Tinguely, D., Koeb, L., Schopf, J., and Le, P. K.,** On the relation between free plasma and saliva amitriptyline and nor-amitriptyline, *Int. Pharmacopsychiatr.*, 17, 136, 1982.
303. **Kragh-Sorenson, P. and Larsen, N. E.,** Factors affecting nortriptyline steady-state kinetics plasma and saliva levels, *Clin. Pharmacol. Ther.*, 28, 796, 1980.
304. **Ben-Aryeh, H., Naon, H., Szurgel, R., Gutman, D., and Hefetz, A.,** Salivary lithium concentration-a tool for monitoring psychiatric patients, *Oral Surg.*, 50, 127, 1980.
305. **deGier, J. J., Hart, B. J., Wilderink, P. F., and Nelemans, F. A.,** Comparison of plasma and saliva levels of diazepam, *Br. J. Clin. Pharmacol.*, 10, 151, 1980.
306. **Hallstrom, C. and Lader, M. H.,** Diazepam and *N*-desmethyl-diazepam concentrations in saliva, plasma and CSF, *Br. J. Clin. Pharmacol.*, 9, 333, 1980.
307. **Giles, H. G., Miller, R., MacLeod, S. M., and Sellers, E. M.,** Diazepam and *N*-desmethyldiazepam in saliva of hospital inpatients, *J. Clin. Pharmacol.*, 20, 71, 1980.
308. **Freeborn, S. F., Calvert, R. T., Black, P., MacFarlane, T., and D'Souza, S. W.,** Salivary and blood pethidine concentrations in the mother and the newborn baby, *Br. J. Obstet. Gynecol.*, 87, 966, 1980.
309. **Newton, R., Broughton, L. J., Lind, M. J., Morrison, P. J., Rogers, H. J., and Bradbrook, I. D.,** Plasma and salivary pharmacokinetics of caffeine in man, *Eur. J. Clin. Pharmacol.*, 21, 45, 1981.
310. **Khanna, N. N., Bada, H. S., and Somani, S. M.,** Use of salivary concentration in prediction of serum caffeine and theophylline concentrations in premature infants, *J. Pediatr.*, 96, 494, 1980.

311. **Mackichan, J. J., Duffer, P. K., and Cohen, M. E.,** Salivary concentrations and plasma protein binding of carbamazepine and carbamazepine 10,11-epoxide in epileptic patients, *Br. J. Clin. Pharmacol.*, 12, 31, 1981.
312. **Lucek, P. J. and Dixon, R.,** Chlordiazepoxide concentrations in saliva and plasma measured by radioimmunoassays, *Res. Commun. Chem. Pathol. Pharmacol.*, 29, 397, 1980.
313. **Luowa, P. V. and Sotanieme, E. A.,** Saliva and plasma clearance of antipyrene as reflectors of liver function, *Eur. J. Drug Metab.*, 6, 261, 1981.
314. **May, P. R., Van Putten, T., Jenden, D. J., Yale, C., and Dixon, W. J.,** Chlorpromazine levels and outcome of treatment in schizophrenics, *Arch. Gen. Psychiatr.*, 38, 202, 1981.
315. **Monaco, F., Piredda, S., Mutani, R., Mastropaolo, C., and Tondi, M.,** Free fraction of valproic acid in tears, saliva, cerebrospinal fluid, *Epilepsia*, 23, 23, 1981.
316. **Nau, H., Jesdinsky, D., and Wittfoht, W.,** Microassay for primidone and metabolites, *J. Chromatogr.*, 11, 71, 1980.
317. **Paxton, J. W.,** Effects of aspirin on salivary and serum phenytoin kinetics, *Clin. Pharmacol. Ther.*, 27, 170, 1980.
318. **Skinner, S. F., Robertson, L. T., Artero, M., and Gerding, R. K.,** Longitudinal study of phenobarbital in serum, cerebrospinal fluid, and saliva in the dog, *Am. J. Vet. Res.*, 41, 600, 1980.
319. **Tanaka, E., Kinoshita, H., Yoshida, T., and Kuroiwa, Y.,** Saliva levels of trimethadione and its metabolite as an index of drug metabolizing activity in rats, *J. Pharmacobiodyn.*, 4, 922, 1981.
320. **Hardee, G. E., Stewart, T., and Capowacchia, A. C.,** Tobacco smoke xenobiotics in mothers milk, *Toxicol. Lett.*, 15, 109, 1983.
321. **Barchowsky, A., Stargel, W. W., Shand, D. G., and Routledge, P. A.,** Saliva concentration of lidocaine and its metabolites in man, *Ther. Drug Monitoring*, 4, 335, 1982.
322. **Guechot, J., Fiet, J., Passa, P., Villette, J. M., Gourmel, B., Tabeau, F., Cathelineau, G., and Dreux, C.,** Physiological and pathological variations in saliva cortisol, *Horm. Res.*, 16, 357, 1982.
323. **Bjornsson, T. D., Meffin, P. J., Peters, F. A., and Blaschke, T. F.,** Saliva and plasma levels and plasma protein binding of clofibric acid in uremic patients, *Clin. Pharmacol. Ther.*, 27, 230, 1980.
324. **Bruun, C., Lambrou, D., Larsen, M. J., Fejerskov, O., and Thylstrup, A.,** Fluoride in mixed human saliva after different topical fluoride treatments and possible relation to caries inhibition, *Community Dent. Oral Epidemiol.*, 10, 124, 1982.
325. **DiGregorio, D. J. and Ruch, E.,** Human whole blood and saliva concentrations of oral and intramuscular promethazine, *J. Pharm. Sci.*, 69, 1457, 1980.
326. **Lena, S. M., Hutchins, P., Wood, C. B., and Turner, P.,** Salivary theophylline estimation in the management of asthma in children, *Postgrad. Med. J.*, 56, 85, 1980.
327. **Ritchel, W. A., Bykadi, G., Norman, E. J., Chixton, R. J., and Dentin, D.,** Salivary elimination of cyclophosphamide in man, *J. Clin. Pharmacol.*, 21, 461, 1981.
328. **Watanabe, J., Urasaki, Y., Nakase, Y., Ueda, H., Iwamoto, K., and Ozeki, S.,** Excretion of indomethacin into saliva following intravenous administration to dogs, *J. Pharmacobiodyn.*, 4, 336, 1981.
329. **Koysooko, R., Ellis, E. F., and Levy, G.,** Relationship between theophylline concentration in plasma and saliva of man, *Clin. Pharmacol. Ther.*, 15, 454, 1974.
330. **Wan, S. H., Martin, S. B., and Azarnoff, D. L.,** Kinetics, salivary excretion of amphetamine isomers, and effect of urinary pH, *Clin. Pharmacol. Ther.*, 23, 585, 1978.
331. **Posti, J.,** Saliva-plasma drug concentration ratios during absorption: theoretical considerations and pharmacokinetic implications, *Pharm. Acta Helv.*, 57, 83, 1982.
332. **Hepner, G. W.,** Breath analysis: gastroenterological applications, *Gastroenterology*, 67, 1250, 1974.
333. **Hepner, G. W. and Vesell, E. S.,** Assessment of aminopyrene metabolism in man by breath analysis after oral administration of [^{14}C]aminopyrene. Effects of phenobarbital, disulfiram and portal cirrhosis, *N. Engl. J. Med.*, 291, 1384, 1974.
334. **Hepner, G. W. and Vessell, E. S.,** Quantitative assessment of hepatic function by breath analysis after oral administration of [^{14}C]aminopyrene, *Ann. Int. Med.*, 83, 632, 1975.
335. **Lauterburg, B. H. and Bircher, J.,** Expiratory measurement of maximal aminopyrine demethylation *in vivo*: effect of phenobarbital, partial hepatectomy, portacaval shunt and bile duct ligation in the rat, *J. Pharmacol. Exp. Ther.*, 196, 501, 1976.
336. **Houston, J. B., Lockwood, G. F., and Taylor, G.,** Use of metabolite exhalation rates as an index of enhanced mixed-function oxidase activity *in vivo*, *Drug Metab. Dispos.*, 9, 449, 1981.
337. **Bircher, J. and Preisig, R.,** Exhalation of isotopic CO_2, in *Methods in Enzymology*, Vol. 77, Jakoby, W. B., Ed., Academic Press, New York, 1981, 3.
338. **Wendel, A. and Dumelin, E. E.,** Hydrocarbon exhalation, *Meth. Enzymol.*, 77, 10, 1981.
339. **Wendel, A., Feuerstein, S., and Kong, K.-H.,** Acute paracetamol intoxication of starved mice leads to lipid peroxidation *in vivo*, *Biochem. Pharmacol.*, 28, 2051, 1979.
340. **Williams, R. T.,** *Detoxication Mechanisms*, Chapman & Hall, London, 1947.

Chapter 5

SPECIES VARIATION IN TARGET ORGAN TOXICITY

J. C. Connelly and J. W. Bridges

TABLE OF CONTENTS

I. INTRODUCTION

The species variation in target organ toxicity exhibited by chemicals may be divided into broad categories. First, for any given compound, there is the possibility of an exaggerated range of susceptibility among species to damage to a particular tissue. Second, there may be a "true" variation in target organ selectivity, e.g., the toxicity of the administered chemical may manifest itself as a hepatic necrosis in species A, and as a renal or pulmonary necrosis in species B. A variant of the latter category is the situation in which a single tissue is affected in the first species, and that tissue plus other(s) are affected in the second (see Section V.A).

At present, many of the readily available examples in the literature reflect differences in sensitivity of species to, most commonly, hepatic lesions; however, there are several fairly well-documented instances of interspecies variation in the site of target organ attack after dosing of toxins. It is probable that confidential results of commercial product testing conceal several more, but it is also likely that in few cases are the reasons for the target organ selectivity completely understood.

II. IMPORTANCE OF SPECIES SELECTIVITY IN TARGET ORGAN TOXICITY

A. Prediction of Toxicity or Carcinogenesis in Man

For practical and ethical reasons, chemicals likely to be encountered by man are tested in animal models, and species variations in sites of toxicity are thus of great significance. The most widely used species for toxicity testing are mice, rats, and dogs, although pigs and nonhuman primates may also be employed. Rats and mice are generally used for carcinogenicity testing. On the grounds that they are particularly sensitive species for the purpose, rabbits are normally used for teratology and eye irritancy studies, and guinea pigs for skin irritancy testing. Screening for neurotoxicity of organophosphate compounds may be carried out in hens, and the potential for cataract induction by dinitro compounds is assessed in ducks. However, it is important to emphasize that in most instances species are selected for toxicity testing not because they are necessarily relevant models for man, but for reasons of ease of breeding and maintenance, particular susceptibility to a toxin or type of toxin, and/or the existence of a large database from which to segregate xenobiotic-induced and spontaneous effects. Heywood[1,2] analyzed toxicity data on about 50 compounds in several species, and concluded that in about 40% of cases the target organs in rodents and nonrodents were dissimilar.

Examples of failures in extrapolation of animal toxicity findings to man are summarized in Tables 1 and 2. Of the well-known Elixir of Sulphanilamide and thalidomide disasters, the former could have been predicted by even the most basic of animal testing, the latter only if the correct species was chosen at the outset. Litchfield[4] claimed a reasonable rate (68%) of success in extrapolating sites of toxicity of six drugs in humans from data obtained in rats and dogs alone, but patients who suffered adverse reactions after treatment with the compounds might have taken a different view. One means of attempting to improve predictability is to consider more species. However, this still does not obviate the possibility that man will exhibit bizarre target organ sensitivity or susceptibility, and carries a danger of rejection of valuable drugs or economically important chemicals because of a unique adverse effect in one of the additional experimental species.

Although this chapter is primarily concerned with toxicity appearing after relatively short periods of compound administration, it must be emphasized that target organs for carcinogens tend to show as great, if not greater, species variation than those for acute toxins (Table 3). In some cases the target organ in humans is predictable, but may be so from findings in only one experimental species; moreover, multiple tissue sites for carcinogenesis are en-

Table 1
SOME FAILURES IN LABORATORY SPECIES TO PREDICT HUMAN TOXICITY[3]

Agent	Toxic outcome	Species
Azauracil	CNS disturbances	All mammals tested
Benzene	Leukemia	All mammals tested
Clioquinol	Eye lesions	All mammals tested
Dinitrophenol	Cataracts	All mammals tested
Naphthylamine	Bladder cancer	Rats and rabbits
Oral contraceptives	Thrombosis	All species tested
Practalol	Skin, eye, intestinal lesions	All species tested
TCDD	Chloracne	All except hairless mouse
Thalidomide	Fetal abnormalities	Most rat strains

Table 2
EFFECTS PREDICTED IN LABORATORY SPECIES BUT NOT OCCURRING IN MAN[3]

Agent	Toxic outcome	Species
Aspirin	Fetal abnormalities	Rats
DDT, dieldrin, phenobarbitone	Liver tumors	Mice
Oral contraceptives	Retinal bleaching	Monkey
Penicillin	Rapidly fatal	Guinea pig

countered more frequently in laboratory animals than in man. In addition, chemicals may be carcinogenic in the same or different organs as in man when administered by routes other than the ones by which human exposure is expected.

B. Development of Selective Toxins

The identification of species differences in target organ susceptibility or specificity may lead to the production of agents capable of selectively destroying organisms for health or economic reasons. For example, during pharmacological screening studies norbormide was found to be a highly selective rat toxin, failing to show toxic signs in a variety of other mammals or birds.[6] Death in rats appeared to be due to irreversible peripheral vasconstriction, which was not observed in resistant species.

Red squill has also been employed as a rodenticide, its efficacy being based on the inability of rodents to vomit and expel the substance, in contrast to species with well-developed emetic centers. However, when absorption in rodents and other mammals is assured by parenteral administration variations in response still persist, death in the former species being due to central nervous system (CNS) effects, and in the latter to cardiac derangements.

The properties of the target organ-selective avicide, 3-chloro-*p*-toluidine, are described below.

The economic importance of understanding species variations in susceptibility to chemicals is emphasized by the case of diazinon. This compound had been used for several years as a fly control around chicken pens with no toxic outcome for the birds. When it was employed for the same purpose in duck pens without further testing, the loss of over 15,000 animals occurred.[7]

C. Development of Target Organ-Selective Drugs

In certain instances target organ variations may be of use in the study and employment of the therapeutic value of drugs. Findings from the testing of antineoplastic agents in

Table 3
TARGET ORGAN TOXICITY FOR CARCINOGENS[5]

	Human		Other species		
Compound	**Target tissue**	**Probable main route of exposure**	**Species**	**Target tissue**	**Route by which exposed**
Cyclophosphamide	Bladder	Oral, injection	Rat	Mammary gland	i.p. injection
			Mouse	Lung, hemopoietic system	s.c. or i.p. injection
				Various sites	oral
Diethylstilbestrol	Uterus, vagina	Oral	Rat	Mammary gland, bladder	s.c. implantation
			Mouse	Mammary gland	oral
				Mammary gland, lymphoreticular, testis	s.c. injection or implantation
			Hamster	Kidney	s.c. injection or implantation
			Monkey	Uterine serosa	s.c. implantation
Melphalan	Hemopoietic system	Oral, injection	Rat	Local, mammary gland	i.p. injection
			Mouse	Lung, lymphosarcoma	i.p. injection
2-Naphthylamine	Bladder	Oral, injection, skin	Mouse	Liver, lung	s.c. injection
			Hamster, dog, monkey	Bladder	oral
Vinyl chloride	Liver, brain,[a] lung[a]	Inhalation, skin	Mouse, rat	Liver, lung, kidney, mammary gland, blood vessels	Inhalation

[a] Denotes indicative evidence.

laboratory animals provide some of the most interesting examples of species differences in both desirable and undesirable target organ toxicity (see below). In addition to their practical utility, such experiments may help to elucidate mechanisms of drug action.

Rarely may a species-selective lesion lead directly to the development of a valuable drug. In early studies of *o′, p′*-DDD, the only experimental animal to demonstrate necrosis and atrophy of the adrenal cortex was the dog; subsequently, man was found to show a similar response, resulting in the successful clinical trial of this insecticide against adrenal cancer.[8]

D. Establishment of Mechanisms of Action of Foreign Compounds

Many toxins must be metabolically activated in order to express their toxicity. If reactive metabolites are formed in the liver, local damage only may occur, or, if the active species are sufficiently stable, they may be released into the bloodstream and affect other tissues. Alternatively, short-lived intermediates may be generated in extrahepatic tissues and exert their toxicity *in situ*. Boyd and colleagues[9] in particular, for example, have attempted to use simple furans showing species-selective target organ toxicity to identify patterns of metabolite formation and transport. Aspects of their excellent work on the furan derivative 4-ipomeanol are discussed later in this chapter.

III. FACTORS COMPLICATING IDENTIFICATION OF TRUE SPECIES VARIATION IN TARGET ORGAN TOXICITY

The nature and extent of toxicity depend upon a large number of factors whose effects in many cases are far from fully documented. Some of the variables are relatively easy to control, others are not. For instance, in the context of investigating target organ-selective toxicity, should we use male or female animals, young or old? Which strain of rat do we select? Generally, such factors assume the greatest importance when considering large differences in target organ susceptibility to dose, particularly those differences which are of such a degree as to be essentially qualitative. It is interesting to note that if 4-ipomeanol had been compared in female mice with female or male rats or rabbits, probably no target organ differences would have been demonstrated. Some variables affecting the toxicity of chemicals are shown in Table 4, with examples where appropriate drawn from compounds discussed elsewhere in this chapter. The question of interaction of compounds is vital. Inducers and inhibitors of drug-metabolizing enzymes particularly may exert profound effects on toxicity. Since the aim in laboratory studies is normally as much as possible to restrict exposure of the animals to the single compound under test, extrapolation of toxicity data to man may be complicated considerably by the fact that humans are exposed continuously to a large number and variety of chemicals, which are potentially capable of interaction with each other or with the test compound.

IV. PRINCIPAL FACTORS GIVING RISE TO SPECIES VARIATION IN TARGET ORGAN SUSCEPTIBILITY AND SELECTIVITY

Even if the factors described in Table 4 are eliminated or controlled, species-dependent target organ selection or degree of vulnerability is evident for a number of drugs and other chemicals.

The sensitivity or organs to damage is influenced by several primary factors, including the distribution and pharmacokinetics of the toxic agent, its metabolic fate (which may be governed by tissue-specific biochemistry), the presence or absence of specific receptors or macromolecules susceptible to damage, and the capacity of the tissue to repair a particular lesion. It follows, therefore, that differences in target organ sensitivity and selectivity among

Table 4
FACTORS POSSIBLY COMPLICATING RECOGNITION OF SPECIES DIFFERENCES IN TARGET ORGAN SUSCEPTIBILITY OR SELECTIVITY

Factor	Comments
Animal-related:	
Strain	Wide variation in paracetamol-induced nephrotoxicity in rat strains; some strains of rats or mice have total absence of particular drug-metabolizing enzymes
Individual differences	Especially important when considering hypersensitivity, allergy, or idiosyncrasies of drug metabolism; possibly important in toxicity of cholestatic agents
Sex, hormonal status	Sex differences in toxicity common in rats and mice; 4-ipomeanol shows sex as well as species differences in mice
Age	Distribution and other properties of drug-metabolizing enzymes change with age; target organ variation in immature and mature male mice with 4-ipomeanol
Preexisting disease	Infection may alter response to chemicals via stress or changes in metabolic status
Environment-related:	
Temperature	May affect toxicity of chemicals, including organophosphate pesticides; body temperature may also be influenced by drugs at various ambient temperatures
Pressure, altitude	Changes in environmental oxygen tension may influence toxicity, particularly of compounds affecting cardiorespiratory function, e.g., increased toxicity of digitalis and red squill at high altitude
Biological rhythms	Circadian rhythms affect response to several drugs; light and dark exposure influence foreign compound metabolism in rat liver (note that some compounds are capable of photoactivation)
Stress	Chemical distribution in body may be affected and toxicity modified; overcrowding can increase toxicity of CNS stimulants in rats
Nutrition	Protein deficiency may increase or decrease toxicity; restriction of overall food intake in normally overfed rats or mice may cause striking reduction in occurrence of tumors
Exposure to other chemicals	Inducers or inhibitors of drug metabolism can increase or decrease toxicity of compuunds, e.g., large shifts in target organ sensitivity with 4-ipomeanol and other furans after 3-MCA pretreatment; increase in CCl_4 hepatotoxicity after induction; potentiation of toxicity of malathion by EPN; species-dependent modification of cyclosphosphamide effects by phenobarbitone
Compound-related:	
Physical form of chemical	May influence selectivity and bioavailability of compound
Impurities	Toxicity may be at least in part related to impurities present in preparation, e.g., 2,4,5-T
Stability of chemical	Degradation prior to absorption may be important, e.g., thalidomide
Route of exposure	Many effects, e.g., oral toxicity of red squill in rodents; modification of toxins including Brown FK, azauracil by gut microflora; hydrolysis of phthalate diesters in gut wall

species may arise from any of several causes, either singly or in combination, including:

1. Variations in specific physiology: possession of appropriate receptors and nature of compound/receptor interactions
2. Anatomical considerations (see discussion of alkyltins below)
3. Differences in ability to absorb toxic compounds, especially when exposed via the oral or percutaneous routes, and differences in distribution
4. Differences in gut microflora population or distribution
5. Qualitative or quantitative differences in the metabolism of toxins. (metabolism is also important in the species-dependent toxicity of many agents that may be placed in the categories listed above)

These factors are discussed, with examples, in the following pages. More detailed case histories are also given for some classes of compounds for which particularly striking species-dependent target organ selectivity has been shown.

A. Physiological Factors

The selective rat toxicant norbormide (see above) causes irreversible constriction of vascular smooth muscle, leading to ischemia and death. The receptor sites for norbormide are apparently different from those for adrenalin, and the compound is toxic to domestic animals and primates only at very high doses. Norbormide receptors in smooth muscle of peripheral blood vessels are thus thought to exist only in the rat.[10]

The heart of the adult rat is notably resistant to the action of cardiac glycosides, with doses of glycoside that are markedly toxic to most species producing little or no effect.[11] Conversely, the rat is much more susceptible to the neurotoxic effects of ouabain. These differences appear to reside in the interaction between the enzyme Na^+, K^+ - ATPase and ouabain. The rat brain enzyme has a much higher affinity for ouabain than does the heart enzyme,[12] and more importantly, the rate of dissociation of the heart Na^+,K^+-ATPase-ouabain complex is very rapid. Species differences in ouabain sensitivity appear to be largely explicable in terms of the stability of its complex with the enzyme.[13] However, in the dog, which exhibits cardiotoxicity, but is less susceptible to ouabain-induced neurotoxicity despite equal sensitivity of the brain and heart enzymes, the blood-brain barrier appears to play a protective role.

Receptor differences have also been proposed to explain the species dependence of the teratogenic effects of thalidomide, but selective metabolic activation now seems to play an important role, as discussed later in this article.

Tetrachlorodibenzo-*p*-dioxin (TCDD) is formed as a trace contaminant in the synthesis of several commercial products, especially trichlorophenols. It is extremely toxic, but the acute oral LD_{50} varies over a 5000-fold range in different species, progressing from hamster (5 mg/kg, least susceptible) to guinea pig (about 1 μg/kg, most susceptible). Considerable qualitative species variation in damage to target organs is also seen (Table 5).

Although hepatomegaly is produced in all species, hepatic lesions range from widespread necrosis in the rabbit,[14] through focal centrilobular damage in mice,[15] to minimal histopathological change in the guinea pig.[16] The last finding may represent a genuine species difference in hepatic vulnerability, but it should be noted that it could be explained by the lethal dose in guinea pigs being insufficient to produce liver damage. Chloracne is the lesion most frequently seen in man;[17] it can be reproduced in rhesus monkeys, hairless mice, and the rabbit ear, but not in rats, hamsters, or guinea pigs.

TCDD produces a loss of lymphoid tissue in all species examined, primarily from the thymus, but also from lymph nodes and spleen; however, there is a very pronounced species difference in the dose required to produce these effects. Apart from urinary tract hyperplasia and wasting, the latter of which is shown by all species but is not in itself apparently lethal[18] thymic involution is the major lesion in guinea pigs. There is little evidence that the consequential loss of immune competence results directly in death.[19]

Polychlorinated aromatic hydrocarbons cause a hyperplastic and hypertrophic lesion of the gastric mucosa in the monkey and cow,[20,21] but do not apparently do so in the mouse, rat, or guinea pig. Various other effects are seen in different species, including edema and ascites in chickens, testicular damage in several species, and pancytopenia, particularly in monkeys (see Allen et al.,[12] and references quoted therein).

Pancytopenia and hemorrhage in the monkey and hepatic necrosis in the rabbit may be lethal, but despite this high degree of toxicity of TCDD, in most species the lesion actually responsible for death is unclear. TCDD is also teratogenic and embryotoxic in rats and mice, and is a promoter of the carcinogenesis of, e.g., diethylnitrosamine.

In man, in addition to chloracne, exposure to TCDD has been claimed to be associated with liver abnormalities, myocardial defects, and other lesions, and with still births and teratogenesis.[22]

The recognition of an hepatic receptor for TCDD has led Poland and Knutson[23] to suggest

Table 5
SPECIES DIFFERENCES IN HISTOPATHOLOGICALLY IDENTIFIABLE LESIONS INDUCED BY TCDD OR RELATED POLYCHLORINATED AROMATIC HYDROCARBONS[23]

	Species						
Lesion	**Mouse**	**Rat**	**Hamster**	**Guinea pig**	**Rabbit**	**Monkey**	**Chicken**
Hepatic damage	+	+ +	+	–	+ + +	+	+
Thymic involution	+	+	+	+	N.D.	+	+
Testicular atrophy	+	+	N.D.	+	N.D.	+	+
Bone marrow effects — pancytopenia	±	N.D.	N.D.	+	N.D.	+ +	+
Chloracne[a]	–	–	–	–	+	+ +	N.D.
Gastric hyperplasia	–	–	–	–	N.D.	+ +	N.D.
Intestinal hyperplasia	N.D.	N.D.	+ +	N.D.	N.D.	+	N.D.
Urinary tract hyperplasia	–	–	N.D.	+ +	N.D.	+ +	N.D.
Focal lung damage	N.D.	+ +	N.D.	N.D.	N.D.	N.D.	N.D.
Edema, hydropericardium	+	–	+	–	N.D.	+	+ +

Note: N.D. = No data located for these species.

[a] One of major manifestations of TCDD toxicity in man.

a general model for the toxicity of halogenated aromatic hydrocarbons that could account for the species differences in many of the specific tissue responses. In this model, some tissues are seen as responding to TCDD with a limited pleiotropic response (e.g., induction of drug metabolizing enzymes), whereas tissues showing "toxic" responses express an additional battery of genes that need not be the same in all tissues. These authors note the lack of sufficient data for the complete understanding of TCDD toxicity, but believe that all of the adverse effects of TCDD could be explained in terms of response to receptor binding.

B. Anatomical Factors

There are very few well-documented examples of anatomical factors as the *principal* cause of species differences in target organ toxicity, although undoubtedly they contribute to many of these phenomena. Dialkyltin compounds, especially dibutyltin salts, are used as stabilizers in the manufacture of plastics, and as antihelminthics in poultry. Dibutyltin salts produce several toxic effects in mammals,[24,25] but the most striking lesion is bile duct damage in susceptible species. In the rat, edema of the pancreas along the course of the bile duct is visible within 12 to 18 hr after an oral dose of 50 mg/kg dibutylin dichloride.[26] When doses are given on successive days, the bile duct becomes dilated and inflamed, and may be partially obstructed. The bile duct may leak and cause a severe peritonitis with patches of necrosis. Liver involvement includes multiple yellow infarcts of various lobes, and the combination of biliary and hepatic damage is lethal. A similar lesion is seen in mice dosed with dibutyltin dichloride, but guinea pigs tolerate repeated oral administration without apparent ill effect. Rabbits are killed by several successive doses of 25 to 50 mg/kg dibutyltin dichloride, although no lesion of the bile duct or gall bladder occurs.

The species selectivity of the lesion appears to be associated with the anatomical relationship between bile duct and pancreas. The pancreas of rabbit or guinea pig lies within loops of the intestine, and is not in direct contact with the bile duct. In rats and mice the lower and major part of the bile duct lies with the pancreas; if pancreatic juice comes into contact with cell debris from the injured bile duct, the hydrolytic enzymes of the pancreatic juice may be activated and so exacerbate the lesion there, or bile escaping from the duct may penetrate the pancreatic acini and cause them to liberate their secretions locally. This interaction between ductular and pancreatic components appears to be necessary for the development of the lesion.

C. Absorption and Distribution

Qualitative and quantitative variations in the biotransformation of chemicals have been much studied and are well established as a cause of species differences in their toxicity. However, it is not always appreciated that differences in absorption, distribution, and excretion of compounds may also contribute significantly to selective toxicity. Moreover, absorption and distribution remain unpredictable, and completely unexpected species effects may occur. Interspecies differences in plasma protein binding are especially marked[27] and the ability to actively transport a foreign compound may vary considerably; either phenomenon may be implicated in target organ selectivity.

The rodenticide zinc phosphide shows marked species variations in toxicity when administered orally. Rats, rabbits, and avian species die rapidly from hepatic and pulmonary lesions. Cats and dogs are more resistant, and the latter may tolerate as much as 1000 mg/kg.[28] Zinc phosphide in itself is not very acutely toxic, but its reaction with mineral acids results in the release of highly toxic phosphine gas. Rats and other species susceptible to zinc phosphide poisoning release gastric hydrochloric acid continuously; cats and dogs are intermittent secretors, and are therefore more resistant.

Many organophosphate compounds, particularly the organophosphate esters, have been

Table 6
MAXIMUM RATES OF PENETRATION OF AN ORGANOPHOSPHATE COMPOUND THROUGH SKIN EXCISED FROM DORSAL THORAX IN VARIOUS SPECIES[a]

Species	Max rate of absorption (μg/cm²/min)
Rat	9.3
Rabbit	9.3
Guinea pig	6.0
Cat	4.4
Goat	4.4
Monkey	4.2
Dog	2.7
Pig	0.3

[a] Data from Adamson and Davies[33] after McCreesh[34].

used as agricultural pesticides, industrial chemicals, and chemical warfare agents. In mammals they show distinct acute and delayed toxicities with a marked degree of variation in species sensitivity. Acutely, poisoning due to organophosphate esters results primarily from phosphorylation and inhibition of acetylcholinesterase. The molecular basis for these effects has been reviewed by Aldridge.[29] In mice, guinea pigs, rats, and rabbits, the acute administration of tri-*o*-cresyl phosphate, one of the most notorious organophosphate esters, is characterized by hyperexcitability, spastic incoordination of muscles, flaccid paralysis, impaired respiration, and death. The immediate cause of death in acute poisoning is generally asphyxiation, arising partly from neuromuscular failure and partly from depression of CNS activity. However, cholinergic neurons are involved in many biological functions, and species differences may result from the manifold consequences of acetylcholinesterase inhibition. In several animal species influence on peripheral nerve function may be the acute lesion leading to death, whereas in primates the effects on the CNS are probably more important. Several phosphate triesters also produce a delayed neurotoxic effect, described as an acute axonopathy, and characterized by axonal degeneration followed by destruction of the myelin sheath.[30] The lesion appears to be related to phosphorylation of an esterase other than acetylcholinesterase,[29] resulting in the onset of ataxia and flaccid paralysis after a latent period.

The reasons for differences in species response to organophosphate esters embrace several factors. Dermal absorption is important in organophosphate toxicity, and the rate of penetration through skin is very variable among species (Table 6). Man is classed among the most susceptible to toxicity via this route, an observation especially significant in exposure of agricultural workers and others to organophosphate pesticides by spraying. Differences in species susceptibility to the delayed neuropathy after oral administration of organophosphates may also be related to extent of absorption in certain instances. Incorporation of emulsifying agents into the dose, which apparently enhance GI absorption of tri-*o*-cresyl phosphate, is required to produce ataxia in dogs. In some nonhuman primates, including the slow loris, progression to irreversible paralysis is shown only after parenteral administration of organophosphates.

Most species show acute toxicity with organophosphate esters, provided that the route of administration allows sufficient access of the compounds to their site of action. However, several species, including rats, mice, rabbits, and guinea pigs, are resistant to delayed neurotoxicity, whereas cats, dogs, cows, sheep, primates, and avians are susceptible. These differences are probably due at least in part to toxicokinetic variables, e.g., rats and mice metabolize and excrete neurotoxic phenylphosphorothioate esters at rates much greater than those observed in susceptible species. However, it is notable that the rat, which is the most resistant of species studied, rapidly abosrbs tri-*o*-cresyl phosphate, and metabolizes it to a cholinergic organophosphate that shows delayed neurotoxity in susceptible species. This has led to the suggestion that the refractory nature of the rat may be due primarily to differences in its neuroanatomy.[31]

An interesting recent finding is that species variation in organophosphate toxicity may not be confined to neuronal effects. Sϕderlund et al.[32] have reported that the flame retardant organophosphate triester, tris(2,3-dibromopropyl)phosphate, binds covalently to rat kidney proteins to a much greater extent than to kidney proteins of mouse, hamster, or guinea pig, and, furthermore, that the rat is the only one of these species in which acute renal tubular necrosis is observed.

Hence, although species selectivity in organophosphate toxicity has been included under the heading of absorption and distribution, primarily because of relevance of percutaneous absorption to poisoning in man, these compounds may in addition exemplify species differences in receptor sensitivity, metabolism, tissue binding, and anatomy.

Some other examples of species variation in toxicity related to absorption, distribution, and excretion of drugs, especially to the distribution of methylglyoxal-*bis*-guanylhydrazone, have been described by Adamson and Davies.[33]

D. Species Variation in Toxicity Related to Gut Flora

The archetypal example of species difference in toxicity related to GI flora, and one of the most striking examples of overprediction by animal screening, is the severe toxicity and mortality produced in the guinea pig, but not other species, by penicillin.[35] These effects are due in part to replacement of the normal intestinal microflora with an overgrowth, or superinfection, of Gram negative coliforms, and are thus secondary in nature, but do indicate the superinfection that is a clinical problem with the use of penicillin in man.

However, the gut microflora of animals are also capable of carrying out many reactions with drugs or other foreign compounds as substrates.[36] Species differences are found both in the nature of these microorganisms, and in their location within the GI tract. For example, the feces of rats and mice contain far more lactobacilli and streptococci than do those of man or various other animals, whereas fewer enterobacteria are isolable from the feces of rabbits or guinea pigs.[37] Furthermore, in man and the dog very few organisms occur in the stomach and proximal small intestine, whereas in coprophagic animals such as the rat and mouse they occur in large numbers along the entire tract from stomach to rectum.

The possible consequences of enterofloral metabolism of foreign compounds include variations in drug metabolism, toxicity, and carcinogenesis among animal species. An example of this is provided by the antineoplastic agent, 6-azauridine, which is absorbed efficiently by several laboratory species, whereas only about 30% is absorbed in man. This unabsorbed azauridine is split by large intestine bacteria to yield 6-azauracil, which is absorbed, and results in CNS toxicity.[38]

Azo compounds are also very commonly used synthetic colorings in foods and pharmaceuticals, and many bacteria show active and nonspecific reduction of these chemicals. The food dye Brown FK, which is a complex mixture of a least six *p*-sulphophenylazo derivatives of 1,3-diaminobenzene and 2,4-diaminotoluene, exhibits species variations in toxicity, apparently as a result of enterfloral metabolism.[39,40] Repeated oral doses of Brown FK cause weight loss and a vacuolar myopathy of heart and skeletal muscle in rats. The same changes are seen in guinea pigs and rabbits. Mice, however, have low susceptibility, and no muscle damage occurs in hamsters.

In the presence of rat gut contents, Brown FK undergoes azo reduction to give sulfanilic acid and a phenzaine thought to be derived from the condensation of polyamines such as

1,2,4-triaminobenzene and 2,4,5-triaminotoluene. Both polyamides produce cardiac and muscular lesions in mice (not susceptible by the oral route) when injected intravenously,[41] while rats given Brown FK intraperitoneally at doses equivalent to the oral doses showed no lesions.[39] The cardiac damage in rats could also be reduced by antibiotic treatment. These results suggest that the species variations reside in the different abilities of gut microflora to produce myotoxic polyamines, rather than a difference in susceptibility of muscle. It has also been proposed that in susceptible species, the intestinal microflora maintain a low redox potential that inhibits oxidation and condensation of the polyamines.[41]

There are many other examples of species differences in the metabolism of compounds by gut microorganisms, some leading to variations in toxicity. A particular concern is the ability of reactions mediated by gut bacteria to produce promoters or co-carcinogens. Such events, if giving rise to species differences in target organ susceptibility and selectivity for carcinogenesis in laboratory animals, will obviously enormously complicate the extrapolation of experimental data to man.

E. Metabolic Factors

Most currently recognized species-dependent target organ effects of toxins are due primarily to the metabolism of these chemicals in animal tissues. Interspecies variations in both the nature and extent of biotransformation of foreign compounds are complex, and although virtually all of the likely pathways of metabolism of a given compound may be predicted, gross differences in the routes utilized by various species may be found.

Many metabolic factors may contribute to species selectivity or susceptibility of target organs to toxins; the following are among the most important.

If the parent compound is itself toxic, then the susceptibility of a tissue will be governed largely (in the presence of appropriate receptors, or macromolecules vulnerable to attack) by the ability of the tissue to detoxify the active moiety.

If the compound is converted to a toxic metabolite, the situation is more complicated.

Qualitative differences in tissue metabolism that may lead to species-dependent toxicity include: (1) the lack of certain foreign compound-metabolizing capabilities in some species; (2) the occurrence of pathways restricted to a single or small group of species; (3) the stability of the active metabolite(s) produced, i.e., whether or not they may be formed in one tissue and transported via the bloodstream to a second target tissue; and (4) substrate specificities of both phase I and phase II foreign compound-metabolizing enzymes (note: in particular, very few interspecies comparisons of the nature of extrahepatic cytochrome(s) P-450 have been carried out).

Quantitatively, tissues of different species may: (1) differ in the amount and activity of phase I and phase II enzymes, and (2) differ in the affinities of these enzymes for the appropriate substrates. The *balance* between activating and deactivating enzymes in tissues of different species, in both the quantitative and qualitative senses, also plays a role in selective toxicity. In addition, whether the parent compound or a metabolite is the ultimate toxin, the possibility that the compound might differentially induce its own (frequently hepatic) metabolism, or that of endogenous compounds, in animal species must be considered. Induction of activating or detoxifying enzymes by a second agent may sometimes cause a striking shift of target organ for a compound already exhibiting a marked degree of species-dependent selective tissue damage (see case history for furans below).

Examples relating to various of the factors describing above are available. For instance, the defective abilities of guinea pigs, cats, and dogs to carry out *N*-hydroxylation of aromatic amines, glucuronidation of small phenols and aromatic acids, or acetylation of primary amino groups, respectively, are well known, and explain the selective toxicity of high doses of the appropriate compounds to these species (Table 7). The anti-allergy drug FPL 52757 (6,8-diethyl-5-hydroxy-4-oxo-4*H*-1-benzopyran-2-carboxylic acid) provides an interesting

Table 7
SPECIES DEFICIENCIES IN METABOLISM OF FOREIGN COMPOUNDS

Species	Compounds	Deficient reaction	Outcome
Cat	Small phenolic compounds, e.g., phenol, 1-naphthol	Glucuronidation	Sulfate rapidly depleted, leading to increased toxicity
Cat	Small aromatic acids, e.g., benzoic acid (also clofibric acid)	Glucuronidation	Glycine rapidly depleted, leading to increased toxicity
Dog	Many primary amino groups, e.g., *p*-aminosalicylic acid	Acetylation	Failure to conjugate exacerbates toxicity
Guinea pig	Aromatic amines, e.g., 2-acetamidofluorene	*N*-hydroxylation	Results in increased toxicity in some cases, but protects against activation of 2-AAF to carcinogenic metabolite
Rat	Aliphatic amines, e.g.,	*N*-hydroxylation	Generally, increased toxicity

example of failure to metabolize a compound leading to toxicity. In most species studied metabolism and elimination of this drug is rapid. However, the dog mainly excretes the unchanged compound, and exhibits marked hepatotoxicity, apparently associated with the accumulation in the bile of high concentrations of FPL 52757, which has a pronounced detergent action.[42] A similar species variation in toxicity has also been reported with another chromone derivative, FPL 57787.[43] Alternatively, several conjugation reactions have been described only in man or other primates.[44] Either situation (pathways uniquely defective or active in certain species) may exacerbate the problem of predicting the metabolism or toxicity of foreign compounds in man.

The diuretic, antihypertensive steroid, spironolactone, illustrates species-selective toxicity due both to differences in cytochrome P-450 populations and to effects on hormone-mediated metabolism. In cortisol-producing species such as the dog and guinea pig, administration of spironolactone causes large and rapid decreases in adrenal microsomal cytochrome P-450 concentration and steroid 17α- and 21-hydroxylase activities, whereas in the adrenals of corticosterone-producing animals, including the rat and the rabbit, cytochrome P-450 levels and rates of 17α- and 21-hydroxylation are increased.[45] Testicular microsomal cytochrome P-450 is also reduced in each of these species by 40 to 90%, the degree of loss in either testis or adrenal being related to the activity of 17α-hydroxylase, which is a microsomal cytochrome P-450-dependent enzyme, and whose activity is high in cortisol-producing but low in corticosterone-producing animals. While the mechanism of cytochrome P-450 loss is uncertain, it appears from the lack of testicular necrosis seen and the ability of progesterone to inhibit spironolactone-mediated cytochrome P-450 destruction, that spironolactone may be a suicide substrate for this cytochrome. A secondary species-selective effect arises from the decline in testosterone production by the testis following dosing with spironolactone. In male rats this leads to liver enlargement[46] and decreased benzo(*α*)pyrene hydroxylase activity,[47] whereas in male guinea pigs there is a considerable increase in the activity of this enzyme. Various explanations involving the control of sex differences in hepatic metabolism by testosterone have been suggested.[47] Administration of spironolactone also results in enlarged and adenomatous thyroids in rats,[46] which has not yet been reported in other species.

As noted above, several explanations for the species-selective teratogenicity of thalidomide have been noted, but current evidence tends to favor a metabolic basis. A wide variety of species, including man, monkeys, rabbits, chicken, and mice are susceptible to thalidomide-induced teratogenicity. Most rat strains, hamsters, and guinea pigs are resistant. Differences in metabolic products in rats and rabbits, particularly the excretion of phenolic metabolites in the latter,[48] suggest that thalidomide may undergo oxidative metabolism via an arene oxide. Gordon and colleagues[49] showed that hepatic preparations from maternal rabbits and from rabbit, monkey, and human fetuses, produced a metabolite of thalidomide that was toxic in human lymphocytes in vitro. The toxicity was enhanced by inhibitors of epoxide hydrolase, and abolished by adding purified epoxide hydrolase to the incubation medium. The target macromolecule for attack by the metabolite is still unknown; however, it may be that the arene oxide is bound to a specific receptor, and results in teratogenicity.

Diesters of *o*-phthalic acid, used extensively as plasticizers, show species variability in hepatic and testicular effects. Treatment of rats with di-(2-ethylhexyl) phthalate (DEHP) for prolonged periods produces liver enlargement with fatty degeneration, and atrophy and focal degeneration of seminiferous tubules in the testis.[50] In contrast, guinea pigs (and dogs) appear to be resistant to the hepatic effects of phthalate esters. Di-*n*-butylphthalate (DBP) also causes rapid and severe seminiferous tubular atrophy in rats and guinea pigs, and focal atrophy alone in mice, but has no effect in hamsters.[51] DEHP likewise has little effect on the hamster testis. The nonsusceptibility of hamsters may be related to their significantly slower intestinal rate of hydrolysis of DEHP to mono-(2-ethylhexyl) phthalate (MEHP), the latter being the form in which most of an oral dose of DEHP is absorbed. However, this is unlikely to be the entire explanation as, although MEHP elicits focal seminiferous tubular atrophy in hamsters, mono-*n*-butylphthalate does not have such an effect. The observed toxicity might be related to a species-dependent depletion of testicular zinc.[51]

Foreign compounds whose administration to animals leads to intrahepatic cholestasis show some interesting species variations, and provide a particular example of difficulties in the prediction of human toxicity from animal data. Although cholestatic lesions have been produced in laboratory species, no model has been developed that duplicates entirely all of the features of human drug-induced cholestasis, and the almost universal lack of a dose-response relationship for these compounds in animals makes extrapolation of the risk of cholestasis associated with therapeutic dosage in man very difficult. The jaundice well known to be produced in man by certain chemicals is extremely difficult to reproduce in animals, and this has been claimed to indicate that drug-induced cholestasis in humans is a hypersensitivity reaction. However, such a conclusion has generally been reached by exclusion rather than the presentation of material proof.

The single compound so far unequivocally demonstrated to induce a dose-dependent and reproducible cholestasis in some laboratory animals, α-naphthylisothiocyanate (ANIT), also exhibits a rather marked species-selective toxicity. In susceptible species, such as the rat, mouse, or guinea pig, acute administration of ANIT invokes hyperbilirubinemia and cessation of bile flow. The hamster is comparatively insensitive to the effects of ANIT, whereas rabbits and dogs do not develop the cholestatic response at all. ANIT has been shown to require biotransformation catalyzed by hepatic microsomal cytochrome P-450 in order to express its toxicity,[52] and it has been demonstrated that a reactive moiety derived from ANIT binds irreversibly to hepatic macromolecules. However, there is no correlation between the extent of this binding in different species and their susceptibility to ANIT-induced cholestasis, and the relative effects of various inhibitors and inducers of cytochrome P-450-mediated metabolism on the covalent binding and toxicity of ANIT are thoroughly inconsistent. (This is in striking contrast to the toxicity of some other compounds such as paracetamol.) ANIT decreases monoxygenase activities in both cholestasis-susceptible and nonsusceptible species before the onset of hyperbilirubinemia in the former[52] and β-naphthylisothiocyanate (BNIT), which does not cause cholestasis in any species tested, also decreases monooxygenase activities. Hence, while metabolism of ANIT appears to lead to a selective hepatotoxicity, no presently understood aspect of that metabolism can be correlated with its species dependence. Many other compounds have been associated with intrahepatic cholestasis in man,[53] some of which unreliably cause the appearance of certain features of cholestasis in animal models, but with no dose-dependence or predictable latent period. The typically low incidence of this form of toxicity obscures species-dependent relationships, and the cholestasis induced by these compounds may indeed be the result of an allergic reaction or drug interaction, but the possibility that variations in biotransformation may be involved in their toxicity is now being more seriously considered.

Aflatoxin B_1 (AFB_1), a mycotoxin produced by molds of the *Aspergillus* genus, is a potent hepatotoxin and hepatocarcinogen in many animal species, although the degree of sensitivity (and organ selectivity in some cases) of different species varies widely. Turkeys, ducklings, cattle, pigs, dogs, cats, rats, rabbits, guinea pigs, and monkeys are classed as susceptible to the acute effects of AFB_1, whereas sheep, mice, hamsters, and chickens are resistant.[54] Among the susceptible species, the necrotic or degenerative lesion may be centrilobular (e.g., pig and guinea pig), periportal (e.g., rat and duckling), or midzonal (e.g., rabbit). In addition, the primary site of biochemical and pathological effects in the mouse appears to be the kidney rather than the liver.[55] Rats, ducklings, and trout are very susceptible to the hepatocarcinogenicity of AFB_1, while sheep, guinea pigs, and mice are resistant. Some strains of mice have in fact been described as completely refractory.[56]

Earlier theories suggest that the species differences in hepatotoxicity and hepatocarcinogenicity of AFB_1 are caused by the production of different metabolites, but activation to AFB_1-2,3-epoxide (AFB_1-8,9-epoxide) is now generally considered necessary for both biological responses. Interspecies variations in AFB_1 effects are thus related to other factors. Rates of initial activation of AFB_1 to the epoxide in different species do not correlate well with the acute toxicity seen, suggesting that the ability to detoxicate the active moiety is a vital consideration. O'Brien et al.[57] used a preparation of quail microsomes, which rapidly produce AFB_1-2,3-epoxide from the parent compound, to compare the capability of cytosolic preparations from several other species to conjugate the epoxide with reduced glutathione. Differences in ability to conjugate ranged from 0 (quail) to 72% (mouse), and correlated exactly with the order of severity of the liver lesion produced when AFB_1 was administered to the appropriate species in vivo. Significantly, a human liver microsomal preparation metabolized AFB_1 to the epoxide rapidly, whereas the post-mitochondrial supernatant from the same liver failed to produce AFB_1-glutathione from quail liver — generated epoxide, indicating that man may have a relatively high susceptibility to the acute effects of AFB_1.

Binding of activated AFB_1 to tissue macromolecules may be important in both its acute toxicity and carcinogenicity, although the pattern is as yet far from clear. Croy and Wogan[56] found that binding of AFB_1- derived material to DNA was far greater in rat liver than in kidney, whereas in the mouse, more binding to renal DNA occurred. These authors correlated their results with the acutely toxic rather than the carcinogenic susceptibilities of rat and mouse tissues to AFB_1, despite the more usual association of DNA binding by reactive metabolites with mutagenicity or carcinogenicity. Cancers of the kidney are seen after the administration of AFB_1 to rats; the common site for AFB_1 carcinogenicity in adult mice is the lung (see Table 3), with little evidence available regarding renal tumors. Ueno et al.[58] have correlated the greater binding of radiolabeled AFB_1 to RNA and protein in guinea pig liver than in rat liver with the higher susceptibility of guinea pigs to AFB_1-induced acute toxicity. Guinea pig hepatic DNA also bound less AFB_1 than did rat liver DNA, which may be associated with the relative resistance of the guinea pig to the hepatocarcinogenicity of AFB_1. However, few animals were used, and the differences in binding to rat and guinea pig macromolecules were small. Interspecies sensitivity to the acute toxicity of AFB_1 now has some mechanistic basis, but a satisfactory explanation for the differences in susceptibility to hepatocarcinogenesis, and the species-dependent sites seen in rat and mouse (Table 3), remains to be found.

Species selectivity involving kidney toxicity is a phenomenon common to several compounds. For example, in susceptible species such as the mouse and hamster, an active metabolite of paracetamol causes hepatic necrosis.[59] Other laboratory animals, including the rat, rabbit, and guinea pig, are relatively resistant. In contrast to the findings with ANIT, the hepatotoxicity correlates with the degree of activation of paracetamol, binding to tissue macromolecules, and glutathione depletion. High doses of paracetamol do not produce detectable histopathological changes in the kidneys of mice, rabbits, or Sprague-Dawley rats,[59] but cause tubular necrosis in male Fischer 344 rats.[60] Man may also show acute renal failure in the absence of fulminant hepatic damage after ingestion of paracetamol.[61] Selection for nephrotoxicity among rat strains (and possibly variations in species susceptibility) may be related to differences in the intrarenal activation of *p*-aminophenol formed from paracetamol by deacetylation.[62]

The selective avicide, 3-chloro-*p*-toluidine (CPT) is acutely highly toxic to most bird species, but has an LD_{50} of 650 to 1500 mg/kg in the rat.[63] In susceptible species, CPT causes necrosis of renal proximal tubules, possibly because the metabolite 2-chloro-4-acetotoluidine (CAT), formed in the liver, is deacetylated in the kidney to CPT, which is then further activated to nephrotoxic metabolites. Renal deacetylase activity is high in the very susceptible starling and chicken, but undetectable in the redtail hawk (a nonsusceptible avian species) and the rat.[64] A high degree of binding of CAT-derived material to starling kidney macromolecules has also been shown,[65] but the relationship of this binding to nephrotoxicity has not been proven. In quail, focal hepatic necrosis is seen in addition to renal necrosis after administration of CPT. CPT causes methemoglobinemia in rats and rabbits, but the causes of death in these and some other species appear to be complex and related possibly to cardiovascular and respiratory factors.[66] The role of metabolism in the development of these lesions is far from clear.

The lung is another organ particularly implicated in species-dependent toxicity. Carbon tetrachloride (CCl_4) is a potent hepatotoxin, although some species, including the rat, have considerable resistance to CCl_4-induced liver necrosis, and many avians are able to survive enormous doses. In rats and guinea pigs CCl_4 also damages pulmonary type II pneumocytes, and in rats and mice causes striking changes in bronchiolar Clara cell morphology.[67] The selective damage to Clara cells, at least, appears to be due to their ability to metabolically activate CCl_4. Renal lesions also occur in male rats after dosing with CCl_4; these are presumed to be caused by renal activation of this chemical, but nephrotoxicity resulting from an active moiety being transported from the liver has not been ruled out.[68] The major pathological changes in man after CCl_4 poisoning are seen in liver and kidney. Various other organs have been described as susceptible to CCl_4-induced damage in laboratory animals, including testes, adrenals, and placenta, but no pattern of species selectivity is yet definable.

1,1-Dichloroethylene (DCE) causes necrosis restricted to Clara cells in mouse lung without any histologically detectable concomitant liver or kidney damage; neither are renal or hepatic cytochrome P-450 and associated monooxygenases affected.[69] However, rats dosed with DCE develop both renal and hepatic toxicity.[70]

Naphthalene selectively produces necrosis of Clara cells and a depression of monooxygenase activities in mouse lung without apparently damaging the liver.[71] Activation of naphthalene appears to be necessary for toxicity, but there is evidence that metabolites formed in the liver may play a part in its pulmonary effects.[72] The administration of 2-methylnaphthalene to mice similarly causes necrosis of Clara cells without affecting the histological picture in either liver or kidney.[73] Liver function in rats, is however, impaired by 2-methylnaphthalene, but only as measured by the clearance of bromosulphthalein, and it is thus impossible to decide unequivocally from studies carried out to date whether 2-methylnaphthalene disturbs physiological mechanisms in mouse liver.

Pyrrolizidine alkaloids are products of various common plants, mostly of the *Crotalaria* and *Senecio* genera, and several are toxic to many animal species. Variations in toxicity may depend to some extent upon the nature of the alkaloids consumed (for example, the alkaloids of *Crotalaria* are associated with lung lesions) as well as upon the animal species affected, although it is generally possible to discuss them collectively without great inaccuracy. Acutely, a confluent, hemorrhagic, typically centrilobular necrosis develops in the liver of susceptible species, including the rat and mouse.[74] Cattle and horses are also very susceptible to this acute lesion, whereas sheep, goats, guinea pigs, and quail are relatively insensitive. A chronic liver lesion characterized by megalocytosis and other changes is also seen in some species after repeated administration of small doses of pyrrolizidine alkaloids. Liver tumors may arise occasionally. In addition, lung lesions follow pyrrolizidine alkaloid exposure in certain animals, e.g., the rat, the primary target being the pulmonary blood vessels. Acute liver and lung damage both appear to result from metabolic activation of the alkaloids, probably to pyrrole derivatives. The lung lesion is found only in those species that develop liver lesions, and the minimum dose levels for producing lung damage are always higher than those producing liver toxicity,[67] suggesting that toxic metabolites are formed primarily (if not exclusively) in the liver and transported to the lung (see Section V.A).

McLean[74] and Huxtable[75] have reviewed most of the current knowledge of pyrrolizidine alkaloid poisoning. There is little consistency in the toxicity induced among domestic animals. Poultry exhibit acute liver damage; chronic lesions occur in dogs, pigs, sheep, and horses. Pigs, sheep, and poultry show lung damage. Horses also suffer from a hemolytic syndrome. Numerous other lesions following the adminstration of various pyrrolizidine alkaloids have been described, including renal damage in rats, but there are insufficient data to support a definite mechanistic basis for these species differences in toxicity.

V. CASE HISTORY — FURANS

A. 4-Ipomeanol

Some of the most striking and well-documented examples of species-dependent target organ selectivity are shown by various substituted furans (see Figure 1). Of these, the most extensively studied has been 4-ipomeanol or 1-(3-furyl)-4-hydroxypentanone,[67,76] one of a group of structurally related compounds produced in sweet potatoes infected with the mold *Fusarium solani*.

Originally linked with the etiology of a severe pulmonary disease in cattle ingesting affected sweet potatoes, 4-ipomeanol has now been demonstrated to cause lung lesions in several rodent species. However, a marked degree of species sensitivity or specificity in target organ damage has also been revealed.

In most laboratory animal species, administration of 4-ipomeanol by oral, intraperitoneal, or intravenous routes at levels of 10 to 60 mg/kg results in lung damage and death within 24 to 36 hr. The most tolerant species so far tested is the hamster, in which the LD_{50} value is approximately 140 mg/kg.[77] Mice, rats, rabbits, hamsters, and guinea pigs dosed with 4-ipomeanol all show a characteristic pulmonary lesion, with necrosis of the nonciliated terminal bronchiolar epithelial (Clara) cells.[78] With minimally toxic doses of 4-ipomeanol, the lesion is localized in the smallest bronchioles, and appears to be the only major pathological change produced. At marginally sublethal doses, cells of the larger airways are affected, and massive pleural effusion and intra-alveolar edema may occur, particularly in the mouse.

When adminstered to some species, 4-ipomeanol damages other organs in addition to the lungs. Thus, whereas only pulmonary lesions occur in rats, rabbits, and guinea pigs, centrilobular hepatic necrosis and occasional renal damage are seen in the hamster. In mice, there are also notable age-related differences in the pattern of target organ toxicity.[79] Doses of 4-ipomeanol near to the LD_{50} reliably produce in adult male mice a renal cortical necrosis as well as pulmonary damage, but young male and female mice are highly resistant to this renal toxicity.

Birds exhibit a marked contrast to rodents in their response to 4-ipomeanol. Doses as

4-Ipomeanol

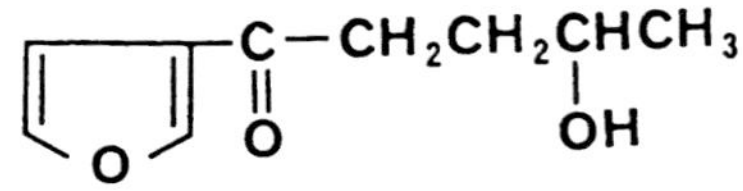

2,3-Benzofuran

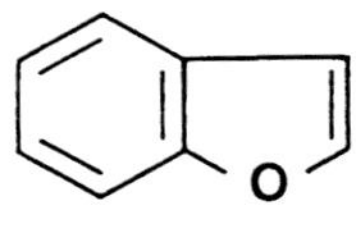

Furosemide

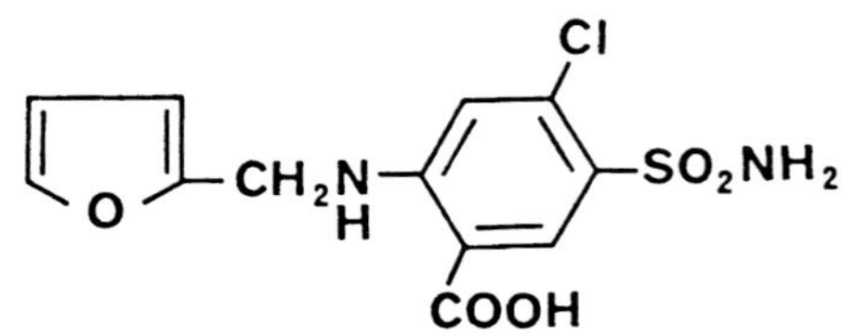

FIGURE 1. Substituted furans showing variations in target organ toxicity in rat and mouse.

high as 75 mg/kg in the rooster and 80 mg/kg in the Japanese quail produce no evidence of cellular necrosis in the lung, or pleural edema, hemorrhage, or effusion. Neither do doses of 4-ipomeanol at the LD_{50} result in detectable renal lesions in either species. However, in both the Japanese quail and rooster, 4-ipomeanol is a potent and dose-dependent hepatotoxin.[80] Species differences in target organ toxicity for 4-ipomeanol are summarized in Table 8.

There is considerable evidence to suggest that metabolic activation in the target tissue is responsible for the selective toxicity of 4-ipomeanol, as follows.

1. The site of 4-ipomeanol-induced lesions within mammalian lung is consistent with the localization of cytochrome P-450, cytochrome *c* reductase, and a major proportion of pulmonary xenobiotic metabolism in Clara cells.[81] These cells are also vulnerable to other metabolically activated toxins and to a number of nitrosamine carcinogens.[82]
2. Investigations using 2-furamide,[83] furosemide,[84] 2-(*N*-ethylcarbamoylhydroxymethyl) furan,[85] 4-ipomeanol, and 3-methylfuran[86,87] imply that metabolic activation of these toxic furans is mediated by cytochromes P-450 localized in their target tissues. Demonstration of the furan moiety as an essential structural site for the activation of toxic furan derivatives,[87] and the involvement of a dihydrofuran 2,3-epoxide in the activation of AFB_1[88] indicate that epoxides may be involved in the metabolism of furans. Glutathione appears to play a role in the detoxication of reactive metabolites of both furosemide and 4-ipomeanol, but only one study has suggested direct modulation by epoxide hydrolase of the covalent binding of furosemide to microsomal protein.[89] Hence, while the possibility of epoxide intermediates in the activation of simple furans is attractive, no definite proof of their formation yet exists.
3. Studies in vivo and in vitro show that without prior metabolism 4-ipomeanol is unable

Table 8
SPECIES DIFFERENCES IN LD_{50} AND TISSUE-SELECTIVE DAMAGE CAUSED BY 4-IPOMEANOL[77,80]

Species	LD_{50} (mg/Kg)	Location of tissue damage: Liver	Kidney	Lung
Mouse (adult male)	42	−	+	+ +
Rat	22	−	−	+ +
Rabbit	40	−	−	+ +
Hamster	140	+ +	+	+ +
Rooster	N.D.	+ +	−	−
Japanese quail	75	+ +	−	−

Note: N.D. = No data.

to bind covalently to tissue macromolecules, although the alkylating metabolites formed are very reactive and short-lived in biological media, and are not directly isolable. Detection of metabolite production has thus been achieved by measuring the covalent binding, primarily to tissue protein, of radiolabeled material derived from 4-ipomeanol.[67,76] The target organs for 4-ipomeanol toxicity in vivo in the different species tested correlate well with the degree and location of covalent binding of the compound in vivo. In rat, rabbit, and guinea pig, the predominant site for covalent binding of 4-ipomeanol is the lung. Small amounts of radiolabel have been detected in rat liver and kidney, but the extent of binding to other tissues was either very low or undetectable.[77,90] The levels of pulmonary alkylation were high in adult male mice, but in this species the extent of renal alkylation was higher still; in contrast, very little 4-ipomeanol was bound to the kidneys of either female mice or immature animals of either sex.[79,91] In hamsters, considerable amounts of 4-ipomeanol were bound in the liver, in addition to the lung. Covalent binding of 4-ipomeanol metabolites in livers of Japanese quail or rooster was nearly three times that in kidney, and more than five times that in eight other tissues.[80] Not only is there close correlation between target tissue and covalent binding of 4-ipomeanol, but autoradiography reveals that there is association of covalently bound radioactivity with those specific cell types that are most vulnerable to 4-ipomeanol toxicity. In kidneys from adult male mice, 4-ipomeanol is located almost entirely in proximal cortical tubular cells, its sites of toxicity, and in all species susceptible to lung damage, binding has been found to occur practically exclusively in Clara cells.[67,76,92]

4. Studies with enzyme inducers and inhibitors provide further support for the role of target tissue metabolism in the etiology of 4-ipomeanol toxicity. The inhibitors pyrazole, piperonyl butoxide, and cobaltous chloride decrease both hepatic and pulmonary covalent binding of 4-ipomeanol in rats, correlating with a decrease in pulmonary toxicity.[67] In mice, alkylation of renal tubular cells is also blocked by piperonyl butoxide.

That generation of toxic metabolites from 4-ipomeanol occurs *in situ* in extrahepatic organs is supported by the effects of inducers of cytochrome P-450-mediated monooxygenase activity. Pretreatment of rats with 3-methylcholanthrene (3-MCA) caused a marked increase in the covalent binding of 4-ipomeanol to liver in vitro without significantly increasing the covalent binding to rat lung preparations.[86] Paralleling these effects, the treatment of rats with 3-MCA caused a marked shift in target organ selectivity in vivo. An increase in covalent binding to the liver was associated with centrilobular necrosis, whereas both pulmonary covalent binding and toxicity were decreased.[93] In "responsive" C57BL/6J mice, pretreatment with 3-MCA followed by 4-ipomeanol frequently resulted in hepatic necrosis and reduced the extent of binding and necrosis in kidney and lung. No corresponding effect was seen in "noninducible" DBA/2J mice.[92] These results strongly support the hypothesis that reactive metabolites of 4-ipomeanol bind in the tissues in which they are generated. They also highlight a point made earlier regarding interactions of compounds capable of inducing or inhibiting enzyme activities with a compound being tested under laboratory conditions for species differences in toxicity.

What, then, are the underlying reasons for the species selectivity in target organ toxicity exhibited by 4-ipomeanol? The striking correlation between quantitative production in different tissues of metabolites capable of binding to those tissues and subsequent damage implies that species variations in local monooxygenase activities per se may determine patterns of toxicity. However, as mentioned previously, other factors governing the activation (or deactivation) pathways may also be important. Of interest here is the finding that although the maximal rates of covalent binding of 4-ipomeanol are similar in rat lung and liver microsomes, the K_m for activation and binding in the liver preparation is 10 times higher than that in lung microsomes.[86]

In view of the rather low levels of cytochrome P-450 in many extrahepatic tissues,[94] the fact that lung microsomes can achieve a higher V_{max} for metabolism and tissue binding of 4-ipomeanol than that of liver microsomes is notable. The results of Boyd and colleagues[86] also suggest that there is poor correlation between the capacity of microsomal preparations from different tissues to metabolize 4-ipomeanol and their total cytochrome P-450 concentration. It is now accepted that in mammalian species the liver at least may contain cytochromes P-450 of multiple type and substrate specificity for xenobiotics.[95,96] Comparatively little is known about multiplicity of lung cytochromes P-450, but the immunological experiments of Guengerich and Mason[97] indicate that there is cross-reactivity between microsomes prepared from untreated, 3-MCA- or phenobarbitone-induced rat lung and IgG antibodies to cytochromes P-450 purified from the livers of phenobarbitone- or 3-MCA-pretreated rats. Two forms of cytochrome P-450 have been demonstrated in rabbit lung,[98] one of which (P-450_I) is indistinguishable on the basis of spectrum, extinction coefficient, molecular weight, amino acid composition, fragmentation with cyanogen bromide, and immunochemistry from the major form of hepatic cytochrome P-450, P-450_{PB}, isolated from rabbits pretreated with phenobarbitone.[99]

P-450_{PB} appears to represent only a minor proportion (5 to 10%) of the hepatic cytochrome concentration in the untreated rabbit, whereas P-450_I comprises some 50% of the total pulmonary cytochrome.[98] The latter cytochrome appears to be responsible for most of the pulmonary metabolism of 4-ipomeanol,[85,100] and may help to explain why active metabolites of 4-ipomeanol achieve toxic intracellular concentrations in the lungs of several species, but not in the liver. If this is the crucial mechanistic factor, the relative proportions of the corresponding cytochromes in hamster hepatic and pulmonary microsomes will be worthy of evaluation. It also implies a striking difference in cytochrome P-450 populations in male and female adult mouse kidney.

The effects of inducers of cytochrome P-450 should also be considered. In mammalian liver phenobarbitone and 3-MCA induce different major forms of the cytochrome, whereas in lung, 3-MCA induces cytochrome P-450 and aryl hydrocarbon hydroxylase activity,[101,102] but phenobarbitone in most species has been found to have no effect. If P-450_{PB} and P-450_I are identical, then this raises an interesting question concerning the control of synthesis of different forms of cytochrome P-450. Furthermore, as described above, pretreatment with 3-MCA switches the target organ toxicity of 4-ipomeanol from lung to liver in the rat, despite the ability of 3-MCA to induce pulmonary as well as hepatic cytochrome P-450 in rats.[97] The decrease in lung binding and toxicity of 4-ipomeanol in 3-MCA-pretreated rats may thus be due to a greatly increased rate of hepatic clearance (and shortened biological half-life) of 4-ipomeanol, sufficient to more than compensate for the increased metabolic capacity of the lung, or 3-MCA may induce cytochromes P-450 of different substrate specificities in the two tissues. If the phenobarbitone form of cytochrome P-450 is normally primarily responsible for the metabolism of 4-ipomeanol, the toxic metabolite generated in 3-MCA-pretreated liver may indeed be different from the one(s) toxic to the lung.

An antibody prepared against purified cytochrome b_5 has been found to almost completely inhibit both NADH- and NADPH-supported pulmonary metabolism and covalent binding of 4-ipomeanol, but to have little effect on its activation in the liver.[103] These results are consistent with a cytochrome P-450-dependent activation of 4-ipomeanol in which transfer of the second electron occurs via cytochrome b_5, and indicate further the desirability of establishing the *nature* of hepatic and pulmonary metabolites of 4-ipomeanol. Interestingly, the NADH/cytochrome b_5/cytochrome b_5 reductase system appears to be able to support to a large extent the cytochrome P-450-dependent renal ω- and (ω-1)-hydroxylation of fatty acids.[104] However, rat kidney microsomes at least are practically incapable of mediating the covalent binding of 4-ipomeanol.

The effects of 4-ipomeanol in some avian species, while reinforcing the hypothesis of target tissue activation of 4-ipomeanol *in situ*, tend to complicate the issue of involvement of different cytochrome P-450 forms in such activation. Rather than terminal airways, lungs in the bird have continuous set of conducting airways, and do not possess cells analogous to the ciliated and nonciliated bronchiolar cells of the mammal.[105] Consistent with this is the shift from pulmonary to hepatic toxicity of 4-ipomeanol in birds. However, there seem to be differences in lung content of cytochrome P-450 among bird species; the total cytochrome P-450 concentration (expressed as nanomoles per milligram protein) in quail lung is comparable with that in rodent lung, whereas this cytochrome is indetectable in the lungs of Rhode Island Red roosters.[106] The total covalent binding of radiolabeled 4-ipomeanol in rat and quail lung is also comparable;[80] nonetheless, no necrosis has been observed in quail lung. Any of several explanations may account for this anomaly. Quail lung may lack the sensitive macromolecule whose binding to activated 4-ipomeanol initiates pulmonary toxicity, i.e., the binding observed is nonspecific and does not lead to organ damage. Alternatively, the pulmonary cytochrome P-450 of quail may mediate the production of 4-ipomeanol metabolites different from those generated in the liver. Third, the lung toxic metabolite may originate in the liver, and its differential toxicity be related to differences in protein sensitivity. Arguing against the last hypothesis is the lack of glutathione depletion in the lung in contrast with the liver, and the findings that other active metabolites generated in bird liver are toxic to the lung when transported there via the blood circulation. Induction studies in quail also show that phenobarbitone has little effect on either the toxicity or covalent binding of 4-ipomeanol in vivo (a situation comparable with that in the rat). However, 3-MCA, too, fails to significantly increase hepatic covalent binding or toxicity of 4-ipomeanol in quail, despite inducing hepatic cytochrome P-450 threefold.

Detoxicating enzymes may also participate in the tissue-selective toxicity of 4-ipomeanol. Endogenous glutathione appears to play a protective role, although the glutathione-dependent threshold for covalent binding and tissue damage seen with some hepatotoxins in mammals[59] is not evident in either rat lung or quail liver.[80] Little is known about the induction or other properties of avian glutathione transferase, but one explanation for the failure of phenobarbitone and 3-MCA to enhance the organ-selective toxicity of 4-ipomeanol in the rat and quail, respectively, would be greater induction of detoxifying pathways than of those responsible for its activation. Phenobarbitone is an inducer of hepatic epoxide hydrolase activity in rat, whereas 3-MCA is not.[107] Neither inducer appears to be very potent in the induction of rat pulmonary epoxide hydrolase activity.[102] The ability of some furans to induce hepatic epoxide hydrolase activity exceeds that of the "classical" inducers[108] and it would be informative to know the effect of 4-ipomeanol on hepatic and pulmonary epoxide hydrolase, and the rate of turnover of any 4-ipomeanol epoxide formed.

Although activation by discrete subspecies of cytochrome P-450 is attractive as a key to the species variation in target organ selectivity of 4-ipomeanol, a definitive answer awaits better characterization of extrahepatic cytochrome P-450 populations, demonstration of the pathways of activation of 4-ipomeanol and other simple substituted furans and, hence, establishment of the balance between routes of activation and detoxication.

B. Benzofuran

Mice pretreated with 2,3-benzofuran (100 mg/kg, single dose) are claimed to exhibit coagulative necrosis of proximal convoluted tubules of the outer renal cortex.[83] At 200 mg/kg, the renal lesion is exacerbated, and slight hepatic damage occurs. In contrast, the administration of benzofuran (500 mg/kg, intraperitoneal injection) to rats produces a lesion ultrastructurally and histologically typical of intrahepatic cholestasis, including single cell necrosis of hepatocytes,[108] with mild kidney effects. Pretreatment with phenobarbitone eliminates the renal toxicity in mice, but increases the severity of both renal (swelling of proximal convoluted tubules and compression of glomeruli) and hepatic lesions in rats. Pre-induction of rats with 3-MCA results in even greater kidney and liver damage, whereas piperonyl butoxide or cobalt chloride practically abolish the effects of benzofuran in mice. It is apparent that benzofuran is activated in both rat and mouse tissues, but the target organ pattern and varying results of pretreatment with inducers or inhibitors are difficult to explain. If benzofuran is transformed to an active metabolite (e.g., a 2,3-epoxide), it might be postulated that in mouse, cobalt chloride or piperonyl butoxide inhibit its formation, and phenobarbitone increases the activity of detoxicating enzymes to a greater extent than it increases monooxygenase activities. Phenobarbitone is not generally a good inducer extrahepatically, and the renal toxicity might be due to active metabolites transported from the liver. However, the lack of hepatotoxicity in the nonpretreated mouse, and the increased hepatotoxicity in the rat after administration of phenobarbitone *plus* benzofuran remain unexplained.

C. Furosemide

Furosemide, a diuretic drug used in the treatment of cardiovascular and renal diseases, causes extensive hepatotoxicity and nephrotoxicity in rodents at high doses. In the mouse, administration of furosemide results in a coagulative necrosis of the renal inner cortex and subcortical areas, midzonal to centrilobular necrosis of liver parenchymal cells after low doses, and centrilobular lesions after higher ones.[83] Intravenous injection of furosemide produces both severe hepatic midzonal and renal lesions, whereas kidney injury is infrequent after intraperitoneal injection. Rats dosed with furosemide develop necrosis of proximal convoluted tubules, but are resistant to hepatic damage. Similarly, hamsters do not exhibit hepatic necrosis.[84] Pretreatment with cobalt chloride blocks furosemide-induced renal necrosis in rats; pretreatment with 3-MCA increases mortality, reduces the incidence of kidney damage, and evokes mild single cell hepatic necrosis.[83] Inhibition of monooxygenase activities with cobalt chloride also lessens renal damage in mice, whereas 3-MCA has relatively little effect. Furosemide is activated by hepatic microsomes to a metabolite that binds to proteins[84] and the results of McMurtry and Mitchell[83] suggest that 2-furamide (and possibly other nephrotoxic furans) may be activated and bind to kidney macromolecules *in situ*. If this is the case, target organ switching and zonal changes in toxicity within tissues after pretreatment with monooxygenase inducers may, as with benzofuran, be explained by differential actions on intoxicating and detoxicating pathways, but the integration of these effects is unclear.

VI. CASE HISTORY — ANTINEOPLASTIC AGENTS

The furans described above represent a class of structurally closely related compounds whose toxicity in different species (especially comparative aspects in birds and mammals) is particularly useful in helping to elucidate mechanisms of toxicity. In contrast, the antineoplastic drugs form a therapeutic class of structurally very dissimilar compounds, whose species-selective target organ toxicity highlights the problems of predicting safe doses, and the occurrence of iatrogenic responses, in man. In some respects too, in designing an anticancer drug one is seeking to produce an agent possessing organ-selective properties, but to suit a situation where the characteristics of that organ may be very little different from surrounding (or other) tissues in which damage must be minimized.

Certain features of species-selection in the design and study of antineoplastic drugs should be considered.

1. Tests are required in sufficient animal species to pick up indicators of serious toxicity, while not excluding almost every potentially valuable drug, bearing in mind points (2) and (3).
2. Antitumor agents are administered to man at or near their toxic dose, hence some toxicity is practically inevitable, and it is especially important that advance warning of likely adverse reactions should be available.
3. Although larger amounts of the antineoplastic drugs are administered to animals in toxicity trials than are employed therapeutically, the high doses used in medicine mean that probably more toxicity data are available for humans in comparison with laboratory species than is the case for many chemicals.
4. Altered toxicity resulting from drug-interaction is important, since combination therapy in the treatment of cancer is very common.

A very large number of human organs or tissues are damaged by anticancer drugs in clinical use, even if only the major toxicities of each agent are taken into account (Table 9). Standard laboratory toxicity testing procedures are very good at predicting some of these, adequate at detecting others, and in certain cases leave much to be desired. Owens[109] considered 21 antineoplastic drugs with a variety of modes of action, and found that whereas the predictability of bone marrow, GI, hepatic, and renal toxicity was good in rodents, dogs, or monkeys, that of nervous system toxicity was questionable, and skin disturbances were not anticipated (Table 10).

Some of these inadequacies seem to have been eliminated by the changing state-of-the-art in toxicity testing. Certain others were probably due to deficiencies of observation in animal studies. It has been pointed out[110] that the unanticipated toxic psychoses and motor and sensory disturbances seen in patients dosed with 6-azauracil have their counterparts in the somnolence, ataxia, and paralysis observed in rodents, dogs, and cats after high doses, and reported in the literature. Similarly, terephthalanilide-induced ophthalmoplegia and *cis*-diamminedichloroplatin (cisplatin)-induced ototoxicity could be detected retrospectively in animals once they had been described in man. Guarino[111] and Heywood[2] consider that for major manifestations of toxicity, there is good predictability between species for anticancer drugs, and the former noted that of 120 reported principal toxicities there were only two instances (hematological in both cases) that occurred in the monkey but not in the dog; i.e., that dog studies, properly carried out and interpreted, provide a very high degree of predictability.

However, apart from the fact that clinical signs of toxicity may differ in animals and man, two other particular problems remain.

One of these is the overprediction sometimes seen in animal studies, especially of renal

Table 9
MAJOR TOXICITIES DETECTED IN HUMAN ORGANS OR TISSUES DURING CLINICAL TRIALS OF ANTINEOPLASTIC AGENTS

Tissue/organ	Drugs producing major toxicity
Bladder	Camptothecin
	Cyclophosphamide
Brain	L-Asparaginase
	Azaribine
	Cycloleucine
Heart	Adriamycin
	Daunorubicin
Intestine	Camptothecin
	Methotrexate
Kidney	Streptozotocin
Liver	6-Mercaptopurine
	Methotrexate
Lung	Bleomycin
	Busulfan
Pancreas	L-Asparaginase
Peripheral nervous system	Vincristine
Testis, ovary	Cyclophosphamide

Table 10
PREDICTABILITY OF HUMAN TOXICITY OF ANTICANCER DRUGS BY ANIMAL TESTING[109]

	Toxicity shown in			
	Rodent	Dog	Monkey	Human
Actinomycin P2				
Bone marrow	±	±	N.D.	+[a]
GI	+	+	N.D.	+
Skin and appendage[b]	–	–	N.D.	+
4-Aminopyrazolo(3,4-D)pyrimidine				
Bone marrow	+	+	N.D.	+
GI	+	+	N.D.	+
Hepatic	+	+	N.D.	+
Carzolamide				
Bone marrow	–	–	N.D.	+[a]
GI	–	+	N.D.	+
Hepatic	±	+	N.D.	+
Nervous	–	–	N.D.	+
2-Chloroethylmethanesulfonate				
Bone marrow	N.D.	+	+	?
GI	N.D.	+	+	+
Hepatic	+	+	N.D.	+
Renal	+	+	N.D.	+
Cyclophosphamide				
Bone marrow	+	+	+	+
GI	+	+	+	+
Skin and appendage[b]	–	–	–	+
Methotrexate				
Bone marrow	+	+	+	+
GI	+	+	+	+
Skin and appendage[b]	–	–	N.D.	+

Table 10 (continued)
PREDICTABILITY OF HUMAN TOXICITY OF ANTICANCER DRUGS BY ANIMAL TESTING[109]

	Toxicity shown in			
	Rodent	Dog	Monkey	Human
Mithramycin				
Bone marrow	−	−	N.D.	+[a]
GI	−	−	N.D.	+
Hepatic	±	+	N.D.	+
Vincaleukoblastine				
Bone marrow	+	+	N.D.	+
GI	+	+	N.D.	+
Nervous	−	−	N.D.	+[c]
Skin and appendage[b]	−	−	N.D.	+

Note: N.D. = No data.

[a] Primarily thrombocytopenia.
[b] Dermatitis and alopecia.
[c] Peripheral neuropathy.

and hepatic damage. Such findings cannot be disregarded, because very high doses of the drug might result in the appropriate signs in man, but Schein[110] observes that there do appear to be genuine species differences in organ sensitivity among dog, monkey, and man. Hence, the adverse reactions predicted by animal screens may appear in man at greater or lesser dose levels, or may be exhibited in a different order relative to the overall toxic outcome of a drug's use. Also, anticipated organ dysfunction may arise in man, but be expressed by changes in different clinical or chemical variables.

The second, and more serious, problem is that of false negatives in animal studies.[112] Despite improved animal screening, important human toxicities are still unpredicted; for example, the pancreatitis associated with L-asparaginase. In some cases, false negatives may reflect the relatively small numbers of large animals (in pragmatic terms, the most successful predictors of antineoplastic drug toxicity) used in testing, compared with the numbers of patients subsequently exposed to the drug. Many of these patients are already receiving other therapeutic agents or have existing organ failure, and may not bear comparison with previously normal, healthy animals. However, some differences almost certainly represent variations in physiology (e.g., the relative resistance of the monkey to thrombocytopenia or GI toxicity) or metabolic capability, and will influence the choice of species chosen for testing.

Rabbits and guinea pigs, for example, are resistant to the toxicity of methotrexate (MTX), dichloromethotrexate (DCM), and aminopterin.[113] These findings correlate with high hepatic activities of aldehyde oxidase, which converts these drugs to 7-hydroxy metabolites (little antifolate activity). However, relative rates of MTX catabolism do not explain all of the species differences in antifolate toxicity. MTX is 13-fold more toxic than DCM in the dog, despite the fact that this species lacks the ability to metabolize DCM, the more potent of the two analogues. Similarly, the slight difference in affinity for aldehyde oxidase cannot explain the much greater toxicity of aminopterin than DCM in the mouse. Many other pathways of metabolism are followed by anticancer drugs,[111] and species differences in their prominence may be important; in particular, the ability of extrahepatic tissues to activate or deactivate chemicals may play a role in local toxicity and the genesis of specific organ lesions. Adaptation also complicates the study of MTX toxicity; certain organs in mice

Table 11
DELAYED TOXIC RESULTS OF ADMINISTRATION OF ANTINEOPLASTIC DRUGS TO MAN

Effects seen	Drugs responsible
Alopecia	ACD, CPA, VBL, VCR
Bone marrow aplasia	BSF
Cardiotoxicity	ADM, BSF, DMC
Cataract	BSF
Hepatic damage	CAB, 6-MP, MTX
Neurotoxicity	Central: Ara-C, MTX
	Peripheral: VNL, VCR
Pigmentation	BSF
Pulmonary fibrosis	BLM, BSF, CPA, MTX
Reproduction	Male: BSF, CAB, CPA, 6-MP, NMU, PCB
	Female: BSF, CAB, CPA, 6-MP, NMU, TTA
Skeletal	MTX
Thrombocytopenia	BCNU, BSF, CCNU
Urinary	CPA, 6-MP

Abbreviations: ACD = actinomycin D; ADM = adriamycin; Ara-C = cytosine arabinoside; BCNU = 1,3-bis(2-chloroethyl)-1-nitrosourea; BLM = bleomycin; BSF = busulfan; CAB = chlorambucil; CCNU = 1-(2-chloroethyl)-3-cyclohexyl-1-nitrosourea; CPA = cyclophosphamide; DMC = daunomycin; 6-MP = 6-mercaptopurine; MTX = methotrexate; NMU = nitrogen mustard; TTA = thiotepa; VBL = vinblastine; VCR = vincristine.

respond to MTX by increasing their rate of dihydrofolate reductase synthesis, an effect which may not occur in other species.[114]

Interspecies variations in antineoplastic drug toxicity are manifold, both quantitatively and qualitatively, and particularly when rodents are considered. MTX hepatotoxicity is much less pronounced in mice than in rats, and the latter also develop lung lesions.[115] In this case, damage to both organs in rats is predictive of human toxicity, although the pulmonary lesion in man is unusual.[116,117] Rat and mouse studies often overpredict hepatic and renal toxicity in man. Conversely, few signs of nervous sytem dysfunction occur in rodents given 5-fluorouracil, whereas the dog is peculiarly sensitive.[118] Humans exhibit neurotoxicity, but it is generally reversible.[119] CCNU (1-(2-chloroethyl)-3-cyclohexyl-1-nitrosourea) produces bone marrow damage in the dog, monkey, and man; neutropenia is particularly severe or lethal in dogs, whereas monkeys die of chronic interstitial nephritis.[120] Cardiopulmonary toxicity also occurs in dogs.[121] Nephrotoxicity,[122] but not pulmonary or hepatic toxicity, has been reported in man. The pulmonary toxicity arising in humans from BCNU (1,3-bis(2-chloroethyl)-1-nitrosourea) therapy is better predicted by dogs than by monkeys.[121] Early deaths in rats after 6-mercaptopurine adminsitration is due to lung damage, an effect not seen in mouse, cat, or dog.[123] Many other species differences in target organ susceptibility or selectivity have been reported for the common antineoplastic drugs,[124] but the mechanisms underlying them, as in most of the examples above, are obscure.

Delayed toxicity (possibly including carcinogenesis) of anticancer drugs is a special problem in underprediction. This has become particularly important with the discovery of cardiomyopathy due to adriamycin and daunomycin,[125] but there are many other examples of anticancer drug-induced delayed toxicities in humans (Table 11). Woolley and Schein[112] claim that animal systems have been generally successful in the prediction of delayed toxicity. However, the tissue affected in animals may not correlate directly with the one affected in man; e.g., delayed hepatotoxicity would be predicted from the administration to dogs of either BCNU or CCNU, whereas in man the delayed reaction is manifested in the bone marrow.

VII. CONCLUSIONS

Quantitative differences in the toxicity (= lethality) of chemicals to animal species, as expressed by LD_{50} values or similar measurements, have been recognized for many years. So too has the necessity to administer sometimes 10- or 100-fold the amount of a compound to a second species to elicit the same degree of organ damage as was seen in the first. Widespread recognition of qualitative species differences in target organ toxicity, of single or multifactorial origin, is a more recent phenomenon. However, more examples are now rapidly coming to light.

The influences determining some of the interspecies variation described in this article can be readily eliminated, or modified by manipulation of the animals models (e.g., where absorption, gut microfloral metabolism, or lack of emetic response largely regulates toxicity, the route of administration can be made parenteral). For better or worse, others cannot, and since they exist, they may as well be exploited.

The use of species-dependent, organ-selective lesions in the understanding of mechanisms of toxicity is probably the keystone of their value, emphasizing the need to follow up unexpected adverse effects either in man or in animals. Boyd and colleagues have pointed the way here with their studies of the lung toxic furans. Such understanding may lead to the more rational design of tissue-selective drugs; serendipity has played a major role in the discovery of many of those known at present. Also, although toxicity testing has improved since Barnes and Denz[126] concluded that extrapolation of animal data to man was a matter of guesswork, current procedures are far from perfect. Heywood[2] has indicated that the simple parameters of toxicity (clinical observation, body weight, organ weight, and histopathological examinations) are the most effective. However, it is likely that toxicity tests of the future, designed to require fewer animals and to allow valuable (and, it is hoped, selective) drugs to achieve clinical use more rapidly, will depend for their accuracy upon some knowledge of the mechanisms of the biological actions of these drugs.

ACKNOWLEDGMENT

The authors would like to thank Ms. Eleanor Beaton for her assistance in unearthing the material for this article.

REFERENCES

1. **Heywood, R.,** Target organ toxicity, *Toxicol. Lett.*, 8, 349, 1981.
2. **Heywood, R.,** Target organ toxicity. II., *Toxicol. Lett.*, 18, 83, 1983.
3. **Bridges, J. W. and Hubbard, S. A.,** Principles, practice, problems and priorities in toxicology, in *Current Approaches to Occupational Health*, Vol. 2, Gardner, A. W., Ed., J. Wright & Sons, Bristol, 1982, 251.
4. **Litchfield, J. T.,** Evaluation of the safety of new drugs by means of tests in animals, *Clin. Pharmacol. Ther.*, 3, 665, 1962.
5. **Tomatis, L., Agthe, C., Bartsch, H., Huff, J., Montesano, R., Saracci, R., Walker, E., and Wilbourn, J.,** Evaluation of the carcinogenicity of chemicals: a review of the monograph programme of the International Agency for Research on Cancer (1971 to 1977), *Cancer Res.*, 38, 877, 1978.
6. **Roszkowski, A. P.,** Comparative toxicity of rodenticides, *Fed. Proc. Fed. Am. Soc. Exp. Biol.*, 26, 1082, 1967.
7. **Dougherty, E.,** Thiophosphate poisoning in white pekin ducks, *Avian Dis.*, 1, 127, 1957.
8. **Bergenstal, D. M., Hertz, R., Lipsett, M. B., and Moy, R. H.,** Chemotherapy of adrenocortical cancer with o', p'-DDD (Rothane), *Ann. Intern. Med.*, 53, 672, 1960.
9. **Boyd, M. R., Grygiel, J. J., and Minchin, R. F.,** Metabolic activation as a basis for organ-selective toxicity, *Clin. Exp. Pharmacol. Physiol.*, 10, 87, 1983.

10. **Roszkowski, A. P.,** The pharmacological properties of norbormide, a selective rat toxicant, *J. Pharmacol. Exp. Ther.*, 149, 288, 1965.
11. **Allen, J. C. and Schwartz, A.,** A possible biochemical explanation for the insensitivity of the rat to cardiac glycosides, *J. Pharmacol. Exp. Ther.*, 168, 42, 1969.
12. **Tobin, T. and Brody, T. M.,** Rates of dissociation of enzyme-ouabain complexes and $K_{0.5}$ values in $(Na^+ + K^+)$ adenosine triphosphatase from different species, *Biochem. Pharmacol.*, 21, 1553, 1972.
13. **Akera, T., Larsen, F. S., and Brody, T. M.,** The effect of ouabain on sodium- and potassium-activated adenosine triphosphatase from the hearts of several mammalian species, *J. Pharmacol. Exp. Ther.*, 170, 17, 1969.
14. **Vos, J. G. and Beems, R. B.,** Dermal toxicity studies of technical polychlorinated biphenyls and fractions thereof in rabbits, *Toxicol. Appl. Pharmacol.*, 19, 617, 1971.
15. **Vos, J. G., Moore, J. A., and Zinkl, J. G.,** Toxicity of 2,3,7,8-tetrachlorodibenzo-p-dioxin (TCDD) in C57B1/6 mice, *Toxicol. Appl. Pharmacol.*, 29, 229, 1974.
16. **McConnell, E. E., Moore, J. A., Haseman, J. K., and Harris, M. W.,** The comparative toxicity of chlorinated dibenzo-p-dioxins in mice and guinea pigs, *Toxicol. Appl. Pharmacol.*, 44, 335, 1978.
17. **Rose, H. A. and Rose, S. P. R.,** Chemical spraying as reported by refugees from South Vietnam, *Science*, 177, 710, 1972.
18. **Gasiewicz, T. A., Holscher, M. A., and Neal, R. A.,** The effects of total parenteral nutrition on the toxicity of 2,3,7,8-tetrachlorodibenzo-*p*-dioxin in the rat, *Toxicol. Appl. Pharmacol.*, 54, 469, 1980.
19. **Greig, J. B., Jones, G., Butler, W. H., and Barnes, J. M.,** Toxic effects of 2,3,7,8-tetrachlorodibenzo-*p*-dioxin, *Food Cosmet. Toxicol.*, 11, 585, 1973.
20. **Allen, J. R. and Norback, D. H.,** Polychlorinated biphenyl- and triphenyl-induced gastric mucosal hyperplasia in primates, *Science*, 179, 498, 1973.
21. **McConnell, E. E., Moore, J. A., Gupta, B. N., Rakes, A. H., Luster, M. I., Goldstein, J. A., Haseman, J. K., and Parker, C. E.,** The chronic toxicity of technical and analytical pentachlorophenol in cattle. I. Clinicopathology, *Toxicol. Appl. Pharmacol.*, 52, 468, 1980.
22. **Allen, J. R., Hargraves, W. A., Hsia, M. T. S., and Lin, F. S. D.,** Comparative toxicology of chlorinated compounds on mammalian species, *Pharmacol. Ther.*, 7, 513, 1979.
23. **Poland, A. and Knutson, J. C.,** 2,3,7,8-Tetrachlorodibenzo-*p*-dioxin and related halogenated hydrocarbons: examination of the mechanism of toxicity, *Annu. Rev. Pharmacol. Toxicol.*, 22, 517, 1982.
24. **Barnes, J. M. and Stoner, H. B.,** Toxic properties of some dialkyl and trialkyl tin salts, *Br. J. Ind. Med.*, 15.
25. **Seinen, W., Vos, J. G., Van Krieken, R., Penninks, A., Brands, R., and Hooykaas, H.,** Toxicity of organotin compounds. III. Suppression of thymus-dependent immunity in rats by di-*n*-butyltindichloride and di-*n*-octyltindichloride, *Toxicol. Appl. Pharmacol.*, 42, 213, 1977.
26. **Barnes, J. M. and Magee, P. N.,** The bilary and hepatic lesions produced experimentally by dibutyltin salts, *J. Pathol. Bact.*, 75, 267, 1958.
27. **Bridges, J. W. and Wilson, A. G. E.,** Drug-serum protein interactions and their biological significance, in *Progess in Drug Metabolism*, Vol. 1, Bridges, J. W. and Chasseaud, L. F., Eds., Wiley Interscience, New York, 1976, 193.
28. **Johnson, H. D. and Voss, E.,** Toxicological studies of zinc phosphide, *J. Am. Pharm. Assoc. Sci. Ed.*, 41, 468, 1952.
29. **Aldridge, W. N.,** Organophosphorus compounds: molecular basis for their biological properties, *Sci. Prog. Oxford*, 67, 131, 1981.
30. **Schaumburg, H. H. and Spencer, P. S.,** Toxic neuropathies, *Neurology*, 29, 429, 1979.
31. **Baron, R. L.,** Delayed neurotoxicity and other consequences of organophosphate esters, *Annu. Rev. Entomol.*, 26, 29, 1981.
32. **Søderlund, E. J., Nelson, S. D., Von Bahr, C., and Dybing, E.,** Species differences in kidney toxicity and metabolic activation of tris (2,3-dibromopropyl) phosphate, *Fundamental Appl. Toxicol.*, 2, 187, 1982.
33. **Adamson, R. H. and Davies, D. S.,** Comparative aspects of absorption, distribution, metabolism and excretion of drugs, in *International Encyclopedia of Pharmacology and Therapeutics*, Michelson, M. J., Ed., Pergamon Press, Oxford, 1973, 851.
34. **McCreesh, A. H.,** Percutaneous toxicity, *Toxicol. Appl. Pharmacol.*, 7(Suppl. 2), 20, 1965.
35. **Hamre, D. M., Rake, G., McKee, C. M., and MacPhillamy, H. B.,** The toxicity of penicillin as prepared for clinical use, *Am. J. Med. Sci.*, 206, 642, 1943.
36. **Scheline, R. R.,** Drug metabolism by the gastrointestinal microflora, in *Extrahepatic Metabolism of Drugs and Other Foreign Compounds*, Gram, T. E., Ed., Specialist Medical and Scientific Books, Jamaica, 1980, 551.
37. **Drasar, B. S., Hill, M. J., and Williams, R. E. O.,** The significance of the gut flora in safety testing of food additives, in *Metabolic Aspects of Food Safety*, Roe, F. J. C., Ed., Blackwell Scientific, Oxford, 1970, 245.

38. **Handschumacher, R. E., Calabresi, P., Welch, A. D., Bono, V., Fallon, H., and Frei, E.,** Summary of current information on 6-azauridine, *Cancer Chemother. Rep.*, 21, 1, 1962.
39. **Grasso, P., Gaunt, I. F., Hall, D. E., Golberg, L., and Batstone, E.,** Studies on Brown FK. III. Administration of high doses to rats and mice, *Food Cosmet. Toxicol.*, 6, 1, 1968.
40. **Grasso, P., Muir, A., Golberg, L., and Batsone, E.,** Studies on brown FK. IV. Cytopathic effects of Brown FK on cardiac and skeletal muscle in rat, *Food Cosmet. Toxicol.*, 6, 13, 1968.
41. **Walker, R., Grasso, P., and Gaunt, I. F.,** Myotoxicity of amine metabolites from Brown FK, *Food. Cosmet. Toxicol.*, 8, 539, 1970.
42. **Eason, C. T., Parke, D. V., Clark, B., and Smith, D. A.,** The mechanism of hepatotoxicity of a chromone carboxylic acid (FPL 52757) in the dog, *Xenobiotica*, 12, 155, 1982.
43. **Smith, D. A.,** Differences in toxicity due to species variation in the metabolism of an oral anti-allergy agent, *Br. J. Pharmacol.*, 66, 422, 1979.
44. **Caldwell, J.,** The current status of attempts to predict species differences in drug metabolism, *Drug. Metab. Rev.*, 12, 221, 1981.
45. **Menard, R. H., Bartter, F. C., and Gillette, J. R.,** Spironolactone and testicular cytochrome P-450: impairment of steroid 21-hydroxylation in the adrenal cortex, *Arch. Biochem. Biophys.*, 173, 395, 1976.
46. **Lumb, G., Newberne, P., Rust, J. H., and Wagner, B.,** Effects in animals of chronic administration of spironolactone — a review, *J. Environ. Pathol. Toxicol.*, 1, 641, 1978.
47. **Menard, R. H., Stripp, B., and Gillette, J. R.,** Spironolactone and testicular cytochrome P-450: decreased testosterone formation in several species and changes in hepatic drug metabolism, *Endocrinology*, 94, 1628, 1974.
48. **Schumacher, H., Smith, R. L., and Williams, R. T.,** The metabolism of thalidomide: the fate of thalidomide and some of its hydrolysis products in various species, *Br. J. Pharmacol.*, 25, 338, 1965.
49. **Gordon, G. B., Spielburg, S. P., Blake, D. A., and Balasubramanian, V.,** Thalidomide teratogenesis: evidence for a toxic arene oxide metabolite, *Proc. Natl. Acad. Sci. U.S.A.*, 78, 2545, 1981.
50. **Thomas, J. A., Darby, T. D., Wallin, R. F., Garvin, P. J., and Martis, L.,** A review of the biological effects of di-(2-ethylhexyl) phthalate, *Toxicol. Appl. Pharmacol.*, 45, 1, 1978.
51. **Gray, T. J. B., Rowland, I. R., Foster, P. M. D., and Gangolli, S. D.,** Species differences in the testicular toxicity of phthalate esters, *Toxicol. Lett.*, 11, 141, 1982.
52. **El-Hawari, A. M. and Plaa, G. L.,** α-Naphthylisothiocyanate (ANIT) hepatotoxicity and irreversible binding to rat liver microsomes, *Biochem. Pharmacol.*, 26, 1857, 1977.
53. **Plaa, G. L. and Priestly, B. G.,** Intrahepatic cholestasis induced by drugs and chemicals, *Pharmacol. Rev.*, 28, 207, 1977.
54. **Newberne, P. M. and Butler, W. H.,** Acute and chronic effects of aflatoxin on the liver of domestic and laboratory animals: a review, *Cancer Res.*, 29, 236, 1969.
55. **Akao, M., Kuroda, K., and Wogan, G. N.,** Aflatoxin B_1. The kidney as a site of action in the mouse, *Life Sci.*, 10, 495, 1971.
56. **Croy, R. G. and Wogan, G. N.,** Quantitative comparison of covalent aflatoxin-DNA adducts formed in rat and mouse livers and kidneys, *J. Natl. Cancer Inst.*, 66, 761, 1981.
57. **O'Brien, K., Moss, E., Judah, D., and Neal, G.,** Metabolic basis of the species difference to aflatoxin B_1 induced hepatotoxicity, *Biochem. Biophys. Res. Commun.*, 114, 813, 1983.
58. **Ueno, I., Friedman, L., and Stone, C. L.,** Species difference in the binding of aflatoxin B_1 to hepatic macromolecules, *Toxicol. Appl. Pharmacol.*, 52, 177, 1980.
59. **Mitchell, J. R., Jollow, D. J., Potter, W. Z., Davis, D. C., Gillette, J. R., and Brodie, B. B.,** Acetaminophen-induced hepatic necrosis. I. Role of drug metabolism, *J. Pharmacol. Exp. Ther.*, 187, 185, 1973.
60. **Mitchell, J. R., McMurtry, R. J., Statham, C. N., and Nelson, S. D.,** Molecular basis for several drug-induced nephropathies, *Am. J. Med.*, 62, 518, 1977.
61. **Cobden, I., Record, C. O., Ward, M. K., and Kerr, D. N. S.,** Paracetamol-induced acute renal failure in the absence of fulminant liver damage, *Br. Med. J.*, 284, 21, 1982.
62. **Newton, J. F., Yoshimoto, M., Bernstein, J., Rush, G. F., and Hook, J. B.,** Acetaminophen nephrotoxicity in the rat. II. Strain differences in nephrotoxicity and metabolism of *p*-aminophenol, a metabolite of acetaminophen, *Toxicol. Appl. Pharmacol.*, 69, 307, 1983.
63. **Schafer, E. W.,** The acute oral toxicity of 369 pesticidal, pharmaceutical and other chemicals to wild birds, *Toxicol. Appl. Pharmacol.*, 21, 315, 1972.
64. **Mull, R. L. and Giri, S. N.,** The role of renal aromatic *N*-deacetylase in selective toxicity of avicide 3-chloro-*p*-toluidine in birds, *Biochim. Biophys. Acta*, 273, 222, 1972.
65. **Giri, S. N., Barger, A. E., Bickford, A. A., and Peoples, S. A.,** Histopathological features and biochemical mechanism of 2-chloro-4-acetotoluidine in starling, *Exp. Mol. Pathol.*, 29, 426, 1978.
66. **Felsenstein, W. C., Smith, R. P., and Gosselin, R. P.,** Toxicologic studies on the avicide, 3-chloro-*p*-toluidine, *Toxicol. Appl. Pharmacol.*, 28, 110, 1974.

67. **Boyd, M. R.,** Biochemical mechanisms in chemical-induced lung injury. Roles of metabolic activation, *CRC Crit. Rev. Toxicol.*, 7, 103, 1980.
68. **Hook, J. B.,** Toxic responses of the kidney, in *Casarett and Doull's Toxicology: The Basic Science of Poisons*, 2nd ed., Doull, J., Klaasen, C. D., and Amdur, M. O., Eds., Macmillan, New York, 1980, chap. 11.
69. **Krijgsheld, K. R., Lowe, M. C., Mimnaugh, E. C., Trush, M. A., Ginsberg, E., and Gram, T. E.,** Lung-selective impairment of cytochrome P-450 dependent monooxygenases and cellular injury by 1,1-dichlorethylene in mice, *Biochem. Biophys. Res. Commun.*, 110, 675, 1983.
70. **Jenkins, L. J. and Andersen, M. E.,** 1,1-Dichloroethylene nephrotoxicity in the rat, *Toxicol. Appl. Pharmacol.*, 46, 131, 1978.
71. **Tong, S. S., Hirokata, Y., Trush, M. A., Mimnaugh, E. G., Ginsberg, E., Lowe, M. C., and Gram, T. E.,** Clara cell damage and inhibition of pulmonary mixed-function oxidase by napththalene, *Biochem. Biophys. Res. Commun.*, 100, 944, 1981.
72. **Warren, D. L., Brown, D. L., and Buckpitt, A. R.,** Evidence for cytochrome P-450 mediated metabolism in the bronchiolar damage by naphthalene, *Chem. Biol. Interact.*, 40, 287, 1982.
73. **Griffin, K. A., Johnson, C. B., Breger, R. K., and Franklin, R. B.,** Pulmonary toxicity, hepatic and extrahepatic metabolism of 2-methylnaphthalene in mice, *Toxicol. Appl. Pharmacol.*, 61, 185, 1981.
74. **McLean, E. K.,** The toxic action of pyrrolizidine (Senecio) alkaloids, *Pharmacol. Rev.*, 22, 429, 1970.
75. **Huxtable, R. J.,** New aspects of the toxicology and pharmacology of pyrrolizidine alkaloids, *Gen. Pharmacol.*, 10, 159, 1979.
76. **Boyd, M. R.,** Biochemical mechanisms in pulmonary toxicity of furan derivatives, in *Reviews in Biochemical Toxicology*, Vol. 2, Hodgson, E., Bend, J. R., and Philpot, R. M., Eds., Elsevier, Amsterdam, 1980, 71.
77. **Dutcher, J. S. and Boyd, M. R.,** Species and strain differences in target organ alkylation and toxicity by 4-ipomeanol: predictive value of covalent binding in studies of target organ toxicities by reactive metabolites, *Biochem. Pharmacol.*, 28, 3367, 1979.
78. **Boyd, M. R.,** Evidence for the Clara cell as a site of cytochrome P-450 dependent mixed-function oxidase activity in the lung, *Nature (London)*, 269, 713, 1977.
79. **Jones, R., Allegra, C., and Boyd, M.,** Predictive value of covalent binding in target organ toxicity: age-related differences in target organ alkylation and toxicity by 4-ipomeanol, *Toxicol. Appl. Pharmacol.*, 48, A129, 1979.
80. **Buckpitt, A. R., Statham, C. N., and Boyd, M. R.,** In vivo studies on the target tissue metabolism, covalent binding, glutathione depletion and toxicity of 4-ipomeanol in birds, species deficient in pulmonary enzymes for metabolic activation, *Toxicol. Appl. Pharmacol.*, 65, 38, 1982.
81. **Philpot, R. M. and Wolf, C. R.,** The properties and distribution of the enzymes of pulmonary cytochrome P-450 dependent monooxygenase systems, in *Reviews in Biochemical Toxicology*, Vol. 3, Hodgson, E., Bend, J. R., and Philpot, R. M., Elsevier/North Holland, New York, 1981, 51.
82. **Reznik-Schuller, H. M. and Hague, B. F., Jr.,** A morphometric study of the pulmonary Clara cell in normal and nitrosohepta-methyleneimine-treated European hamsters, *Exp. Pathol.*, 18, 366, 1980.
83. **McMurtry, R. J. and Mitchell, J. R.,** Renal and hepatic necrosis after metabolic activation of 2-substituted furans and thiophenes, including furosemide and cephaloridine, *Toxicol. Appl. Pharmacol.*, 42, 285, 1977.
84. **Mitchell, J. R., Nelson, W. L., Potter, W. Z., Sasame, H. A., and Jollow, D. J.,** Metabolic activation of furosemide to a chemically reactive hepatotoxic metabolite, *J. Pharmacol. Exp. Ther.*, 199, 41, 1976.
85. **Guengerich, F. P.,** Studies on the activation of a model furan compound — toxicity and covalent binding of 2-(*N*-ethylcarbamoyl-hydroxymethyl) furan, *Biochem. Pharmacol.*, 26, 1909, 1977.
86. **Boyd, M. R., Burka, L. T., Wilson, B. J., and Sasame, H. A.,** In vitro studies on the metabolic activation of the pulmonary toxin, 4-ipomeanol, by rat lung and liver microsomes, *J. Pharmacol. Exp. Ther.*, 207, 677, 1978.
87. **Boyd, M. R., Statham, C. N., Franklin, R. B., and Mitchell, J. R.,** Pulmonary bronchiolar alkylation and necrosis by 3-methylfuran, a naturally occurring potential atmospheric contaminant, *Nature (London)*, 272, 270, 1978.
88. **Swenson, D. H., Lin, J., Miller, E. C., and Miller, J. A.,** Aflatoxin B_1-2,3-oxide as a probable intermediate in the covalent binding of aflatoxin B_1 and B_2 to rat liver DNA and ribosomal RNA in vivo, *Cancer Res.*, 37, 172, 1977.
89. **Wirth, P. J., Bettis, C. J., and Nelson, W. L.,** Microsomal metabolism of furosemide. Evidence for the nature of the reactive intermediate involved in covalent binding, *Mol. Pharmacol.*, 12, 759, 1976.
90. **Buckpitt, A. R. and Boyd, M. R.,** Xenobiotic metabolism in birds, species lacking pulmonary Clara cells, *Pharmacologist*, 20, 181, 1978.
91. **Longo, N. S. and Boyd, M. R.,** Sex differences in renal alkylation and toxicity by 4-ipomeanol in the mouse, paper presented at 19th Annu. Meet. of the Society of Toxicology, Washington, D.C., 1980.
92. **Boyd, M. R. and Dutcher, J. S.,** Renal toxicity due to reactive metabolites formed in situ in the kidney: investigations with 4-ipomeanol in the mouse, *J. Pharmacol. Exp. Ther.*, 216, 640, 1981.

93. **Boyd, M. R. and Burka, L. T.,** In vivo studies on the relationship between target organ alkylation and the pulmonary toxicity of a chemically reactive metabolite of 4-ipomeanol, *J. Pharmacol. Exp. Ther.*, 207, 687, 1978.
94. **Connelly, J. C. and Bridges, J. W.,** The distribution and role of cytochrome P-450 in extrahepatic organs, in *Progress in Drug Metabolism,* Vol. 5, Bridges, J. W. and Chasseaud, L. F., Eds., Wiley Interscience, New York, 1980, chap. 1.
95. **Guengerich, F. P.,** Isolation and purification of cytochrome P-450 and the existence of multiple forms, *Pharmacol. Ther.*, 6, 99, 1979.
96. **Lu, A. Y. H. and West, S. B.,** Multiplicity of mammalian microsomal cytochromes P-450, *Pharmacol. Rev.*, 31, 277, 1980.
97. **Guengerich, F. P. and Mason, P. S.,** Immunological comparison of hepatic and extrahepatic cytochromes P-450, *Mol. Pharmacol.*, 15, 154, 1979.
98. **Wolf, C. R., Szutowski, M. M., Ball, L. M., and Philpot, R. M.,** The rabbit pulmonary monooxygenase system: characteristics and activities of two forms of pulmonary cytochrome P-450, *Chem. Biol. Interact.*, 21, 29, 1978.
99. **Serabjit-Singh, C. J., Wolf, C. R., and Philpot, R. M.,** The rabbit pulmonary monooxygenase system. Immunochemical and biochemical characterisation of enzyme components, *J. Biol. Chem.*, 254, 9901, 1979.
100. **Serabjit-Singh, C. J., Wolf, C. R., Philpot, R. M., and Plopper, C. G.,** Cytochrome P-450: localization in rabbit lung, *Science,* 207, 1469, 1980.
101. **Hulshoff, A., Lubawy, W. C., and Kostenbauder, H. B.,** The effect of some cigarette smoke constitutents and other compounds on the metabolism of benzo(a)pyrene in rabbit lung 9000 g supernatant, *Xenobiotica,* 8, 711, 1978.
102. **Dansette, P. M., Alexandrov, K., Azerad, R., and Frayssinet, Ch.,** The effect of some mixed function oxidase inducers on aryl hydrocarbon hydroxylase and epoxide hydrolase in nuclei and microsomes from rat liver and lung. The effect of cigarette smoke, *Eur. J. Cancer,* 15, 915, 1979.
103. **Sasame, H. A., Gillette, J. R., and Boyd, M. R.,** Effects of anti-NADPH cytochrome *c* reductase and anti-cytochrome b_5 antibodies on the hepatic and pulmonary microsomal metabolism and covalent binding of the pulmonary toxin 4-ipomeanol, *Biochem. Biophys. Res. Commun.*, 84, 389, 1978.
104. **Ellin, A. and Orrenius, S.,** Fatty acid hydroxylation in rat kidney cortex microsomes, *Mol. Cell. Biochem.*, 8, 69, 1975.
105. **Fedde, M. R.,** Respiration, in *Avian Physiology,* Sturkie, P. D., Ed., Springer-Verlag, New York, 1976, chap. 6.
106. **Buckpitt, A. R. and Boyd, M. R.,** Relationship between metabolism and toxicity of xenobiotics in avian species, in *Progress in Drug Metabolism,* Vol. 7, Bridges, J. W. and Chasseaud, L. F., Eds., Wiley Interscience, New York, 1983, chap. 6.
106. **Bresnick, E., Mukhtar, H., Stoming, T. A., Dansette, P. M., and Jerina, D. M.,** Effects of phenobarbital and 3-methylcholanthrene administration on epoxide hydrolase levels in liver microsomes, *Biochem. Pharmacol.*, 26, 891, 1977.
108. **Connelly, J. C.,** The Toxicity of Some Oxygen Heterocycles, Ph.D. thesis, University of Surrey, Surrey, England, 1983.
109. **Owens, A. H.,** Predicting anticancer drug effects in man from laboratory animal studies, *J. Chronic Dis.*, 15, 223, 1962.
110. **Schein, P. S.,** The prediction of clinical toxicities of anticancer drugs, in *Pharmacological Basis of Cancer Chemotherapy,* Williams & Wilkins, Baltimore, 1975, 383.
111. **Guarino, A. M.,** Pharmacologic and toxicologic studies of anticancer drugs: of sharks, mice and men (and dogs and monkeys), *Methods Cancer Res.*, 17, 91, 1979.
112. **Woolley, P. V. and Schein, P. S.,** Clinical pharmacology and phase I trial design, *Methods Cancer Res.*, 17, 177, 1979.
113. **Johns, D. G. and Valerino, D. M.,** Metabolism of folate antagonists, *Ann. N.Y. Acad. Sci.*, 186, 378, 1971.
114. **Zaharko, D. S., Dedrick, R. L., Young, D. M., and Peale, A. L.,** Tolerance of long-term methotrexate infusions by mice, *Biochem. Pharmacol.*, 25, 1317, 1976.
115. **Custer, R. P., Freeman-Narrod, M., and Narrod, S. A.,** Hepatotoxicity in Wistar rats following chronic methotrexate administration: a model of human reaction, *J. Natl. Cancer Inst.*, 58, 1011, 1977.
116. **Chabner, B. A., Myers, C. E., Coleman, C. N., and Johns, D. G.,** The clinical pharmacology of antineoplastic agents, *N. Engl. J. Med.*, 292, 1107, 1975.
117. **Gutin, P. H., Green, M. R., Bleyer, W. A., Bauer, V. L., Wiernik, P. H., and Walker, M. D.,** Methotrexate peneumonitis induced by intrathecal methotrexate therapy, *Cancer,* 38, 1529, 1976.
118. **Harvey, H. J., MacEwen, E. G., and Hayes, A. A.,** Neurotoxicosis associated with use of 5-fluorouracil in five dogs and one cat, *J. Am. Vet. Med. Assoc.*, 171, 277, 1977.

119. **Weiss, H. D., Walker, M. D., and Wiernik, P. H.,** Neurotoxicity of commonly used antineoplastic agents, *N. Eng. J. Med.*, 291, 75, 1974.
120. **Schaeppi, U., Fleischman, R. W., Phelan, R. S., Ethier, M. F., and Luthra, Y. K.,** CCNU (NSC-79037). Preclinical toxicologic evaluation of a single intravenous infusion in dogs and monkeys, *Cancer Chemother. Rep.*, 5(Part 3), 53, 1974.
121. **Oliverio, V. T.,** Toxicology and pharmacology of the nitrosoureas, *Cancer Chemother Rep.*, 4(part 3), 13, 1973.
122. **Silver, H. K. B. and Morton, D. L.,** CCNU nephrotoxicity following sustained remission in oat cell carcinoma, *Cancer Treat. Rep.*, 63, 226, 1979.
123. **Philips, F. S., Sternberg, S. S., Hamilton, L., and Clarke, D. A.,** The toxic effects of 6-mercaptopurine and related compounds, *Ann. N.Y. Acad. Sci.*, 60, 283, 1954.
124. International Agency for Research on Cancer, *IARC Monographs on the Evaluation of the Carcinogenic Risk of Chemicals to Humans. Some Antineoplastic and Immunosuppressive Agents,* Vol. 26, IARC, Lyon. 1981.
125. **Gilladoga, A. C., Manuel, C., Tan, C. T. C., Wollner, N., Sternberg, S. S., and Murphy, M. L.,** The cardiotoxicity of adriamyacin and daunomycin in children, *Cancer,* 37, 1070, 1976.
126. **Barnes, J. M. and Denz, F. A.,** Experimental methods used in determining chronic toxicity: a critical review, *Pharmacol. Rev.*, 6, 191, 1954.

Chapter 6

THE IMPORTANCE OF TISSUE-SPECIFIC BIOCHEMISTRY IN THE DETERMINATION OF TARGET ORGAN TOXICITY

Steven D. Aust

TABLE OF CONTENTS

I. INTRODUCTION

In this chapter an attempt will be made to discuss the importance of tissue-specific biochemistry in determining the target organ for a toxic chemical. This will not be a review of the biochemical basis for tissue-specific toxicities for a list of toxins or even one toxin, although examples will be given. Therefore, the list of cited literature is relatively short. Instead, this will be a discussion of how tissue-specific biochemistry can determine the target organ for a toxin. It is, of course, very important to determine the biochemical mechanism of action for all toxins in order to be able to make quick and accurate diagnoses, to be able to estimate risks, to develop therapeutics, etc. Frequently, however, toxins show tissue specificity and it is important to know why, generally for the same reasons given above.

The sensitivity of a particular tissue to the toxicity of a chemical can be influenced by the concentration of the toxin in the tissue, the activity of deactivating enzymes or activating enzymes and deactivating enzymes in the case of pro-toxins, the activity of enzymes involved in defense or repair mechanisms, the physical structure of the tissue relative to the mechanism of action of the toxin, or other factors which may be considered tissue-specific biochemistry. With a few exceptions, only the latter will be addressed here. The importance of tissue distribution, activation, deactivation, and other defense mechanisms, as well as excretion will be dealt with in other chapters. However, it might be well to note at this time each of these factors may have a tissue-specific biochemical basis. For example, the tissue distribution of paraquat is so dramatically influenced by the fact that it is actively transported into the lung such that it becomes a lung-specific toxin.[1-4] This is tissue-specific biochemistry, but it is relative only to tissue distribution of the toxin.

Although this subject is called tissue-specific it is important to realize that the specificity may not be absolute. The mechanism of toxicity may involve a biochemical process which is only found in a specific tissue, most likely related to some unique function of the tissue. However, in many cases the toxin may affect a biochemical process which is present in many tissues but in a certain tissue the process may be "rate-limiting" and therefore any interference with its capacity in that tissue may result in a tissue-specific toxicity. The former is a rather simple concept. A toxin that interacts with a specific receptor or enzyme which is uniquely located in a tissue (probably because of some unique function of the tissue), would most likely show toxicity in that tissue. Most biochemical processes are present in almost all tissues. How then can a toxin which affects a common biochemical process be a tissue-specific toxin? The question may be answered by asking how critical is the biochemical process which is affected by the toxin to the function of the tissue, or how much can a biochemical process in a tissue be negatively affected before the tissue shows signs of toxicity? This may be termed the rate-limiting concept, i.e., the activity of some enzyme or biochemical pathway determines the functional capacity of a tissue, or a biochemical pathway may just be sufficient to support the function of some tissue.

A biochemical process may be rate-limiting because it is involved in the primary function of the tissue, and any inhibition of this activity would result in the function not being adequate, or it may be only a minor biochemical pathway in the tissue but of such limited capacity that any perturbation of that capacity by a toxin would result in impaired function of the tissue. It is interesting to note that certain cellular processes which may be considered protective or "house-cleaning" may have considerable reserve, i.e., they are not rate-limiting until the activity or reserve is greatly depressed. This seems to be the case with glutathione and the glutathione reductase plus glutathione peroxidase system.[5] Reduced glutathione levels, or the activity of glutathione reductase, may be decreased considerably before the protection against oxidative stress is rate-limiting and the tissue suffers damage.

A biochemical pathway may also be rate-limiting because of some unusual metabolic demand of a tissue. The oxygen demand of neuronal tissue is so high, for example, that toxins which affect the delivery of oxygen or oxidative phosphorylation (i.e., carbon monoxide, cyanide, azide, etc.) are frequently neurotoxins. An analogy would be glucose metabolism by the brain. Substances which inhibit glucose metabolism (α-chlorhydrin, 1-amino-3-chloro-2-propanol, and 6-chloro-6-deoxyglucose) are also frequently neurotoxic.

II. TISSUE-SPECIFIC BIOCHEMISTRY

Tissue-specific biochemistry would definitely determine the target organ for any toxin affecting a biochemical process which only occurred in that tissue. There are many examples of biochemical processes which occur in only one tissue or one type of tissue. These are generally related to some specific function the tissue is designed to accomplish. Examples include the biochemistry related to neurotransmitters, neuromodulators, and ions in neuronal tissues, the biosynthesis of surfactant in the lung, the biochemistry involved in tissue-specific hormone synthesis (i.e., the adrenals, pituitary, pancreas, testes, ovaries, etc.), the biochemistry involved in hormone action in specific tissues (i.e., the testes, ovaries, mammary gland, etc.), active transport systems in the kidney, the biochemistry involved in bile synthesis in the liver, etc.

Toxicities resulting from tissue-specific biochemistry can be related to the two families of biochemicals involved in communication between cells: hormones and neurotransmitters. Hormones are a variety of chemicals which are carried in the circulation and tissues which are meant to respond to them have receptors for the hormones. Therefore toxins which somehow interfere with the function of the hormone receptors would most likely show toxicities only in those tissues containing the receptors. Neurotransmitters, on the other hand, are much less varied in chemical structure and only effective in cells which are anatomically close to the neuronal cells which release the transmitters. Thus, toxins affecting the biochemistry of neurotransmitter release, recognition, hydrolysis, etc., may affect any of the tissues in which the neurotransmitter is used. The tissue specificity of toxicity would therefore depend more on the tissue concentration of toxin and the relative activity or importance of the affected neuronal function in the tissue. Moreover, neurotransmitter levels can be easily affected by precursor availability. Therefore, interference with precursor availability can affect neurotransmitter availability; however, grossly unphysiological conditions are usually required to affect the synthesis of steroids, peptides, or other hormones.

Regions of the brain which show high rates of glucose utilization have been found to be very susceptible to chemicals which indirectly stimulate metabolism or which inhibit glucose utilization. Metabolism in these regions must be near rate-limiting so both further demands for energy and the inhibition of glucose metabolism produce toxicities. Kainic acid is an example of a toxin which is specific to neuronal tissue because it mimics, even more potently, the excitatory neurotransmitter glutamic acid.[6,7] Kainic acid is thought to be toxic by virtue of causing periods of sustained, repetitive discharge which can lead to neuronal cell death. Kainic acid must be an excellent ligand for the neuronal receptor for glutamic acid. When given systemically the toxic effects of kainic acid are seen primarily in the limbic system, notably the hippocampus and amygdala. Interestingly, Collins et al.[8] reported that glucose utilization in the areas of the brain affected by kainic acid increases about fourfold, followed by very low rates of glucose utilization in these areas when neuronal damage and cell loss is seen. It is interesting to speculate that the "excitotoxic" effects of kainic acid or other substances which show an excitotoxic effect (i.e., cell death due to sustained, repetitive discharge) may in fact do so by creating a situation whereby some biochemical process, in this case perhaps energy production, becomes rate-limiting, resulting in cell death. In this case the tissue-specific toxicity is initially caused by the occurrence of a specific receptor in the tissue. However, the mechanism of toxicity is perhaps related to increasing oxygen demand to a point that it cannot be maintained and cell death occurs.

III. TISSUE-SPECIFIC, RATE-LIMITING BIOCHEMISTRY

The second way for a biochemical pathway to determine the target organ for a toxin is for the affected pathway to be rate-limiting for the tissue. A biochemical pathway may be rate-limiting for a tissue because it is intimately involved in the major function of the tissue, or because it is so low in activity that the tissue has no reserve. In the first case any perturbation of the biochemical pathway will result in impaired primary function of the tissue. The question then can be asked as to the relative importance of the function of the tissue to the well-being of the animal. Whether a toxicity has occurred probably also depends somewhat upon one's definition of toxicity. In the second case, any perturbation of the biochemical pathway may dramatically affect the tissue because of the lack of reserve activity. The biochemical pathway does not have to be directly related to the major function of the tissue if it is only peripherally required for the major tissue function and there is little reserve capacity. A toxin affecting such activity would eventually cause an impairment of the major function of the tissue.

It should be noted that there may not in fact be any real differences between this concept and the former. The first is really only the extremes of the second; i.e., a biochemical process can certainly be made rate-limiting by a toxin and if the process only occurs in a single tissue that tissue will be the target tissue. Also, if the biochemical function was the primary function of the tissue, the tissue would most likely be the target organ for any toxin affecting that specific biochemical function. For example, toxins which affect the acetylcholine receptor would be specific for cholinergic neurons as the receptor is required for the function of these neurons. Rarely would one ask about their relative importance to the function of the tissue, although the question could be asked perhaps as to its relative importance in various cholinergic tissues.

The concept of rate-limiting biochemistry in determining target organ toxicities can easily be applied to toxins which affect DNA, RNA, or protein synthesis. The synthesis of DNA, RNA, and protein is generally rate-limiting in tissues with high proliferative or turnover rates. These would include blood forming tissues, skin, the mucous membranes (as in the GI tract), and the gonads. In most tissues of adults, cell division is the exception rather than the rule. Thus, toxins which affect processes of cell division or growth would be relatively ineffective in most tissues of adults. In contrast, the dividing cells of proliferating tissues or tissues with high cell turnover rates would in fact be susceptible to toxins which affect these processes. Likewise, these tissues would be sensitive to toxins which affect a biochemical process involved in providing large amounts of some constituent for the cells. Such is the case for heme synthesis for blood forming organs.

This concept of affecting biochemistry which is rate-limiting is often used in the field of cancer chemotherapy. In various tumor cells certain biochemical processes may be more rate-limiting than in normal tissue. Chemicals which interact with DNA, RNA, or protein synthesis are often effective as chemotherapeutic agents because these processes are more active in proliferating tumor cells and therefore these biochemistries are more likely to be rate-limiting, i.e., the tumor cells will be more sensitive to the chemical. Certain tumors are more dependent upon glycolytic enzymes than the corresponding nontumor tissue. Therefore, chemicals which interfere with glycolysis may be more toxic to tumor cells because this pathway is more likely to be rate-limiting in tumor cells.

Another example of how toxicities can be tissue-specific when a biochemical process or pathway is critical or rate-limiting is in certain teratogeneses. Embryogenesis is a very precisely programed sequence of events. Different biochemical processes involved in cell differentiation or proliferation for the organs (tissues) of the fetus occur at a precise time or stage of development; i.e., various biochemical processes or a biochemical reaction may become rate-limiting for organogenesis of different organs or tissues at various times during embryogenesis. Interference with any process at a time when that process is rate-limiting may result in teratogenesis in a specific tissue. An excellent demonstration of this phenomenon was provided by Harbison and Becker[9] using diphenylhydantoin. A variety of malformations in different tissues was caused by a single chemical simply by administering the chemical at various times of gestation. The time at which the chemical was administered was correlated with the known sequence of organogenesis; i.e., the biochemical function affected by the chemical was rate-limiting at a specific time during the development of the affected tissue. If the chemical was present in the fetus at the time when the biochemical process could be rate-limiting for the tissue the chemical was teratogenic for that tissue. At any other time the tissue would not be the target tissue because the biochemical process affected by the chemical was not rate-limiting.

Blood-forming tissues require a high rate of proliferation to produce adequate quantities of blood cells. This makes these tissues, and therefore blood cells, especially vulnerable to chemicals acting mainly on DNA and mitosis. Any effect of a chemical on this biochemistry will quickly be observed as a case of dyserythropoiesis. The hematopoietic organs will also require ample supplies of metabolites, vitamins, and energy to produce the blood cells. Substances which are needed in special abundance, such as heme, will also be required in relatively large quantities and any toxin which affects the biosynthetic pathways for this type of precursor could easily produce dyserythropoiesis. Erythrocytes are especially handicapped because they lose their nucleus and mitochondria; they must survive without the ability to resynthesize enzymes or membrane proteins and can only generate a small amount of ATP and NADPH. The ATP is primarily used for active transport and the NADPH for reduction of methemoglobin by glutathione reductase. These requirements can be major yet the biochemical pathways to produce these substances are minor. They are therefore quite susceptible to becoming rate-limiting. Any toxin which affects these processes would be apt to show toxicity to the tissue.

There are a number of chemicals which exhibit striking tissue-specific toxicity. Sometimes the chemical is of such high value as to render the tissue-specific toxicity of primary interest. Such is the case for the antineoplastic adriamycin which unfortunately produces debilitating side effects in the heart.[10,11] Consequently, considerable effort has been directed toward determining the biochemical basis for the cardiotoxicity of adriamycin.

It is generally thought that the mechanisms of adriamycin toxicity or even its antitumor effects are due to the oxidation-reduction activity of the adriamycin-iron complex.[12] The autoxidation of reduced adriamycin chelated iron results in the monovalent reduction of oxygen to superoxide. Superoxide is then thought to promote a number of radical-dependent, destructive processes such as lipid peroxidation or DNA strand scission. The mechanism by which superoxide may initiate these processes has been proposed to be via an iron-catalyzed generation of the very reactive hydroxyl radical or by the formation of some type of iron-oxygen complex capable of initiating the processes.

Liver is particularly rich in enzymes capable of reducing adriamycin chelated iron, but adriamycin is not toxic to the liver. Two hypotheses have been investigated as to the biochemical basis of the cardiotoxicity of adriamycin. First, adriamycin has fairly high affinity for membranes and binding is best to negatively charged complex lipids such as phosphatidyl serine and cardiolipin. Goormaghtigh et al.[13] have argued that the specificity for binding of adriamycin to these lipids and their presence in cardiac tissue may be the biochemical basis for the cardiotoxicity of adriamycin. Again, as was the case for paraquat, this is really the biochemical basis for a unique tissue distribution rather than an effect on a rate-limiting biochemical process. Others have proposed that the cardiotoxicity of adriamycin is related to a deficiency in a biochemical defense mechanism. Doroshow et al.[14] have pointed out that cardiac tissue contained considerably less superoxide dismutase and catalase activity than does liver. Doroshow et al.[14] also found that cardiac glutathione peroxidase activity was actually depressed by adriamycin treatment. This would be an example of tissue-specific toxicity resulting from the lack of good reserve capacity in a protection system; in other words, the protection mechanism becomes rate-limiting because of a lack of reserve activity or supply of protective biochemical. Babson et al.[15] have clearly shown that adriamycin can have little effect on lipid peroxidation in hepatocytes until the "reserve" of cellular reduced glutathione is depressed.

VanHelden and Wiil[16] compared the binding of adriamycin to chromatin from heart and skeletal muscle of chick embryo cells and found that in both instances adriamycin interacted with the DNA. Inhibition of DNA and RNA synthesis was much more pronounced in the heart muscle. This may be interpreted as a biochemical difference affecting the tissue specificity of this toxin, but it must be remembered that inhibition of DNA synthesis may not be a major factor in the cardiac toxicity of adriamycin due to the general lack of mitosis in adult myocytes. Since the replication of heart muscle cells is extremely limited, they may be very sensitive to the inhibition of protein synthesis by adriamycin; protein synthesis may be rate-limiting and any effect on protein synthesis in this tissue may be expressed as exquisite sensitivity or tissue-specific toxicity.

There is considerable interest in the different tissues affected by TCDD in different species. TCDD is 2,3,7,8-tetrachlorodibenzo-*p*-dioxin, perhaps the most toxic of a number of congeners of polyhalogenated aromatic hydrocarbons.[17] The mechanism of toxicity for TCDD and similar compounds is unknown but there is a striking relationship between the toxicity of a compound of this nature and its ability to serve as a ligand for a cellular receptor involved in the induction of a number of enzymes including microsomal aryl hydrocarbon hydroxylase activity. These chemicals all cause anorexia and thymic atrophy in all species, but effects in other tissues are highly variable. In addition, there is an extremely wide difference in toxicity in various species. Even more interesting, species which show widely different susceptibility to toxicity can also show toxicity in different tissues. For example, the guinea pig is very susceptible to the toxicity of TCDD whereas the hamster is very resistant and the rat is somewhere in between. The rat shows dramatic hepatotoxicity, the hamster very little, and the guinea pig none.[17] Since the toxicity of TCDD has been proposed to be mediated by, or at least well correlated with what is actually called the TCDD receptor because of its high affinity for TCDD, there have been attempts to compare the amounts of receptor, and its relative affinity for TCDD, in various tissues[18] and various species.[19] This research was prompted mainly by the discovery of strains of mice which are responsive and nonresponsive to the toxicity of TCDD. The nonresponsive strains appear to have a receptor which has lowered affinity for TCDD, pehaps via mutation, and decreased sensitivity to the enzyme inductive and toxic effects of TCDD as compared to the responsive strains.[20] However, no correlations could be found as to the amount of receptor or its affinity for TCDD and either tissue- or species-specific toxicities. There is little but correlative data to suggest that TCDD toxicity is in fact mediated by the receptor. It would appear that a more fruitful approach would be to determine which biochemical pathway(s) are affected by TCDD and whether these pathways are tissue-specific or rate-limiting in affected tissues or species, or how the effect on a pathway may universally affect one function (i.e., anorexia) in all species but at different doses. Obviously, considerable research will be conducted on TCDD for it is extremely toxic and shows extreme differences in species and tissue toxicity. The ultimate questions are what are the effects of TCDD in humans, and how sensitive are humans to the various toxic effects of TCDD.

In summary, tissue-specific biochemistry can certainly determine the target organ for a number of toxins. Naturally toxin, especially active toxin, concentration must be carefully considered. This makes the subject very complex, especially for those toxins which are subject to both activation and inactivation reactions. Even in these cases, tissue-specific biochemistry is important. There are many examples of tissue-specific biochemistry which are exclusive, i.e., they only occur in certain tissues. It is not overly difficult to determine how this may result in determining the target organ for a toxin affecting such processes. The picture is complicated considerably for the consideration of toxins that affect a biochemical pathway which may become rate-limiting. It is somewhat simpler in those tissues in which the biochemical pathway is involved in the major function of the tissue. It is more difficult when the biochemical pathway is not involved in the major function of the tissue, but is quite susceptible to becoming rate-limiting in the presence of a toxin.

REFERENCES

1. **Haley, T. J.,** Review of the toxicology of paraquat (1,1′-dimethyl-4,4′-bipyridinium chloride), *Clin. Toxicol.*, 14, 1, 1979.
2. **Rose, M. S., Smith, L. L., and Wyatt, I.,** Evidence for energy-dependent accumulation of paraquat into rat lung, *Nature (London)*, 252, 314, 1974.
3. **Waddell, W. J. and Marlowe, C.,** Tissue and cellular disposition of paraquat in mice, *Toxicol. Appl. Pharmacol.*, 56, 127, 1980.
4. **Lock, E. A., Smith, L. L., and Rose, M. S.,** Inhibition of paraquat accumulation in rat lung slices by a component of rat plasma and a variety of drugs and endogenous amines, *Biochem. Pharmacol.*, 25, 1769, 1976.
5. **Sies, H., Brigelius, R., Wefers, H., Miller, A., and Cadenas, E.,** Cellular redox changes and response to drugs and toxic agents, *Fundamental Appl. Toxicol.*, 3, 200, 1983.
6. **Olney, J. W., Rhee, V., and Ho, O. L.,** Kainic acid: a powerful neurotoxin analogue of glutamate, *Brain Res.* 77, 507, 1974.
7. **McGeer, E. G., Innanen, V. T., and McGeer, P. L.,** Evidence on the cellular of adenylcyclase in the neostriatum, *Brain Res.*, 118, 356, 1976.
8. **Collins, R. C., McLean, M., and Olney, J.,** Cerebral metabolic response to systemic kainic acid: 14-C-deoxyglucose studies, *Life Sci.*, 27, 855, 1980.
9. **Harbison, R. D. and Becker, B. A.,** Relation of dosage and time of administration of diphenylhydantoin to its teratogenic effect in mice, *Teratology*, 2, 305, 1969.
10. **Myers, C. E.,** Anthracyclines, in *Pharmacologic Principles of Cancer Treatment,* Chabner, B. A., Ed., W. B. Saunders, Philadelphia, 1982, 416.
11. **Myers, C. E.,** Cardiac toxicity, in *Cancer Principles and Practice of Oncology,* DeVita, V. T., Hellman, S., Rosenberg, S. A., Eds., J. B. Lippincott, Philadelphia, 1982, 1701.
12. **Gianni, L., Corden, B. J., and Myers, C. E.,** The biochemical basis of anthracycline toxicity and antitumor activity, *Rev. Biochem. Toxicol.*, 5, 1, 1983.
13. **Goormaghtigh, E., Chatelain, P., Caspers, J., and Ruysshaert, J. M.,** Evidence of a complex between adriamycin derivatives and cardiolipin: possible role in cardiotoxicity, *Biochem. Pharmacol.*, 29, 3003, 1980.
14. **Doroshow, J. H., Locker, G. Y., and Myers, C. E.,** Enzymatic defenses of the mouse heart against reactive oxygen metabolites, *J. Clin. Invest.*, 65, 128, 1980.
15. **Babson, J. R., Abell, N. S., and Reed, D. J.,** Protective role of the glutathione redox cycle against adriamcyin-mediated toxicity in isolated hepatocytes, *Biochem. Pharmacol.*, 30, 2299, 1981.
16. **VanHelden, P. D. and Wiil, I. J. F.,** Effects of adriamycin on heart and skeletal muscle chromatin, *Biochem. Pharmacol.*, 31, 973, 1982.
17. **Poland, A. and Knutson, J. C.,** 2,3,7,8-Tetrachlorodibenzo-*p*-dioxin and related halogenated aromatic hydrocarbons: examination of the mechanism of toxicity, *Ann. Rev. Pharmacol. Toxicol.*, 22, 517, 1982.
18. **Carlstedt-Duke, J. M. B.,** Tissue distribution of the receptor for 2,3,7,8-tetrachlorodibenzo-*p*-dioxin in the rat, *Cancer Res.*, 39, 3172, 1979.
19. **Gasiewicz, T. A. and Rucci, G.,** Cytosolic receptor for 2,3,7,8-tetrachlorodibenzo-*p*-dioxin: evidence for a homologous nature among various mammalian species, *Mol. Pharmacol.*, in press.
20. **Poland, A., Glover, E., and Kende, A. S.,** Stereospecific, high affinity binding of 2,3,7,8-tetrachlorodibenzo-*p*-dioxin by hepatic cytosol, *J. Biol. Chem.*, 251, 436, 1976.

Chapter 7

ROLE OF KEY DEFENSE SYSTEMS IN TARGET ORGAN TOXICITY

Toshihisa Ishikawa, Theodorus P. M. Akerboom, and Helmut Sies

TABLE OF CONTENTS

I. INTRODUCTION

The formulation of an integrated view on the role of key defense systems in determining target organ toxicity must deal with some crucial problems. Enumeration of the activity contents of enzymes in organs or cell populations in organs, and of the pool size of low molecular weight compounds with antioxidant capacity is now possible in sufficient detail; this forms part of the present chapter. However, there is little doubt that the problem is much more complex, because it is probable that toxicity may also be expressed when the antioxidant systems are not used to full capacity, but when there are topological or logistical problems in addition, in spite of apparently well-dimensioned protection systems.

The present chapter will serve a quite basic purpose, namely to give in a condensed form the framework of the defense systems available against so-called oxidative or reductive stress in major organs. This field is in rapid development, and three very recent survey publications may suffice for allowing cross-referencing.[1-3]

More refined treatments of the topic must take into account the intra- and intercellular gradients of potentially dangerous compounds and metabolites thereof, considering complicated interorgan metabolite activation/deactivation and transport/elimination networks.[4] Further, the secondary responses of the organism, e.g., the cellular responses of chemokinesis and chemotaxis, and even hormonal and other responses (e.g., arachidonate cascade) to toxicants, deserve consideration at a more elaborate level of understanding. Nevertheless, since the antioxidant capacity of the enzymatic and nonenzymatic systems discussed in this chapter has been well established in experimental studies, the basic role of these defense systems in dealing with toxic challenge is of interest for evaluating target organ toxicity.

II. KEY DEFENSE MECHANISMS — OCCURRENCE AND TOPOLOGY

A. Cytochrome P-450 and Epoxide Hydrolase

Cytochrome P-450, the well-characterized terminal oxidase of the microsomal electron transport system, metabolizes not only endogenous substrates but also various exogenous organic compounds. Lipophilic compounds are converted to more polar products, and therefore their subsequent excretion is facilitated. In this respect, cytochrome P-450 serves as a defense against toxic foreign compounds. Cytochrome P-450 is abundant in the liver and at a relatively lower content in the kidney. Other organs, such as lung, heart, and brain contain low levels of cytochrome P-450.[5] In general, the microsomal P-450 of extrahepatic tissue is immunologically similar to the major hepatic microsomal P-450 in the rat. 7-Ethoxycoumarin *O*-deethylase activity is induced by 3-methylcholanthrene (3-MCA) in liver, lung, kidney, and brain. Benzo[*a*]pyrene hydroxylase activity is induced by phenobarbital in liver and testis, and by 3-MCA in liver, lung, kidney, and heart (Table 1). The intraorgan distribution of cytochrome P-450 and of its related enzymes within various organs has been analyzed in detail by immunochemical methods,[6,7] showing that the cytochrome P-450 system is present at a higher content in the centrilobular area of the liver, and is localized in the Clara cells of the lung. Distal tubules and collecting ducts of the kidney, and absorptive cells of the villi of the small intestine exhibit high staining for cytochrome b_5 which researchers believe transfers the second electron to cytochrome P-450.[8] An autoradiographic study with ipomeanol fixation has suggested that sebaceous cells of the skin have cytochrome P-450-dependent xenobiotic metabolism.[9]

Apparently paradoxically, cytochrome P-450 converts some foreign compounds into carcinogenic, mutagenic, or otherwise chemically reactive metabolites, including free radicals and epoxides, which may give rise to deleterious effects on the tissue. This metabolic activation by cytochrome P-450 is exemplified by hepatic and pulmonary necrosis caused by acetaminophen, bromobenzene, 4-ipomeanol, and benzo[*a*]pyrene;[10] reactive metabolites formed from oxidative metabolism are bound to cellular components irreversibly.[11]

Table 1
ORGAN DISTRIBUTION OF CYTOCHROME P-450 DEPENDENT MONOOXYGENASE ACTIVITY AND MICROSOMAL EPOXIDE HYDROLASE IN RAT TISSUES

Tissue	Induction	Cytochrome P-450[5] (mU/g w/w) 7-Hydroxycoumarin	Cytochrome P-450[5] (mU/g w/w) 3-Hydroxybenzo[*a*]pyrene	Microsomal epoxide hydrolase[17] (mU/g w/w)
Liver	None	19.0	3.7	113.0
	PB	27.2	18.6	
	MC	251.0	47.6	
Kidney	None	0.62	0.05	6.9
	PB	0.26	0.05	
	MC	9.68	7.26	
Heart	None	0.023	0.004	0.03
	PB	0.036	0.007	
	MC	0.023	0.011	
Lung	None	0.012	0.007	0.76
	PB	0.064	0.003	
	MC	0.600	0.139	
Brain	None	0.004	0.001	0.23
	PB	0.006	0.004	
	MC	0.107	0.002	
Testis	None	0.072	0.02	8.7
	PB	0.254	0.17	
	MC	0.241	0.05	
Spleen	None			1.9
Adrenal	None			7.9

Note: Data are recalculated based on the amounts of microsomal protein in 1 g tissue presented in Reference 21. 1 U corresponds to 1 μmol product/min. PB: phenobarbital; MC: 3-methylcholanthrene.

Epoxide hydrolase may support the detoxication function of cytochrome P-450, catalyzing reactive epoxides formed via the monooxygenase reaction to trans-1,2-dihydrodiols, which are in general chemically more stable. The function and properties of this enzyme have been described in recent reviews.[12,13] While epoxide hydrolase activity is found in both cytosolic and microsomal fractions of various mammalian organs,[14] hydrolase activity of the human liver is predominant in the microsomal fraction.[12] The existence of epoxide hydrolase on the surface of the smooth and rough endoplasmic reticulum as well as the nuclear envelope has been recognized by immuno-electron-microscopy.[15] In addition, semiquantitative immunohistochemical analysis has revealed that the intralobular distribution of epoxide hydrolase within livers of untreated rats is essentially identical to that determined for cytochrome P-450 MC-B.[16] These observations suggest that microsomal epoxide hydrolase is located primarily in the endoplasmic reticulum in close association with the cytochrome P-450 system.

Table 1 represents microsomal epoxide hydrolase activity toward benzo[*a*]pyrene-4,5-oxide in rat tissues. The liver has the highest activity of the tissues, and testis and kidney are the organs with the next highest activities. Although continuously exposed to environmental foreign compounds, the lung exhibits a fairly low activity of the enzyme as compared with the activity for liver, which may be a reason for the susceptibility of the lung to some xenobiotics such as benzo[*a*]pyrene. Epoxide hydrolase was also found in the skin.[18]

B. Conjugation Reactions

Conjugation reactions present important defense mechanisms against endogenously gen-

Table 2
ORGAN DISTRIBUTION OF GSH, GSH PEROXIDASES, GSSG REDUCTASE, AND GSH S-TRANSFERASES IN RAT AND HUMAN TISSUES

Tissue	GSH[22-30b] (μmol/g w/w)	GSH peroxidases[22-30b] Se	non-Se	GSSG reductase[22-30b]	GSH S-transferases[31] Rat	Human
Liver	5.6	269	143	67	710	120
Kidney	2.8	146	66	146	86	130
Heart	1.1	225	0	9	26	46
Lung	1.5	105	27	41	29	34
Brain	1.8	33	12		66	91
Eye lens	5.1	70		10	12	
Adipose tissue	3.0	53	15	20		
Stomach	7—8	88	13	17	40	50
Intestine	3—4	8	8	35	29	29
Erythrocytes	1.9	26	0	7	3	1
Skin	4.5[a]				13	2

Note: The final 5 columns are measured in mU/mg protein, protein in 105,000 × *g* supernatant. 1 U corresponds to 1 μmol product/min.

[a] μmol/g protein.[30b]

erated reactive intermediates and numerous xenobiotics by preventing these compounds from possible interaction with components in living cells.

1. Glutathione S-Transferases

This is a family of multifunctional proteins catalyzing the reaction of glutathione (GSH) with various electrophilic compounds such as benzo[*a*]pyrene-4,5-oxide, bromosulfophthalein, ethacrynic acid, and numerous halo- and nitrobenzene derivatives, to form the corresponding GSH thioethers.[19,20] The GSH conjugates thus formed are subsequently eliminated from the cell into bile or are further degraded enzymatically to mercapturic acids to be excreted in the urine.

GSH S-transferase activity is distributed in almost all organs and tissues (Table 2). Liver, the major xenobiotic-metabolizing organ, bears the largest capacity for the GSH conjugation reaction. GSH S-transferases are localized predominantly in the cytosol, amounting to about 10% of total soluble proteins in rat liver.[19] About 7% of total GSH S-transferase activity is present in the mitochondrial matrix space.[32] Microsomal GSH S-transferases, which are not identical to the cytosolic, have also been reported.[33,34] Kidney has a specific activity similar to that of human liver, whereas the testis is the organ with the next highest activity in rat. It is well established that GSH S-transferase activity is carried by several isozymes with overlapping substrate specificity and that there is a variation in the occurrence of the isozymes among organs.[35]

GSH S-transferases are distributed nonuniformly throughout the organ. Intestinal mucosa, which is the first tissue exposed to many foreign compounds, exhibits a gradient of decreasing GSH S-transferase activity from villi to crypt cells and along the intestinal tract.[36] In rat kidney, ligandin (transferases 1-1 and/or 1-2) is localized in the proximal renal tubular epithelium.[37] For liver GSH S-transferases, it has been demonstrated by immunochemical staining that they are present in greater amounts within the centrilobular region than within midzonal and periportal regions.[38] Similar distributions have been observed for cytochrome P-450 isozymes, cytochrome P-450 reductase, cytochrome b_5, and epoxide hydrolase,[39] suggesting that GSH S-transferase may operate efficiently on highly reactive metabolites generated from monooxygenase reactions in the liver. However, this high efficiency would

Table 3
ORGAN DISTRIBUTION OF UDP-GLUCURONYLTRANSFERASES AND ARYL SULFOTRANSFERASE IN RAT AND HUMAN TISSUES

Tissue	UDP-glucuronyltransferase (rat)[43] Group 1	UDP-glucuronyltransferase (rat)[43] Group 2	Aryl sulfotransferase (human)[44]
Liver	2,400	270	41.9
Kidney	810	<3	3.7
Lung	100	<1	
Testis	410	<2	
Spleen	140	<4	2.8
Intestine	340	39	8.4
Adrenal			8.9
Thyroid			1.0

Note: Substrates for UDP-glucuronyltransferases groups 1 and 2, and aryl sulfotransferase are 1-naphthol, morphine, and phenol, respectively. 1 U corresponds to 1 μmol product/min. All columns are measured in mU/g w/w.

become a potential trigger for a rapid depletion of GSH in the centrilobular region upon an excess dose of substrates for cytochrome P-450 or GSH S-transferases, since this region contains less GSH.[40] This may explain the cause of centrilobular necrosis.

2. *UDP-Glucuronyltransferases*

These catalyze the transfer of the glucuronyl moiety of UDP-glucuronic acid to a variety of endogenous and exogenous compounds bearing hydroxyl, carboxyl, amino, imino, or sulfhydryl groups.[41] The transferase activity is tightly associated with the endoplasmic reticulum in the cell, and multiple forms of the enzyme have been found.[42] Rat liver glucuronyltransferase activities have been classified on the basis of their inducibility by two prototype xenobiotic inducers, 3-MCA and phenobarbital.[42] Table 3 presents activities of groups 1 and 2 in various rat organs. Group 1 activity, which is chiefly inducible by 3-MCA, is found in various tissues including liver, kidney, small intestine, lung, skin, testis, and spleen. In contrast to this, group 2 activity, which is inducible by phenobarbital, is detectable only in liver and intestine.

Regarding intraorgan distribution of glucuronyltransferases, it has been suggested from pharmacokinetics that the enzyme active with harmol (7-hydroxy-1-methyl-9H-pyrido-(3,4b)-indole) is localized within hepatocytes in the midzonal or centrilobular regions.[45] High activity of glucuronyltransferase was found in fractions enriched with Clara cells of rat lung.[46]

3. *Sulfotransferases*

Compounds which contain one or more hydroxyl groups are particularly good substrates for sulfate conjugation. Sulfotransferases catalyzing the conjugation of these compounds, are a group of proteins classified into four enzymes, i.e., aryl, hydroxysteroid, estrone, and bile-salt sulfotransferases.[47] 3-Phosphoadenosine 5-phosphosulfate (PAPS) is required as an activated sulfate donor, which is synthesized from ATP and inorganic sulfate by the sequential action of ATP sulfurylase and adenosine 5-phosphosulfate kinase. In vivo, sulfate conjugation reactions are regulated by several factors such as the supply of inorganic sulfate and the capacity for synthesis of PAPS residues. The capacity of sulfate conjugation is in general low; however, the physiological significance of sulfotransferases as a defense system results from their fairly high affinity for various phenolic compounds as compared with that for glucuronyltransferases.

Table 4
ORGAN DISTRIBUTION OF SOD AND CATALASE IN HUMAN TISSUES[52-54]

	CuZn SOD[a]		Mn SOD[a]		Catalase[b]	
Tissue	(U/g w/w)	(U/mg protein[c])	(U/g w/w)	(U/mg protein[c])	(kU/g w/w)	(kU/mg protein[c])
Liver	2,400	32.0	600	7.9	98.0	1.4
Kidney	530	9.3	390	7.4	13.0	0.37
Heart	310	8.0	440	11.0	0.9	0.03
Lung	210	4.0	100	1.9	7.7	0.20
Brain	270	14.0	82	3.7	0.2	0.01
Pancreas	230	5.0	200	4.9	4.9	0.11
Muscle	260	5.8	62	1.3	1.3	0.03
Spleen	300	3.5	92	1.2	3.9	0.03
Uterus	430		96			
Testis	220		72			
Adipose tissue	24	5.0	11	2.4	1.5	0.42
Erythrocytes	530	1.5	0	0.0	384	1.15

[a] Data are represented in the unit defined in Reference 54a.
[b] 1 U corresponds to the rate catalyzing 1% of H_2O_2 present in the assay system (15 m*M*) per minute.
[c] Protein in 20,000 × *g* supernatant.

The liver has the highest activity of sulfate conjugation. Indirect evidence indicates that aryl sulfotransferase is predominantly localized within the periportal regions of the liver lobule[45] being different from glucuronyltransferases. At high substrate concentrations, however, the hepatic sulfate conjugation is readily saturated and glucuronyltransferases take over the conjugation of phenolic compounds.[48] Intestine has a relatively high ability for sulfate conjugation among the extrahepatic organs (Table 3). For human lung, sulfate conjugation is considered to be the major conjugation mechanism for phenolic substances. Monohydroxybenzo[*a*]pyrenes or 1-naphthol are metabolized by the short-term organ culture of human lung predominatly to their respective sulfate conjugates with little or no glucuronide conjugate being formed, in contrast to rat lung which primarily forms glucuronide conjugates.[49]

C. Superoxide Dismutases (SOD)

The superoxide anion radical is generated in the living cell in enzymatic and nonenzymatic modes under physiological conditions. Whereas the dismutation reaction can proceed non-enzymatically, in the cell enzymatic catalysis by superoxide dismutases plays a predominant role.[50] Superoxide dismutases are present in all cell types of mammalian organs, and the existence of two different types has been well established. In rat liver CuZn-SOD is located in the cytosolic compartment, and Mn-SOD in the mitochondrial matrix. In human liver the Mn-SOD is also present in the cytosol. The mitochondrial intermembrane space is equipped with the CuZn-enzyme.[51]

The tissue distribution of both the CuZn- and the Mn-SOD activities are presented in Table 4. The different tissues have a similar SOD activity, except for liver having a much higher activity and adipose tissue a lower activity. However, if the activity is related to the amount of protein, the erythrocytes and lung show the lowest specific activities. In many tissues the Mn-SOD activity is almost as high as the CuZn-SOD activity, and it correlates with the amount of mitochondria, which are absent in erythrocytes. Recently, the occurrence of a third type of dismutase, EC-SOD, also containing Cu and Zn, has been reported.[55] In plasma and lymph EC-SOD dominates, and in cerebrospinal fluid CuZn-SOD is the major enzyme. The contribution of other serum metalloproteins showing dismutase activity such as ceruloplasmin, transferrin, or ferritin is low.

The intraorgan distribution of SOD has been studied using immunohistochemical techniques and has revealed large differences in activities within the various tissues.[56] In liver, SOD is more prominent in the centrilobular hepatocytes than in the periportal. In kidney, SOD activity appears to be confined to the thick ascending segment of Henle's loop, whereas other portions of the nephron, both cortical and medullary, including distal, proximal convoluted, and collecting tubules lacked immunoreactivity. In lung, the enzyme is located mainly in the bronchiolar ciliated epithelial cells, whereas in the alveolar type I and II cells a lower activity was observed.

D. Catalase

Catalase and GSH peroxidase are the major enzymes involved in the reduction of hydrogen peroxide. The relative contribution of both enzymes to the elimination of intracellularly generated hydrogen peroxide has been subject of many studies[57] and appears to be related to their kinetic properties and the spatial compartmentation.[57,58] Catalase is present in practically all mammalian cells and was shown to be functionally associated with the peroxisomes.[59] Catalase activities are highest in liver and kidney (Table 4), correlating with the cellular content of peroxisomes as identified by histochemical techniques. However, in species other than rat, e.g., guinea pig or Rhesus monkey, considerable catalase activities have also been found in the soluble cytosol of liver.[60] A high activity of the enzyme is also found in erythrocytes, while lung and pancreas show intermediate values. Very low activities are found in heart and brain. In human liver, no striking differences between centrilobular and periportal zones were observed.[61] Interestingly, lipofuscin granules, visualized with the same technique, appear to be more abundant in the centrilobular region. In lung tissue highest activities are observed in the Clara cells and the type II pneumocytes.[62] In heart and possibly skeletal muscle,[63] the small amount of catalase is confined to microperoxisomes, situated between adjacent myofibrils, and often closely related to mitochondria.[64,65] Intramitochondrial localization of catalase has been reported,[66] but it still is a matter of controversy.

E. GSH and GSH Peroxidases

GSH is linked to the cellular defense mechanisms in multiple ways.[67,68] It could offer protection by direct nonenzymatic reaction with radicals, for example, during radiation under hypoxic conditions,[69] but more knowledge is required in this area. Enzymatic reduction of hydroperoxides by GSH involves GSH peroxidases in the intact cell.[70] GSH transferases catalzye the conjugation of GSH with a variety of xenobiotics (see Section II.B) possessing toxicity directly or after microsomal metabolic activation (see Section II.A).

Mammalian cells are equipped with high concentrations of glutathione (Table 2). Highest concentrations are observed in liver, a central organ in detoxication, and in tissues which are primarily exposed to environmental stress, such as the skin, eyes, and the GI tract. Intracellularly, GSH is found both in the cytosolic compartment and in the mitochondrial matrix, and is effectively kept in the reduced state by glutathione reductase. This flavoprotein uses NADPH as the reductant, and is also present in both compartments (see Section III.B).

The "glutathione status"[71] describes the relative proportions of the different GSH species including GSH-mixed disulfides. There exists a relationship between the amount of glutathione disulfide (GSSG) and GS-S protein,[72] a reaction which is probably catalyzed by thioltransferase.[73] The formation of GS-S protein may play a role in metabolic changes caused by GSH disulfide, stimulation of flux through the pentose phosphate pathway,[74] inhibition of protein synthesis, and influence on tubulin polymerization.[71]

Uneven distribution of GSH within tissues has been shown with histochemical techniques. GSH is about twice as high in the pericentral zone as in the periportal.[40] In lung, GSH was seen especially in the interalveolar epithelial cells.[74a] However, after isolation of different lung cell types, Clara and type II alveolar cells were found to contain 2.6 and 5.5 nmol

GSH per milligram of protein, respectively.[74b] In the kidney, GSH staining was localized in the proximal convoluted tubules in the cortex, and was also prominent in the periphery of the eye lens.[74a]

Se-dependent GSH peroxidase is capable of reducing both H_2O_2 and organic hydroperoxides.[70] Non-Se-dependent activities, identified as the GSH S-transferases 1-2 and 2-2 (transferases B and AA),[75,76] metabolize organic hydroperoxides but not H_2O_2. These enzymes are not able to reduce membrane lipid hydroperoxides to lipid alcohols,[77] unless phospholipase A is present.[78,79] Thus, the physiological significance of these peroxidases to control lipid peroxidation in vivo is still uncertain. Of interest in this respect is the recent report of a microsomal GSH peroxidase activity in heart and liver, which appears to be capable of reducing hydroperoxide derivatives of phospholipids.[80]

GSH-dependent protection against lipid peroxidation is exerted by factors present in the cytosol.[81,82] Also in the microsomes such a factor, different from the cytosolic, has been described.[83-86] The microsomal factor is probably of protein nature and seems to act by scavenging free radicals. Its physiological importance as a defense against free radical injury remains to be elucidated.

Detoxication of intracellularly formed H_2O_2, substantially increased by certain drugs causing redox cycling, involves the Se-dependent GSH peroxidase, exhibiting high activities in the liver, kidney, and heart of the rat (Table 2). In the human, low activities have been reported in heart and lung.[87] Hepatic Se-dependent GSH peroxidase has a greater preponderance in the periportal zone compared to the centrilobular zone.[88] In lung, high activities are found in the alveolar macrophages and in type II pneumocytes.[89]

D. Vitamins E, C, and A

SOD, catalase, and GSH peroxidase are located in the hydrophilic milieu in the cell, to serve as the first defense prior to production of more reactive oxygen species such as the hydroxyl radical ($OH^{\cdot}$). However, these enzymes are inefficient in breaking chain reactions in the lipid peroxidation process which has been initiated in the biomembrane space. The physiological role of vitamin E as a lipid-soluble antioxidant is chain-breaking by removal of chain propagators such as lipid peroxy radical ($ROO^{\cdot}$).[90] Vitamin E reacts with organic peroxy radicals with a rate constant of 10^5 to 10^6 *M*/sec, several orders of magnitude faster than polyunsaturated fatty acids.[90] The lung contains a relatively high level of vitamin E in the heavy mitochondria and endoplasmic reticulum, which may result from the need for antioxidant protection because of the high oxygen tension in the tissue (Table 5). The heart, which is continuously exposed to freshly oxygenated blood from the lung, exhibits the next highest values for mitochondrial and microsomal vitamin E levels. In vitro, microsomes from these organs show a high resistance to NADPH or ascorbate/Fe^{2+}-induced lipid peroxidation, being due to the higher ratio of vitamin E to polyunsaturated fatty acids.[93] On the other hand, it appears noteworthy that membranes of rat adrenocortical cells and human erythrocytes carry specific binding sites for vitamin E.[94,95] The physiological role of these binding sites remains to be solved.

While the vitamin E radical is formed via the reaction with the lipid peroxy radicals, it is continuously reduced to vitamin E by electron donors in vivo. Vitamin C is known to form a powerful synergistic antioxidant combination with vitamin E by reducing the vitamin E radical.[91,96-98] Furthermore, vitamin C reacts with the superoxide radical,[99] activated electrophiles, and various alkylating agents.[100] The vitamin C content in human organs is about ten times higher than vitamin E content (Table 5), corresponding to a concentration of more than 1 m*M* in the cell. Eye lens shows a relatively high level of vitamin C. For the lung about 50% of the pulmonary vitamin C was found in the fluid lining of the air spaces,[101] which may serve an important protective role in the respiratory tract.[102]

In addition to vitamins E and C, the antioxidant activity of carotenoids including vitamin

Table 5
ORGAN DISTRIBUTION OF VITAMINS E AND C IN RAT AND HUMAN TISSUES

Tissue	Vitamin E (rat) Total[92] (nmol/g w/w)	Nuclear[a]	Mitochondria[a] Heavy	Light	Microsome[a]	Microsome[93] (nmol/mg PUFA[b])	Vitamin C (human)[94] (nmol/g w/w)
Liver	18	0.07	0.63	0.28	0.19	1.6	570—910
Kidney	20	0.19	0.16	0.26	0.26	1.6	280—850
Heart	20	0.58	0.60	0.67	0.86	4.4	280—850
Lung	23	0.42	1.40	0.60	1.30	10.0	400
Brain						1.2	740—850
Muscle							170—280
Spleen							570—850
Eye lens							1,400—1,800

[a] Values are nmol/mg protein. From Reference 92.
[b] Peroxidizable polyunsaturated fatty acids.

Table 6
ANTIOXIDANTS IN HUMAN PLASMA[107a]

Antioxidant	Concentration (μ*M*)
Vitamin E	10—40
Vitamin C	40—140
Normal adult	40—140
+ 2 g supplement	100—160
Carotenoids	2
Urate	160—450

A has been suggested because of its quenching capacity to singlet molecular oxygen[103,104] and of effective inhibition of superoxide radical production in polymorphonuclear leukocytes stimulated with tumor promoters.[105] Liver is the major storage organ for vitamin A of mammals. Vitamin A is present mainly in the Kupffer cells as retinyl ester which is found in the soluble fraction apparently in association with a very low-density lipoprotein.[106]

Furthermore, such nonenzymatic antioxidants are also found in plasma (Table 6). It would be of interest that urate prevents lipid peroxidation of the erythrocyte membrane as effectively as vitamin C at concentrations below those normally found in plasma. Evidence has suggested that urate may act as an antioxidant to scavenge reactive oxygen species such as hydroxyl radical and singlet oxygen.[107,107a,108]

III. ANCILLARY REACTIONS

In order for some of the defense systems to operate, ancillary reactions are obligatory. Those are of different types, including (1) transformation reactions required to prepare substrates for detoxication reactions, (2) regenerative reactions such as GSH regeneration from GSSG, involving both GSSG reductase and continuous NADPH supply, and (3) transport systems, including the intracellular transfer by ligandin and the export from cells via excretion mechanisms, e.g., into bile. In the following, an example of both (1) and (2) will be given, and excretion mechanisms will be treated elsewhere in this volume.[109]

A. NADPH:Quinone Oxidoreductase

This enzyme, also known as DT-diaphorase, catalyzes an essential ancillary reaction if viewed in the context of quinone detoxication. As stated in Section II.B, the conjugation reactions catalyzed by UDP-glucuronyltransferases and sulfotransferases require hydroxyl groups, so that the quinone must be reduced prior to becoming a substrate for glucuronidation or sulfation. The importance of the enzyme catalyzing this two-electron reduction[110] under conditions of oxidative stress lies in the circumvention of two successive one-electron steps of reduction of quinones. Such a pathway would proceed via the semiquinone, and autoxidation may well give rise to reactive oxygen species produced from the superoxide anion radical formed; this pathway also is thought to be responsible for the mutagenicity of quinones.[111] The protective function of this enzyme has been demonstrated with perfused liver by the detection of low-level chemiluminescence elicited with a model quinone, menadione, where the photoemission detected from the liver surface was substantially augmented upon inhibition of NADPH:quinone reductase by dicoumarol.[112] Conversely, when the enzyme activity was induced in mice by pretreatment with the anticarcinogenic antioxidant 2,(3)-*tert.*-butyl-4-hydroxy-anisole (BHA),[113] photoemission elicited by menadione was substantially suppressed.[114] Several other parameters of cytotoxcity were much more pronounced when the enzyme was inhibited, e.g., trypan blue exclusion, Ca^{2+} distribution, or blebbing of isolated hepatocytes.[115]

B. GSSG Reductase and NADPH Supply

As stated in Sectin II.E, the continuous operation of GSH peroxidase requires the regeneration of GSH from GSSG by the reductase. Thus, if the available activity is diminished, this defense system may become inoperative. This is illustrated, for example, by the alkylation of the enzyme by nitrosoureas, e.g., 1,3-bis(2-chloroethyl)-1-nitrosourea (BCNU).[116] Cells become more susceptible to oxidative challenge when GSSG reductase has been inactivated in this way,[118] and the therapeutic action of such nitrosoureas against tumor cells is thought to be related to this effect. Recently, a competitive inhibition of GSSG reductase from human erythrocytes by GSH S-conjugates was shown, using both kinetic and X-ray crystallographic analysis.[118] Furthermore, target organ toxicity may be influenced by nutritional factors; since the enzyme uses FAD as prosthetic group, riboflavin deficiency might aggravate a loss of GSH regenerating capacity.[120]

Obviously, the supply of NADPH can also become crucial in different organs. The various sources of NADPH in cells, as well as the subcellular distribution, were discussed in detail.[58]

IV. CONCLUDING REMARKS

The aforementioned capacities of key defense systems in mammalian organs have been shown to play a role in target organ toxicity. Clearly the scope of the problem is even larger. For example, there is a modulating role of these defense systems relating to toxic effects of ionizing radiation. Furthermore, quite important in the medical field is the use of free radical-generating drugs with antiparasitic action which utilizes the very low defense capacity of some of the protozoic parasites such as trypanosomes or plasmodia.[121-123]

This chapter attempted to provide an elementary introduction into this active area, which of course, also extends to pathobiochemical states. For instance, the fortification of an antioxidant defense system in inflammatory states such as by injection of SOD or the supplementation of selenium to boost GSH peroxidase activity have even found clinical applications.

ACKNOWLEDGMENTS

Work coming from the authors' laboratory was supported by Deutsche Forschungsgemeinschaft, Schwerpunktsprogramm "Mechanismen toxischer Wirkungen von Fremdstoffen", by Ministerium für Wissenschaft und Forschung, Nordrhein-Westfalen, and by the National Foundation of Cancer Research, Washington, D.C. The help by Dr. G. Steger in preparing this manuscript is gratefully acknowledged.

REFERENCES

1. **Packer, L., Ed.,** *Oxygen Radicals in Biological Systems, Methods in Enzymology,* Vol. 105, Academic Press, Orlando, Fla., 1984.
2. **Bors, W., Saran, M., and Tait, D., Eds.,** *Oxygen Radicals in Chemistry and Biology,* Walter de Gruyter, Berlin, 1984.
3. **Sies, H., Ed.,** *Oxidative Stress,* Academic Press, London, 1985.
4. **Cohen, G. M.,** Metabolic basis of target organ toxicity, in *Biological Basis of Detoxication,* Caldwell, J. and Jacoby, W. B., Eds., Academic Press, New York, 1983, 325.
5. **Guengerich, F. P. and Mason, P. S.,** Immunological comparison of hepatic and extrahepatic cytochrome P-450, *Mol. Pharmacol.,* 15, 154, 1979.
6. **Dees, J. H., Coe, L. D., Yasukochi, Y., and Masters, B. S.,** Immunofluorescence of NADPH-cytochrome *c* (P-450) reductase in rat and minipig tissues injected with phenobarbital, *Science,* 208, 1473, 1980.
7. **Baron, J. and Kawabata, T. T.,** Intratissue distribution of activating and detoxication enzymes, in *Biological Basis of Detoxication,* Caldwell, J. and Jacoby, W. B., Eds., Academic Press, New York, 1983, 105.
8. **Tavassoli, M., Ozols, J., Sugimoto, G., Cox, K. H., and Müller-Eberhard, U.,** Localization of cytochrome b_5 in rat organs and tissues by immunohistochemistry, *Biochem. Biophys. Res. Commun.,* 72, 281, 1976.
9. **Fouts, J. R.,** The metabolism of xenobiotics by isolated pulmonary and skin cells, *Trends Pharmacol. Sci.,* 3, 164, 1982.
10. **Boyd, M. R.,** Biochemical mechanisms in chemical induced lung injury: roles of metabolic activation, *CRC Crit. Rev. Toxicol.,* 103, 1980.
11. **Jollow, D. J., Mitchell, J. R., Potter, W. Z., Davis, D. C., Gillette, J. R., and Brodie, B. B.,** Acetaminophen-induced hepatic necrosis. II. Role of covalent binding in vivo, *J. Pharmacol. Exp. Ther.,* 187, 195, 1973.
12. **Oesch, F.,** Microsomal epoxide hydrolase, in *Enzymatic Basis of Detoxication,* Vol. 1, Jakoby, W. B., Ed., Academic Press, New York, 1980, 277.
13. **Lu, A. Y. H. and Miwa, G. T.,** Molecular properties and biological function of microsomal epoxide hydrolase, *Ann. Rev. Pharmacol. Toxicol.,* 20, 513, 1980.
14. **Gill, S. S. and Hammock, B. D.,** Distribution and properties of a mammalian soluble epoxide hydrolase, *Biochem. Pharmacol.,* 29, 389, 1980.
15. **Waechter, F., Bentley, P., Germann, M., Oesch, F., and Stäubli, W.,** Immuno-electron-microscopic studies on the subcellular distribution of rat liver epoxide hydrolase and the effect of phenobarbital and 2-acetamidofluorene treatment, *Biochem. J.,* 202, 677, 1982.
16. **Kawabata, T. T., Guengerich, F. P., and Baron, J.,** An immunohistochemical study on the localization and distribution of epoxide hydrolase within livers of untreated rats, *Mol. Pharmacol.,* 20, 709, 1981.
17. **Oesch, F., Glatt, H., and Schmassmann, H.,** The apparent ubiquity of epoxide hydrolase in rat organs, *Biochem. Pharmacol.,* 26, 603, 1977.
18. **Oesch, F., Schmassmann, H., and Bentley, P.,** Specifity of human, rat, and mouse skin epoxide hydrolase towards K-region epoxides of polycyclic hydrocarbons, *Biochem. Pharmacol.,* 27, 17, 1978.
19. **Jakoby, W. B. and Habig, W. H.,** Glutathione transferases, in *Enzymatic Basis of Detoxication,* Vol. 2, Jakoby, W. B., Eds., Academic Press, New York, 1980, 63.
20. **Boyland, E. and Chasseaud, L. F.,** The role of glutathione and glutathione S-transferases in mercapturic acid biosynthesis, *Adv. Enzymol.,* 32, 173, 1969.
21. **DePierre, J. W. and Morgenstern, R.,** Comparison of the distribution of microsomal and cytosolic glutathione S-transferase activities in different organs of the rat, *Biochem. Pharmacol.,* 32, 721, 1983.

22. **Akerboom, T. P. M. and Sies, H.,** Assay of glutathione, glutathione disulfide and glutathione mixed disulfides in biological samples, in *Detoxication and Drug Metabolism: Conjugation and Related Systems,* Vol. 77, Jakoby, W. B., Ed., Academic Press, New York, 1981, 373.
23. **Sies, H., Wendel, A., and Burk, R. F.,** Se and non-Se glutathione peroxidases: enzymology and cell physiology, in *Oxidases and Related Redox Systems,* King, T. E., Mason, H. S., and Morrison, M., Eds., Pergamon Press, Oxford, 1982, 169.
24. **Chow, C. K., Reddy, K., and Tappel, A. L.,** Effect of dietary vitamin E on the activities of the glutathione peroxidase system in rat tissues, *J. Nutr.,* 103, 618, 1973.
25. **Jenkinson, S. G., Lawrence, R. A., Burk, R. F., and Gregory, P. E.,** Non-selenium-dependent glutathione peroxidase activity in rat lung: association with lung glutathione S-transferase activity and the effects of hyperoxia, *Toxicol. Appl. Pharmacol.,* 68, 399, 1983.
26. **Bhuyan, K. C., Bhuyan, D. K., and Podos, S. M.,** Cataract induced by selenium in rat. I. Effect on the lenticular protein and thiols, *IRCS Med. Sci.,* 9, 194, 1981.
27. **Fecondo, J. V. and Augusteyn, R. C.,** Superoxide dismutase, catalase and glutathione peroxidase in the human cataractous lens, *Exp. Eye Res.,* 36, 15, 1983.
28. **Boyd, S. C., Sasame, H. A., and Boyd, M. R.,** High concentrations of glutathione in glandular stomach: possible implications for carcinogenesis, *Science,* 205, 1010, 1979.
29. **Reddy, K. and Tappel, A. L.,** Effect of dietary selenium and autoxidized lipids on the glutathione peroxidase system of gastrointestinal tract and other tissues in the rat, *J. Nutr.,* 104, 1069, 1974.
30. **Ishikawa, T. and Sies, H.,** Cardiac transport of glutathione disulfide and S-conjugate. Studies with isolated perfused rat heart during hydroperoxide metabolism, *J. Biol. Chem.,* 259, 3838, 1984.
30a. **Beutler, E. and Srivastava, S. K.,** GSH metabolism of the lens, in *Glutathione,* Flohé, L., Benöhr, H. C., Sies, H., Waller, H. D., and Wendel, A. Eds., Georg Thieme, Stuttgart, 1974, 201.
30b. **Summer, K.-H. and Göggelmann, W.,** 1-Chloro-2,4-dinitrobenzene depletes glutathione in rat skin and is mutagenic in *Salmonella typhimurium, Mutat. Res.,* 77, 91, 1980.
31. **Baars, A. J., Mukhtar, H., Zoetemelk, C. E. M., Jansen, M., and Breimer, D. D.,** Glutathione S-transferase activity in rat and human tissue and organs, *Comp. Biochem. Physiol.,* 70c, 285, 1981.
32. **Wahlländer, A., Soboll, S., and Sies, H.,** Hepatic mitochondrial and cytosolic glutathione content and the subcellular distribution of GSH-S-transferases, *FEBS Lett.,* 97, 138, 1979.
33. **Morgenstern, R., Guthenberg, C., and DePierre, J. W.,** Microsomal glutathione S-transferase, purification, initial characterization and demonstration that is not identical to the cytosolic glutathione S-transferase A, B, and C, *Eur. J. Biochem.,* 128, 243, 1982.
34. **Friedberg, T., Bentley, P., Stasiecki, P., Glatt, H. R., Raphael, D., and Oesch, F.,** The identification, solubilization, and characterization of microsome-associated glutathione S-transferases, *J. Biol. Chem.,* 254, 12028, 1979.
35. **Mannervik, B., Guthenberg, C., Jensson, H., Warholm, M., and Alin, P.,** Isozymes of glutathione S-transferases in rat and human tissues, in *Functions of Glutathione: Biochemical, Physiological, Toxicological, and Clinical Aspects,* Larsson, A., Orrenius, S., Holmgren, A., and Mannervik, B., Eds., Raven Press, New York, 1983, 75.
36. **Pinkus, L. M., Ketley, J. N., and Jakoby, W. B.,** The glutathione S-transferase as a possible detoxification system of rat intestinal epithelium, *Biochem. Pharmacol.,* 26, 2359, 1977.
37. **Fleischner, G. M., Robbins, J. B., and Arias, I. M.,** Cellular localization of ligandin in rat, hamster, and man, *Biochem. Biophys. Res. Commun.,* 74, 992, 1977.
38. **Redick, J. A., Jakoby, W. B., and Baron, J.,** Immunohistochemical localization of glutathione S-transferases in livers of untreated rats, *J. Biol. Chem.,* 257, 15200, 1982.
39. **Redick, J. A., Kawabata, T. T., Guengerich, F. P., Krieter, P. A., Shires, T. K., and Baron, J.,** Distributions of monooxygenase components and epoxide hydratase within the livers of untreated male rats, *Life Sci.,* 27, 2465, 1980.
40. **Smith, M. T., Loveridge, N., Wills, E. D., and Chayen, J.,** The distribution of glutathione in rat liver lobule, *Biochem. J.,* 182, 103, 1979.
41. **Kasper, C. B. and Henton, D.,** Glucuronidation, in *Enzymatic Basis of Detoxication,* Vol. 2, Jakoby, W. B., Ed., Academic Press, New York, 1980, 3.
42. **Bock, K. W., Burchell, B., Dutton, G. J., Hänninen, O., Mulder, G. J., Owens, I. S., Siest, G., and Tephly, T. R.,** UDP-glucuronosyltransferase activities. Guidelines for consistent interim terminology and assay conditions, *Biochem. Pharmacol.,* 32, 953, 1983.
43. **Bock, K. M., von Clausbruch, U. C., Kaufmann, R., Lilienblum, W., Oesch, F., Pfeil, H., and Platt, K. L.,** Functional heterogeneity of UDP-glucuronyltransferase in rat tissues, *Biochem. Pharmacol.,* 29, 495, 1980.
44. **Beström, H. and Wengle, B.,** Studies on ester sulfates, *Acta Endocrinol.,* 56, 691, 1967.
45. **Pang, K. S., Koster, H., Helsema, I. C. M., Scholtens, E., and Mulder, G. J.,** Aberrant pharmacokinetics of harmol in the perfused rat liver preparation: sulfate and glucuronide conjugations, *J. Pharm. Exp. Ther.,* 219, 134, 1981.

46. **Johns, K. G., Holland, J. F., Foureman, G. L., Bend, J. R., and Fouts, J. R.,** Xenobiotic metabolism in Clara cells and alveolar type II cells isolated from lungs of rats treated with β-naphthoflavone, *J. Pharm. Exp. Ther.*, 225, 316, 1983.
47. **Jakoby, W. B., Sekura, R. D., Lyon, E. S., Marcus, C. J., and Wang, J.-L.,** Sulfotransferases, in *Enzymatic Basis of Detoxication*, Vol. 2, Jakoby, W. B., Ed., Academic Press, New York, 1980, 199.
48. **Reinke, L. A., Belinsky, S. A., Evans, R. K., Kauffman, F. C., and Thurman, R. G.,** Conjugation of p-nitrophenol in the perfused rat liver: the effect of substrate concentration and carbohydrate reserves, *J. Pharm. Exp. Ther.*, 217, 863, 1981.
49. **Mehta, R. and Cohen, G. M.,** Major differences in the extent of conjugation with glucuronic acid and sulfate in human peripheral lung, *Biochem. Pharmacol.*, 28, 2479, 1979.
50. **Fridovich, I.,** Superoxide radical: an endogenous toxicant, *Ann. Rev. Pharmacol. Toxicol.*, 23, 239, 1983.
51. **Weisiger, R. A. and Fridovich, I.,** Mitochondrial superoxide dismutase. Site of synthesis and intramitochondrial localization, *J. Biol. Chem.*, 248, 4793, 1973.
52. **Marklund, S. L., Westman, N. G., Lundgren, E., and Roos, G.,** Copper- and zinc-containing superoxide dismutase, manganese-containing superoxide dismutase, catalase, and glutathione peroxidase in normal and neoplastic human cell lines and normal human tissues, *Cancer Res.*, 42, 1955, 1982.
53. **Westman, N. G. and Marklund, S. L.,** Copper- and zinc-containing dismutase and manganese-containing superoxide dismutase in human tissues and human malignant tumors, *Cancer Res.*, 41, 2962, 1981.
54. **Marklund, S.,** Distribution of CuZn superoxide dismutase and Mn superoxide dismutase in human tissue and extracellular fluids, *Acta Physiol. Scand. Suppl.*, 492, 19, 1980.
54a. **McCord, J. M. and Fridovich, I.,** Superoxide dismutase: an enzymic function for erythrocuprein (hemocuprein), *J. Biol. Chem.*, 22, 6049, 1969.
55. **Marklund, S. L.,** Mammalian superoxide dismutases, in *Oxygen Radicals in Chemistry and Biology*, Bors, W., Saran, M., and Tait, D., Eds., Walter de Gruyter, Berlin, 1984.
56. **Thaete, L. G., Crouch, R. K., Schulte, B. A., and Spicer, S. S.,** The immunolocalization of copper-zinc superoxide dismutase in canine tissues, *J. Histochem. Cytochem.*, 31, 1399, 1983.
57. **Chance, B., Sies, H., and Boveris, A.,** Hydroperoxide metabolism in mammalian organs, *Physiol. Rev.*, 50, 527, 1979.
58. **Orrenius, S. and Sies, H.,** Compartmentation of detoxication reactions, in *Metabolic Compartmentation*, Sies, H., Ed., Academic Press, London, 1982, 485.
59. **DeDuve, C. and Baudhuin, P.,** Peroxisomes (microbodies and related particles), *Physiol. Rev.*, 46, 323, 1966.
60. **Masters, C. J. and Holmes, R.,** Peroxisomes: new aspects of cell physiology and biochemistry, *Physiol. Rev.*, 57, 816, 1977.
61. **Roels, F. and Goldfischer, S.,** Cytochemistry of human catalase. The demonstration of hepatic and renal peroxisomes by a high temperature procedure, *J. Histochem. Cytochem.*, 27, 1471, 1979.
62. **Goldenberg, H., Huettinger, M., Kollner, U., Kramar, R., and Pavelka, M.,** Catalase positive particles from pig lung, *Histochemistry*, 56, 253, 1978.
63. **Christie, K. N. and Stoward, P. J.,** Catalase in skeletal muscle fibers, *J. Histochem. Cytochem.*, 27, 814, 1979.
64. **Herzog, V. and Fahimi, H. D.,** Microbodies (peroxisomes) containing catalase in myocardium: morphological and biochemical evidence, *Science*, 185, 271, 1974.
65. **Hand, A. R.,** Peroxisomes (microbodies) in striated muscle cells, *J. Histochem. Cytochem.*, 22, 207, 1974.
66. **Nohl, H. and Hegner, D.,** Evidence for the existence of catalase in the matrix space of rat-heart mitochondria, *FEBS Lett.*, 89, 126, 1978.
67. **Larsson, A., Orrenius, S., Holmgren, A., and Mannervik, B., Eds.,** *Functions of Glutathione: Biochemical, Physiological, Toxicological, and Clinical Aspects*, Raven Press, New York, 1983.
68. **Meister, A. and Anderson, M.,** Glutathione, *Ann. Rev. Biochem.*, 52, 711, 1983.
69. **Revesz, L. and Malaisse, E. P.,** Significance of cellular glutathione in radioprotection and repair of radiation damage, in *Functions of Glutathione: Biochemical, Physiological, Toxicological, and Clinical Aspects*, Larsson, A., Orrenius, S., Holmgren, A., and Mannervik, B., Eds., Raven Press, New York, 1983, 163.
70. **Flohé, L.,** Glutathione peroxidase brought into focus, in *Free Radicals in Biology*, Vol. 5, Pryor, W. A., Ed., Academic Press, New York, 1982, 223.
71. **Kosower, N. S. and Kosower, E. M.,** The glutathione status of cells, *Int. Rev. Cytol.*, 54, 109, 1978.
72. **Brigelius, R., Muckel, Ch., Akerboom, T. P. M., and Sies, H.,** Identification and quantification of glutathione in hepatic protein mixed disulfides and its relationship to glutathione disulfide, *Biochem. Pharmacol.*, 32, 2529, 1983.
73. **Mannervik, B., Axelsson, K., Sundewall, A.-C., and Holmgren, A.,** Relative contributions of thioltransferase- and thioredoxin-dependent systems in reduction of low-molecular-mass and protein disulfides, *Biochem. J.*, 213, 519, 1983.

74. **Brigelius, R.,** Glutathione oxidation and activation of pentose phosphate cycle during hydroperoxide metabolism: a comparison of livers from fed and fasted rats, *Hoppe Seyler's Z. Physiol. Chem.*, 364, 989, 1983.
74a. **Asghar, K., Reddy, B. G., and Krishna, G.,** Histochemical localization of glutathione in tissues, *J. Histochem. Cytochem.*, 23, 774, 1975.
74b. **Devereux, T. R., Jones, K. G., Bend, J. R., Fouts, J. R., Statham, C. N., and Boyd, M. R.,** *In vitro* metabolic activation of the pulmonary toxin, 4-ipomeanol, in nonciliated bronchiolar epithelial (Clara) and alveolar type II cells isolated from rabbit lung, *J. Pharmacol. Exp. Ther.*, 220, 223, 1981.
75. **Lawrence, R. and Burk, R. F.,** Glutathione peroxidase activity in selenium-deficient rat liver, *Biochem. Biophys. Res. Commun.*, 71, 952, 1976.
76. **Prohaska, J. R.,** The glutathione peroxidase activity of glutathione S-transferases, *Biochim. Biophys. Acta*, 611, 87, 1980.
77. **McCay, P. B., Gibson, D. D., Fong, K.-L., and Roger, K.,** Effect of glutathione peroxidase activity on lipid peroxidation in biological membranes, *Biochim. Biophys. Acta*, 431, 459, 1976.
78. **Grossmann, A. and Wendel, A.,** Non-reactivity of the selenoenzyme glutathione peroxidase with enzymatically hydroperoxidized phospholipids, *Eur. J. Biochem.*, 135, 549, 1983.
79. **Sevanian, A., Muakkassah-Kelly, S. F., and Montestruque, S.,** The influence of phospholipase A2 and glutathione peroxidase on the elimination of membrane lipid peroxides, *Arch. Biochem. Biophys.*, 223, 441, 1983.
80. **Ursini, F., Maiorino, M., Valente, M., Ferri, L., and Gregolin, C.,** Purification from pig liver of a protein which protects liposomes and biomembranes from peroxidative degradation and exhibits glutathione peroxidase activity on phospatidylcholine hydroperoxides, *Biochim. Biophys. Acta*, 710, 197, 1982.
81. **Burk, R. F., Trumble, M. J., and Lawrence, R. A.,** Rat hepatic cytosolic glutathione-dependent enzyme protection against lipid peroxidation in the NADPH-microsomal lipid peroxidation system, *Biochim. Biophys. Acta*, 618, 35, 1980.
82. **Gibson, D. D., Hornbrook, K. R., and McCay, P. B.,** Glutathione-dependent inhibition of lipid peroxidation by a soluble, heat-labile factor in animal tissues, *Biochim. Biophys. Acta*, 620, 572, 1980.
83. **Christophersen, B. O.,** The inhibitory effect of reduced glutathione on the lipid peroxidation of the microsomal fraction and mitochondria, *Biochem. J.*, 106, 515, 1968.
84. **Burk, R. F.,** Glutathione-dependent protection by rat liver microsomal protein against lipid peroxidation, *Biochim. Biophys. Acta*, 757, 21, 1983.
85. **Burk, R. F., Patel, K., and Lane, J. M.,** Reduced glutathione protection against rat liver microsomal injury by carbon tetrachloride; dependence on O_2, *Biochem. J.*, 215, 441, 1983.
86. **Haenen, G. R. M. M. and Bast, A.,** Protection against lipid peroxidation by a microsomal glutathione-dependent labile factor, *FEBS Lett.*, 159, 24, 1983.
87. **Carmagnol, F., Sinet, P. M., and Jerome, H.,** Selenium-dependent and non-selenium-dependent glutathione peroxidases in human tissue extracts, *Biochim. Biophys. Acta*, 759, 49, 1983.
88. **Yoshimura, S., Komatsu, N., and Watanabe, K.,** Purification and immunohistochemical localization of rat liver glutathione peroxidase, *Biochim. Biophys. Acta*, 621, 130, 1980.
89. **Forman, H. J. and Fisher, A. B.,** Antioxidant enzymes of rat granular pneumocytes: constitutive levels and effect of hyperoxia, *Lab. Invest.*, 45, 1, 1981.
90. **Porter, R. and Whelan, J., Eds.,** *Biology of Vitamin E: Ciba Foundation Symposium 101*, Pitman, London, 1983.
91. **Packer, J. E., Slater, T. F., and Willson, R. L.,** Direct observation of a free radical interaction between vitamin E and vitamin C, *Nature (London)*, 278, 737, 1979.
92. **Taylor, S. L., Lamden, M. P., and Tappel, A. L.,** Sensitive fluorometric method for tissue tocopherol analysis, *Lipids*, 11, 530, 1976.
93. **Kornbrust, D. J. and Mavis, R. D.,** Relative susceptibility of microsomes from lung, heart, liver, kidney, brain and testes to lipid peroxidation: correlation with vitamin E content, *Lipids*, 15, 315, 1979.
94. **Kitabchi, A. E., Wimelasena, J., and Barker, J.,** Specific receptor sites for α-tocopherol in purified isolated adrenocortical cell membrane, *Biochem. Biophys. Res. Commun.*, 96, 1739, 1980.
95. **Wimalasena, J., Davis, M., and Kitabchi, A. E.,** Characterization and solubilization of the specific binding sites for d-α-tocopherol from human erythrocyte membranes, *Biochem. Pharmacol.*, 31, 3455, 1982.
96. **Tappel, A. L.,** Vitamin E as the biological lipid antioxidant, *Vitam. Horm.*, 20, 493, 1962.
97. **Leung, H.-W., Vang, M. J., and Mavis, R. D.,** The cooperative interaction between vitamin E and vitamin C in suppression of peroxidation of membrane phospholipids, *Biochim. Biophys. Acta*, 664, 266, 1981.
98. **Niki, E., Yamamoto, Y., and Kamiya, Y.,** Oxidation of phosphatidylcholine and its inhibition by vitamin E and vitamin C, in *Oxygen Radicals in Chemistry and Biology*, Bors, W., Saran, M., and Tait, D., Eds., Walter de Gruyter, Berlin, 1984, 273.

99. **Nishikimi, M.,** Oxidation of ascorbic acid with superoxide anion generated by the xanthine-xanthine oxidase system, *Biochem. Biophys. Res. Commun.*, 63, 1975.
100. **Edgar, J. A.,** Ascorbic acid and biological alkylating agents, *Nature (London)*, 248, 136, 1974.
101. **Willis, R. J. and Kratzing, C. C.,** Ascorbic acid in rat lung, *Biochem. Biophys. Res. Commun.*, 59, 1250, 1974.
102. **Minchin, R. F. and Boyd, M. R.,** Localization of metabolic activation and deactivation systems in the lung: significance to the pulmonary toxicity of xenobiotics, *Ann. Rev. Pharmacol. Toxicol.*, 23, 217, 1983.
103. **Kensler, T. W. and Trush, M. A.,** Inhibition of phorbol ester-stimulated chemiluminescence in human polymorphonuclear leukocytes by retinoic acid and 5,6-epoxyretinoic acid, *Cancer Res.*, 41, 216, 1981.
104. **Packer, J. E., Mahood, J. S., Mora-Arellano, V. O., Slater, T. F., Willson, R. L., and Wolfenden, B. S.,** Free radicals and singlet oxygen scavengers: reaction of a peroxy-radical with β-carotene, diphenyl furan and 1,4-diazobicyclo(2,2,2)-octane, *Biochem. Biophys. Res. Commun.*, 98, 901, 1981.
105. **Witz, G., Goldstein, B. D., Amoruso, M., Stone, D. S., and Troll, W.,** Retinoid inhibition of superoxide anion radical production by human polymorphonuclear leukocytes stimulated with tumor promotors, *Biochem. Biophys. Res. Commun.*, 97, 883, 1980.
106. **Olson, J. A.,** Some aspects of vitamin A metabolism, *Vitam. Horm.*, 26, 1, 1968.
107. **Proctor, P.,** Similar functions of uric acid and ascorbate in man?, *Nature (London)*, 228, 868, 1970.
107a. **Ames, B. N., Cathcart, R., Schwiers, E., and Hochstein, P.,** Uric acid provides an antioxidant defense in humans against oxidant- and radical-caused aging and cancer: a hypothesis, *Proc. Natl. Acad. Sci. U.S.A.*, 78, 6858, 1981.
108. **Deby, C., Deby-Dupont, G., Noel, F.-X., and Lavergne, L.,** *In vitro* and *in vivo* arachidonic acid conversions into biologically active derivatives are enhanced by uric acid, *Biochem. Pharmacol.*, 30, 2243, 1981.
109. **Levine, W. G.,** Role of excretion, in Target Organ Toxicity, Vol. I, Cohen, G. M., Ed., CRC Press, Boca Raton, Fla., 1986.
110. **Smith, M. T., Evans, C. G., Thor, H., and Orrenius, S.,** Quinone-induced oxidative injury to cells and tissues, in *Oxidative Stress*, Sies, H., Ed., Academic Press, London, 1985, 91.
111. **Chesis, P. L., Levin, D. E., Smith, M. T., Ernster, L., and Ames, B. N.,** Mutagenicity of quinones: pathway of metabolic activation and detoxification, *Proc. Natl. Acad. Sci. U.S.A.*, 81, 1696, 1984.
112. **Wefers, H. and Sies, H.,** Hepatic low-level chemiluminescence during redox cycling of menadione and the menadione-glutathione conjugate: relation to glutathione and NAD(P)H:quinone reductase (DT-diaphorase) activity, *Arch. Biochem. Biophys.*, 224, 568, 1983.
113. **Benson, A. M., Hunkeler, M. J., and Talalay, P.,** Increase of NAD(P)H:quinone reductase by dietary antioxidants: possible role in protection against carcinogenesis and toxicity, *Proc. Natl. Acad. Sci. U.S.A.*, 77, 5216, 1980.
114. **Wefers, H., Komai, T., Talalay, P., and Sies, H.,** Protection against reactive oxygen species by NAD(P)H:quinone reductase induced by dietary antioxidant butylated hydroxyanisole (BHA): decreased hepatic low-level chemiluminescence during quinone redox cycling, *FEBS Lett.*, 169, 63, 1984.
115. **Thor, H., Smith, M. T., Hartzell, P., Bellomo, G., Jewell, S. A., and Orrenius, S.,** The metabolism of menadione (2-methyl-1,4-naphthoquinone) by isolated hepatocytes: a study of the implications of oxidative stress in intact cells, *J. Biol. Chem.*, 257, 12319, 1982.
116. **Babson, J. R. and Reed, D. J.,** Inactivation of glutathione reductase by 2-chloroethyl nitrosourea-derived isocyanates, *Biochem. Biophys. Res. Commun.*, 83, 754, 1978.
117. **Babson, J. R., Abell, N. S., and Reed, D. J.,** Protective role of glutathione redox cycle against adriamycin-mediated toxicity in isolated hepatocytes, *Biochem. Pharmacol.*, 30, 2299, 1981.
118. **Eklöw, L., Moldeús, P., Orrenius, S.,** Oxidation of glutathione during hydroperoxide metabolism: a study using isolated hepatocytes and the glutathione reductase inhibitor 1,3-bis(2-chloroethyl)-1-nitrosourea, *Eur. J. Biochem.*, 138, 459, 1984.
119. **Bilzer, M., Krauth-Siegel, R. L., Schirmer, R. H., Akerboom, T. P. M., Sies, H., and Schulz, G. E.,** Interaction of a glutathione S-conjugate with glutathione reductase: kinetic and X-ray crystallographic studies, *Eur. J. Biochem.*, 138, 373, 1984.
120. **Beutler, E.,** Glutathione reductase, in *Glutathione*, Flohé, L., Benöhr, H. C., Sies, H., Waller, H. D., and Wendel, A., Eds., Georg Thieme, Stuttgart, 1974, 109.
121. **Docampo, R. and Moreno, S. N. J.,** Free radical intermediates in the antiparasitic action of drugs and phagocytic cells, in *Free Radicals in Biology*, Vol. 6, Pryor, W. A., Ed., Academic Press, New York, 1983, 243.
122. **Docampo, R. and Moreno, S. N. J.,** Free radical metabolites in the mode of action of chemotherapeutic agents and phagocytic cells in *Trypanosoma cruzi*, *Rev. Infect. Dis.*, 6, 223, 1984.
123. **Clark, I. A. and Cowden, W. B.,** Antimalarials, in *Oxidative Stress*, Sies, H., Ed., Academic Press, London, 1985, 131.

Chapter 8

THE LIVER AS A TARGET ORGAN FOR TOXICITY

John A. Timbrell

TABLE OF CONTENTS

I. INTRODUCTION

The liver is exposed to many potentially toxic substances and is frequently a target organ in experimental animal studies. By its position, structure, and biochemistry, the liver is especially vulnerable to damage from toxic compounds. However, in man, liver damage is less common and only 9% of adverse drug reactions affect the liver.[1]

Substances taken into the body via the GI tract will be absorbed into the hepatic-portal blood system and from there are carried to the liver. After the GI mucosa and blood, the liver is the first tissue to be exposed (Figure 1). The liver is more susceptible to injury because as a portal of entry to the systemic circulation and the other tissues, it is exposed to higher concentrations of foreign substances prior to their dilution in the systemic circulation.

Once within the liver, foreign compounds readily come into intimate contact with liver cells. The sinusoidal structure of the liver facilitates this and consequently foreign compounds are readily taken up from the blood into hepatocytes. The result of this is that some foreign compounds, such as the drug propranolol[2] may be extensively extracted in the "first pass" through the liver.

The large and diverse metabolic capabilities of the liver enable it to metabolize compounds and although this is generally a detoxication process it may result in toxic injury.

The liver also has an extensive role in intermediary metabolism and interference with this by foreign compounds may also make the liver a target organ. Secretion of bile by the liver may also be a factor both via the biliary excretion of foreign compounds, which may be saturated and give rise to high intrahepatic concentrations, and also by direct damage to the biliary system. Enterohepatic recirculation may prolong exposure of hepatocytes to toxic compounds following uptake from the GI tract and reabsorption into the portal vein.

The liver cells, however, have extraordinary regenerative powers and are able to tolerate increased metabolic demands,[3] and the homogeneous structure of the liver allows it to withstand considerable injury.

A. Basic Structure of the Liver

The liver, the largest gland in the body, is about 2% of the body weight of man,[3] and consists of several lobes, and lies in the peritoneal cavity.

It is served by two blood suplies, the portal vein and the hepatic artery (Figure 1). The portal vein brings blood from the capillaries draining the alimentary tract, spleen, and pancreas and accounts for 75% of the hepatic blood.[3] Portal blood is rich in nutrients, but low in oxygen.[3] The hepatic artery brings oxygenated blood which mixes with the portal blood as it enters the liver.

After passage through the liver, the blood drains via the hepatic vein into the inferior vena cava. The liver receives about 25% of the cardiac output, approximately 1.0 to 1.30 mℓ/min/g.[3] The incoming blood is separated from the outgoing blood by the sinusoids which are in intimate contact with the hepatic parenchymal cells. Thus, the portal and hepatic vein do not meet but each ramifies into venules spreading through the liver like fingers with the space between filled with sinusoids. The relation of the sinusoids to the portal vein, central vein, and hepatic artery is shown in Figure 2. This arrangement allows the liver to function efficiently in the removal of substances from the incoming portal blood.

The sinusoids are specialized capillaries which have a discontinous basement membrane lined with Kupffer and endothelial cells. The latter have fenestrations (2-μm diameter) and clusters of these, known as sieve plates, are found particularly near the portal vein end of the sinusoid.[4] These fenestrations allow large mol wt compounds to pass out of the sinusoid.

The liver is a sponge like cellular mass, mainly composed of parenchymal cells interspersed with a connective tissue framework for support. The parenchyme is arranged in plates one or two cells deep, separated by sinusoids, and arranged into lobules. The lobule is a functional

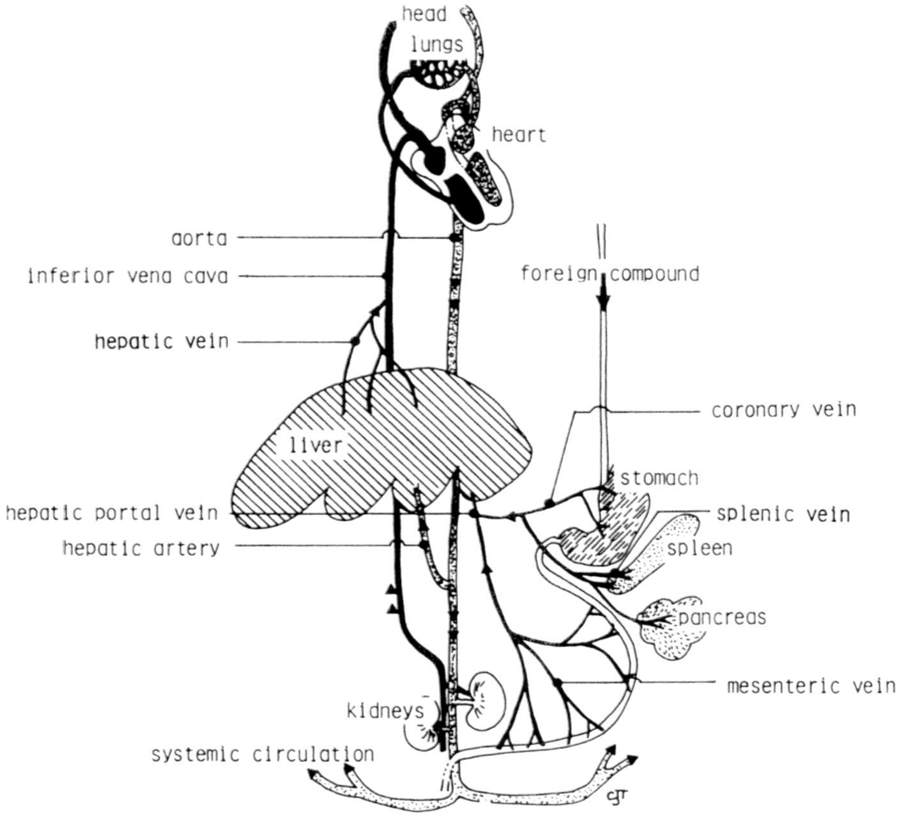

FIGURE 1. Vasculature supplying and draining the liver and its relationship to the systemic circulation.

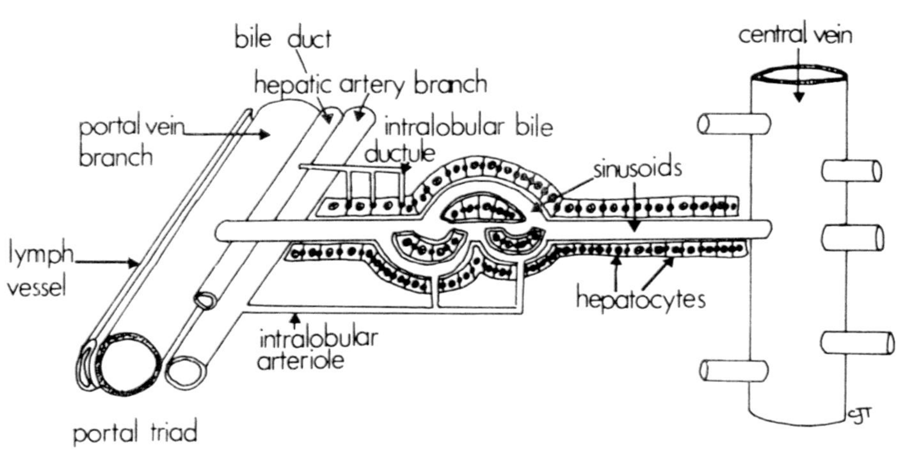

FIGURE 2. Schematic representation of the arrangement and relationship of vessels and sinusoids within the liver. The central vein drains into the hepatic vein.

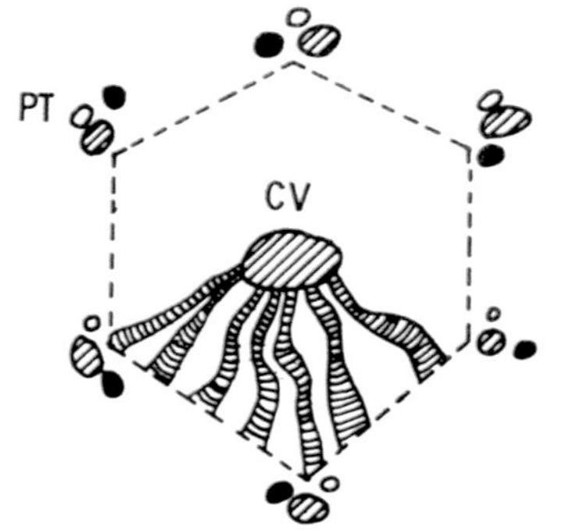

FIGURE 3. Classical liver lobule. PT is the portal tract consisting of branches of the portal vein, hepatic artery, and bile duct. CV is a branch of the central vein. Blood drains from the portal tract toward the central vein at the center. Bile flows towards the portal tract. (From Timbrell, J. A., *Principles of Biochemical Toxicology*, Taylor & Francis, London, 1982. With permission.)

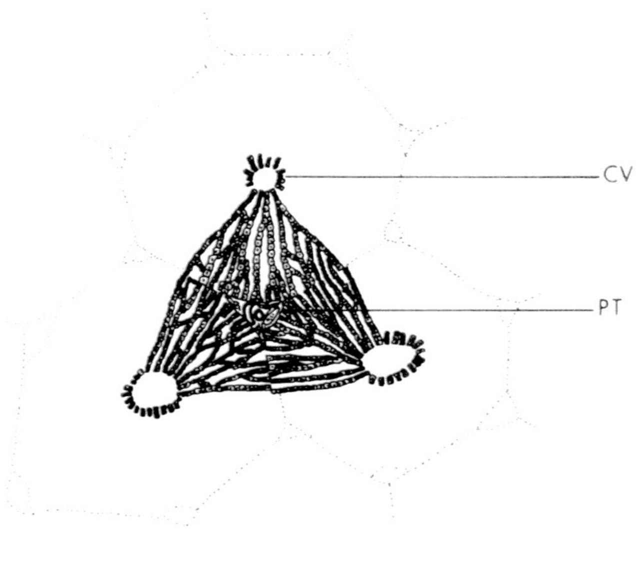

FIGURE 4. Portal lobule. PT and CV are as defined in Figure 3. Blood drains from the portal tract at the center to the central veins at the periphery of the lobule. Bile flows toward the portal tract.

as well as a structural concept and consequently there are three different descriptions of lobules: classical, portal, and hepatic acinus.[3]

The classical lobule describes a hexagonal section of tissue with portal tracts as boundaries (Figure 3). The center of the lobule is the hepatic venule to which blood flows. Bile flows toward the periphery of the lobule. This description emphasizes the endocrine function of the liver and may be helpful for describing centrilobular necrosis.

The portal lobule is a triangular arrangement around the portal tract with a central vein at each corner. This emphasizes the secretion of bile which flows toward the center (Figure 4).

The simple liver acinus is the most recent and probably the most useful concept.[3,5] It is a microcirculatory hepatic unit (Figure 5), bounded by two portal tracts and two terminal

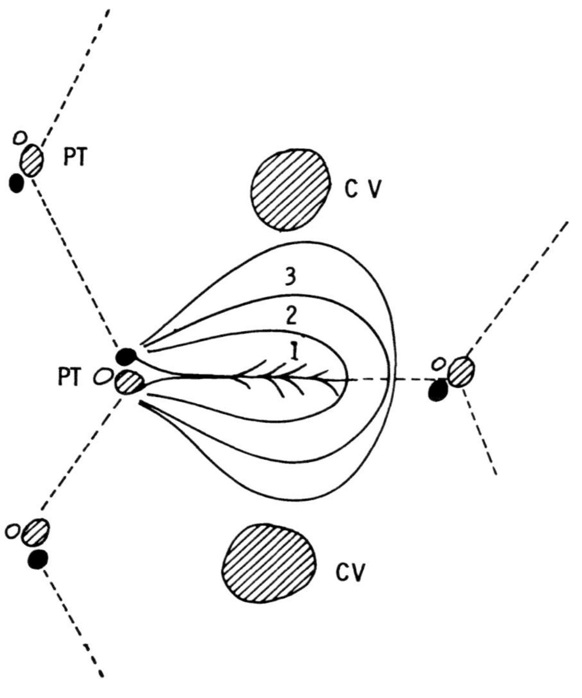

FIGURE 5. Schematic representation of a hepatic acinus. PT and CV are defined in Figure 3. 1, 2, and 3 represent the zones draining the terminal afferent blood vessel. Blood flows from the portal tract toward the periphery of the acinus and bile flows toward the portal tract. (From Timbrell, J. A., *Principles of Biochemical Toxicology*, Taylor & Francis, London, 1982. With permission.)

hepatic venules or central veins. Blood flows from the center to the periphery and bile flows in the opposite direction. The description emphasizes metabolic zonation. There are three zones of cells grouped around the axis and distinguished by the microcirculation. Zone 1 is the area which gains blood from the afferent vessels still rich in nutrients and oxygen. The cells in this zone are the most resistant to damage.[6]

Zone 3 is the area around the central vein or terminal hepatic venule. The cells in this area receive blood that has lost much of the nutrients and oxygen. The cells of this zone, and particularly those in the region where several acini meet, the perivenular ring, are the most sensitive to damage.[6] Zone 2 is the transition zone between zones 1 and 3.

The acinus is also a secretory unit, the bile it produces flowing into the terminal bile ductules in the portal tract.

Hepatic blood flow is an important factor affecting the metabolic activity of the liver.[6] The extraction of foreign compounds from sinusoidal blood will be influenced by blood flow, which is regulated by the hepatic arterioles. Thus, hepatic arterioles may open or close and so allow portal blood into the sinusoids or oxygenated arterial blood.[6] Foreign substances which interfere with the microcirculation may therefore influence hepatic blood flow and consequently alter metabolism and possibly toxicity.[6] The hepatotoxicity of phalloidin may involve intravascular pressure changes.[6] Phenobarbital is well known to enlarge the hepatic vascular network and increase hepatic blood flow.[6] Increased hepatic blood flow will tend to increase the hepatic extraction of foreign compounds. Reduction in liver blood flow due to the swelling of cells blocking the sinusoids may be an early change which is important in liver damage.[6]

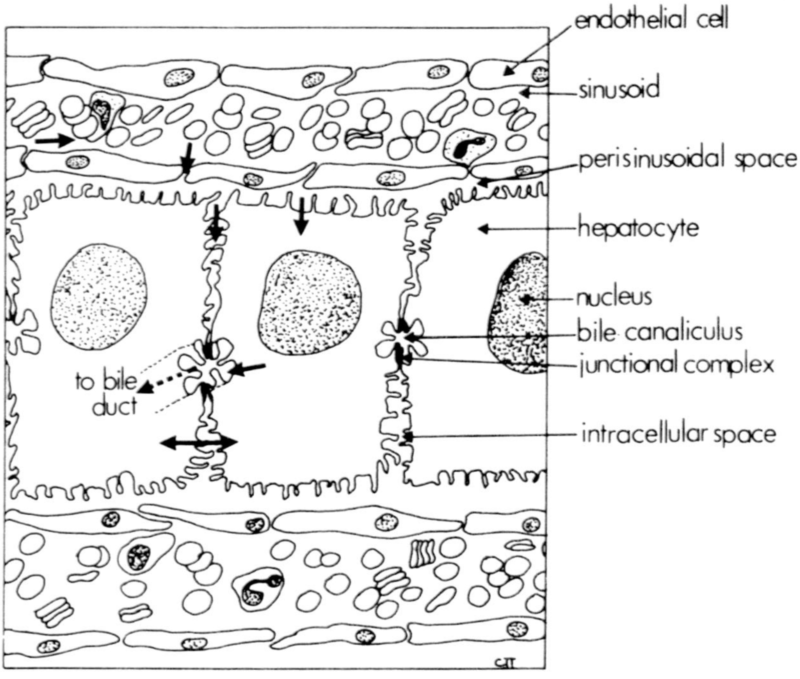

FIGURE 6. Diagrammatic representation of the arrangement of hepatocytes within the liver and the relationship to the sinusoids.

B. Liver Cells

Parenchymal cells constitute the major part of the liver, being about 80% of the cell population by volume.[3] They are arranged in layers or plates, so that they border the sinusoids, the bile canaliculi, and abutt other hepatocytes (Figure 6). The location of the hepatocytes within the lobule influence their function (see below).

The remaining cells are sinusoidal endothelial cells, Kupffer cells, fat storing cells, and pit cells.[7]

C. Localization of Metabolic Activity

Hepatocytes in different areas of the liver may have differences in subcellular structure and metabolic capabilities[8] which may be an important determinant of the localization of toxic effects.

Thus, cells in zone 1, periportal hepatocytes, are more aerobic and are particularly equipped for the tricarboxylic acid cycle, gluconeogenesis,[6] β-oxidation of fatty acids,[6,8] amino acid catabolism, and bile acid excretion.[8] As regards the biotransformation of foreign compounds, glucuronosyl transferase activity is also higher in periportal hepatocytes.[6] Protein synthesis and formation of plasma protein also occur in these hepatocytes,[6] although there may be single cell specialization even within periportal hepatocytes.[8]

Zone 3 (perivenous) hepatocytes have a greater amount of smooth endoplasmic reticulum[8] than periportal hepatocytes, and associated with this cytochrome P-450 and NADPH-cytochrome P-450 reductase activity is higher. Lipid synthesis is also greater in zone 3 hepatocytes.[6,8] Strangely, protective mechanisms such as GSH peroxidase[8,9] and GSH itself[10] are predominant in zone 1 hepatocytes despite predominance of the microsomal monooxygenases in zone 3. This may explain the common occurrence of fatty change in zone 3, where there is less protection from lipid hydroperoxides.[8] As far as other liver cells are concerned, the

Table 1
TYPES OF HEPATIC DAMAGE

1. Interference with bilirubin uptake, excretion, and conjugation
2. Cytotoxic injury
3. Cholestatic injury
4. Mixed cholestatic and cytotoxic injury
5. Fatty liver (steatosis)
6. Cirrhosis
7. Phospholipidosis
8. Liver tumors
9. Vascular lesions
10. Chronic active hepatitis
11. Subacute hepatic necrosis

activity of cytochrome P-450 is low in nonparenchymal cells and the activities of GSH reductase and peroxidase and catalase are also lower than in hepatocytes.[11]

Acetylation was reported to occur in Kupffer cells,[12] but more recently, acetyltransferase activity was not demonstrated in nonparencymal cells.[13]

D. Function of the Liver

The liver serves three basic functions: it takes up nutrients and exogenous substances from the portal blood, it synthesizes a variety of compounds, and it secretes bile.[3]

The uptake of nutrients may be followed by redistribution, synthesis, or catabolism. Thus, triglycerides may be removed from the bloodstream and also secreted in combination with a carrier. The liver is particularly important in the synthesis of plasma proteins, such as albumin and fibrinogen.

Toxicologically, the function of biotransformation is clearly important. In general, foreign compounds are metabolized in the liver to less toxic, more polar, and water-soluble products, which can be more readily excreted and therefore rapidly removed from the body. However, as will be seen later, this in some cases leads to an increase rather than a decrease in toxicity.

All of these functions of the liver: bile production, biotransformation, and synthesis and redistribution of nutrients and intermediary metabolites may be involved in the production of hepatic lesions. Thus, the great metabolic capabilities of the liver make it both a target and an organ able to protect itself.

Fortunately, the liver has a large reserve functional capacity and a significant proportion of the organ may be damaged without a major effect on the organism as a whole.[3] This is partly due to the structural organization, which unlike that in the CNS for instance, is repeated and not localized in one formal, vulnerable, irreplaceable area.

II. TYPES OF HEPATIC DAMAGE

The nature of the damage sustained by the liver reflects its structure and function. The response of the liver to injury is fairly limited; viewed simply, liver injury is either due to the metabolic capabilities of the hepatocyte or involves the secretion of bile. This interference with the normal biochemistry of the cell may lead to direct hepatocyte damage, hepatic tumors, and accumulation of lipids or phospholipids. Interference with bile secretion may give rise to cholestasis and cholestatic injury. These types of injury may follow chronic or acute exposure to hepatotoxins and may affect all or only part of the liver. The range of the types of liver damage is illustrated in Table 1.[14] In some cases several may occur together, for instance, liver tumors or cirrhosis may follow on from early cytotoxic injury.

The various types are discussed briefly in the following subsections.

A. Cytotoxic Damage

Direct damage to the liver parenchyma leading to necrosis is probably the most common form of liver damage. It may follow acute or chronic exposure to a hepatotoxin. Within this classification there are several types of lesion. Necrosis may be zonal, (i.e., primarily affecting one of the three zones), diffuse, or massive.[14] Many compounds which require microsomal enzyme-mediated activation for toxicity such as carbon tetrachloride[15] and paracetamol,[16] cause centrilobular (zone 3) necrosis. Other compounds such as allyl alcohol[17] and cocaine[18] may cause periportal (zone 1) necrosis. Midzonal (zone 2) necrosis is less common. Diffuse necrosis may result from generalized metabolic disruption, as after galactosamine administration.[19]

B. Cholestatic Damage

This type may range from interference with bilirubin transport or conjugation, to bile stasis and bile duct damage.[14]

Interference with bilirubin transport and conjugation casued by compounds such as rifampicin[20] and flavaspidic acid[21] results in hyperbilirubinemia but no hepatic lesion. Similarly, bile stasis such as that due to icterogenin[22] and certain steroids[23] may not cause any parenchymal or bile duct damage — simply bilirubin casts in the canalicular spaces. This is known as canalicular damage.[14] In other cases, bile stasis may be accompanied by damage to the parenchyma, such as with chlorpromazine,[14] this being known as hepatocanalicular injury. Direct damage to the bile ducts (cholangio-destruction), such as caused by α-naphthylisothiocyanate may also be included in this classification.[23]

C. Fatty Liver (Steatosis)

This may be regarded as a type of cytotoxicity, although it may be a reversible change which does not necessarily lead to cell damage. Steatosis or the accumulation of lipid inside the hepatocyte may accompany necrosis as after carbon tetrachloride administration,[15] or it may occur alone as with hydrazine.[24] A large number of toxic compounds may produce steatosis by one of several mechanisms.[14]

D. Phospholipidosis

This lesion is different from steatosis in involving phospholipids and giving rise to lamellated crystalloid inclusions. This type of lesion, which may occur in tissues other than the liver, is caused by chronic treatment with the coronary vasodilator drug 4,4′-diethylaminoethoxyhexestrol.[25]

E. Cirrhosis

This is a chronic lesion which may occasionally follow from acute cytotoxic or cholestatic damage or more commonly from chronic exposure to a hepatotoxin. Thus, a macronodular cirrhosis may follow acute cytotoxicity injury and primary biliary cirrhosis from cholestatic damage.[14]

F. Liver Tumors

Both benign and malignant liver tumors may result from exposure to hepatotoxins.[14] For example, liver adenoma has been associated with the use of contraceptive steroids,[26] and several compounds may cause malignant carcinomas and sarcomas.[14]

Aflatoxin B_1 can produce hepatocellular carcinomas[27] and exposure to vinyl chloride is responsible for the development of angiosarcomas.[28]

G. Vascular Lesions

Hepatic damage may follow or include damage to the vascular network of the liver.[14]

Thus, the pyrrolizidine alkaloids cause hepatotoxicity which includes damage to the small hepatic venous system.[29] Hepatic vein occlusion due to thrombosis may lead to liver damage, as has been associated with use of the contraceptive pill.[30] Such occlusions may lead to necrosis and congestive cirrhosis.

III. CYTOTOXIC DAMAGE

A. Hepatic Necrosis — General Mechanisms

A variety of foreign compounds may cause hepatic necrosis, some via reactive metabolite formation, others via different mechanisms.

The chain of events, which leads from the initial insult to the death of the cell, is at present poorly understood. However, certain initial events and intracellular changes occurring in hepatocytes appear to be common for a number of compounds. Some of these will be discussed in relation to carbon tetrachloride hepatotoxicity. Studies of compounds, which do not cause necrosis but interfere with cellular processes, have revealed that some of these processes are not directly involved in the development of necrosis. Thus, extensive inhibition of protein synthesis by agents such as puromycin may cause fatty liver but not necrosis.[31] Depletion of ATP by compounds such as ethionine also may lead to fatty liver but not necrosis.[32] Ethionine also causes a decrease in RNA synthesis[32] and cellular K^+ and an increase in Na^+.[33] Effects on the nuclear and cytoplasmic architecture may also be observed. However, these important perturbations are clearly not sufficient to cause liver cell necrosis.

Another event, which at one time had been thought to be causally involved in necrosis, is mitochondrial injury.[34] However, although damage certainly occurs after exposure to hepatotoxic compounds with loss of respiratory control and uncoupling of oxidative phosphorylation, this seems to be a late rather than an early event.[34] Thus, damage to other organelles such as the endoplasmic reticulum may be seen earlier, and evidence of other cellular derangements may be found before mitochondrial function is altered. One of the major expected effects of mitochondrial damage would be depression of ATP levels, but as already indicated this may occur after ethionine without loss of cell viability.[32]

Another potential target is the lysosome, where release of the hydrolytic enzymes could conceivably cause necrosis. Again, however, rupture of the lysosomes seems generally to be a late event.[34]

More recently, attention has been focused on cellular membrane distrubances and more particularly the role of calcium.[34] Clearly changes in intracellular calcium homeostasis are involved in cellular injury. However, these changes may occur because of damage to the cell membrane and leakage of extracellular calcium into the cell or by changing the intracellular calcium concentration from within.[34] Thus, liver cell Ca^{2+} increases in isolated cells and in vivo during hepatic damage, but chlorpromazine stops this and protects against cytotoxicity.[35,36]

Other cellular events known to take place in toxic liver necrosis with certain hepatotoxins are lipid peroxidation[15,37] and the covalent binding of reactive intermediates to cellular macromolecules.[38] These will be discussed in relation to carbon tetrachloride hepatotoxicity, but a few salient points will be emphasized here.

The peroxidative breakdown of membrane lipids following free radical attack has been suggested as one of the major factors in cellular damage (Figure 7).[37,39] However, more recently data has been put forward to indicate this event may account for initial localized damage to the endoplasmic reticulum, but that peroxidation products are not produced in sufficient quantities to damage distant parts of the cell.[40]

The same criticism may apply to the role of covalent binding of metabolites to cellular macromolecules in cellular damage. This undoubtedly occurs but its role in cell death is unclear. Thus, with the hepatotoxin acetylhydrazine, covalent binding to protein occurs in

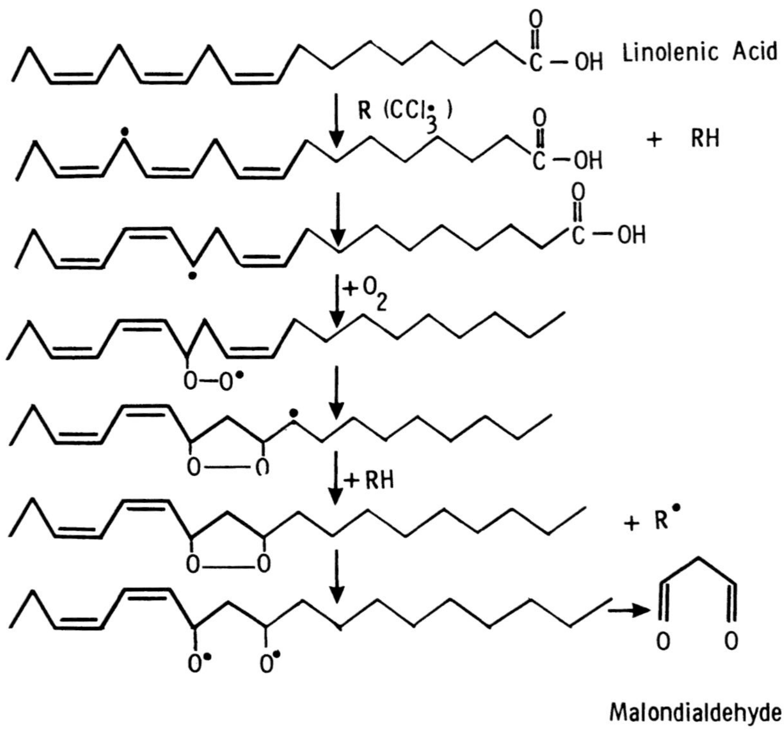

FIGURE 7. Peroxidative destruction of a polyunsaturated lipid initiated by a free radical attack such as by the trichloromethyl radical ($CCl_3\cdot$). (From Timbrell, J. A., *Principles of Biochemical Toxicology*, Taylor & Francis, London, 1982. With permission.)

many tissues of the animal as well as the liver, yet only the liver is damaged.[41] However, this covalent binding of an acetylhydrazine metabolite is only increased in the liver, the target organ, by pretreatments which increase hepatotoxicity (Figure 8).[41] It is probable, therefore, that only a specific part of the covalent binding of hepatotoxins to proteins is involved in the development of hepatic damage.

Other types of liver damage may also be described which result from chronic administration of hepatotoxins or the long-term development of an acute lesion such as subacute hepatic necrosis, fibrosis, and chronic active hepatitis. Also, in some cases several types of lesions may occur together such as venous occlusion and necrosis, necrosis and steatosis, or cholestatic and cytotoxic damage.

Examples of some of these types of liver injury due to toxic compounds will be examined. These examples will attempt to highlight the reason for the liver being the target organ and, wherever known, the mechanisms underlying this organ-specific toxicity.

B. Specific Examples

1. Carbon Tetrachoride

Carbon tetrachloride is one of the most intensely studied hepatotoxins[15] and one which illustrates a number of important principles relating to the liver as a target organ.

Carbon tetrachloride is a simple, chemically stable molecule, a lipid soluble solvent which has wide uses. After various routes of administration (oral, intraperitoneal, subcutaneous, and inhalation), the major toxic effect is damage to the liver.[15] Although lung and renal toxicity have been described,[42] the liver is the major target organ. The type of damage caused is centrilobular (zone 3) hepatic necrosis and fatty liver (steatosis).[14,43]

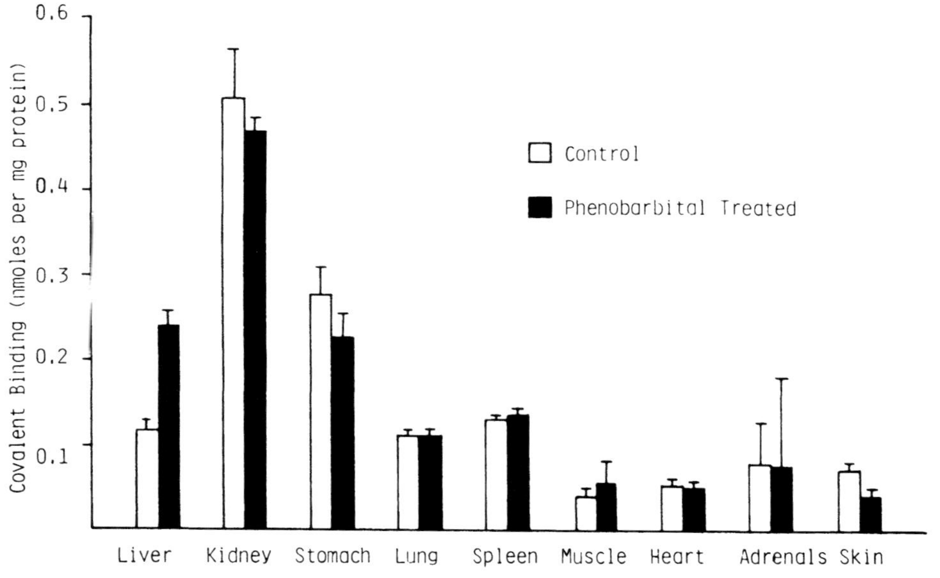

FIGURE 8. Covalent binding of radiolabeled acetylhydrazine to protein from various tissues of control and phenobarbital pretreated rats. (From Woodward, K. N. and Timbrell, J. A., *Toxicology*, 30, 65, 1984. With permission.)

Carbon tetrachloride is widely distributed throughout the body, as would be predicted from its lipid solubility and small molecular size. The fact that this compound damages the liver in particular, therefore, is significant, as all the other major organs will be exposed to similar concentrations of the compound.

The liver is the target organ because of its metabolic capabilities, as the hepatotoxicity of carbon tetrachloride is the result of metabolic activation.[15] Although carbon tetrachloride was originally thought resistant to metabolic attack, it has been recognized for some years that metabolism via the microsomal enzyme system yields a reactive metabolite.[15] As will be discussed, this metabolite reacts with cellular constitutents and the biochemical and pathological consequences can be detected. Although the whole chain of events has yet to be elucidated many important insights into mechanisms of toxicity have been gained from the study of carbon tetrachloride.

When administered to experimental animals of various species, carbon tetrachloride reproducibly causes hepatic damage in a dose-related manner.[14] However, the chicken is not susceptible, seemingly because this species does not metabolize the compound[15] which reinforces the necessary role of metabolism in the hepatotoxicity. Low doses result in fatty liver and destruction of cytochrome P-450.[15] Higher doses also cause centrilobular hepatic necrosis[15,43] and chronic administration leads to cirrhosis.[44] The hepatotoxicity is increased by pretreatment of animals with certain microsomal enzyme inducers such as phenobarbital and decreased by inhibition of these enzymes.[13,45] The centrilobular region is probably the target because of higher levels of the cytochrome P-450 responsible for the activation in this region[46] and possibly because of lower levels of protective substances such as GSH.[10]

The initial metabolic transformation is almost certainly production of the trichloromethyl free radical (Figure 9) catalyzed by cytochrome P-450.[15,37] This free radical is reactive and may undergo a number of reactions (Figure 9). Abstraction of a hydrogen atom from a suitable donor yields chloroform, which may be detected as a metabolite. The donor of the hydrogen atom, which may be an unsaturated fatty acid, may then become a radical and undergo a variety of reactions. Alternatively, the trichloromethyl radical may react covalently

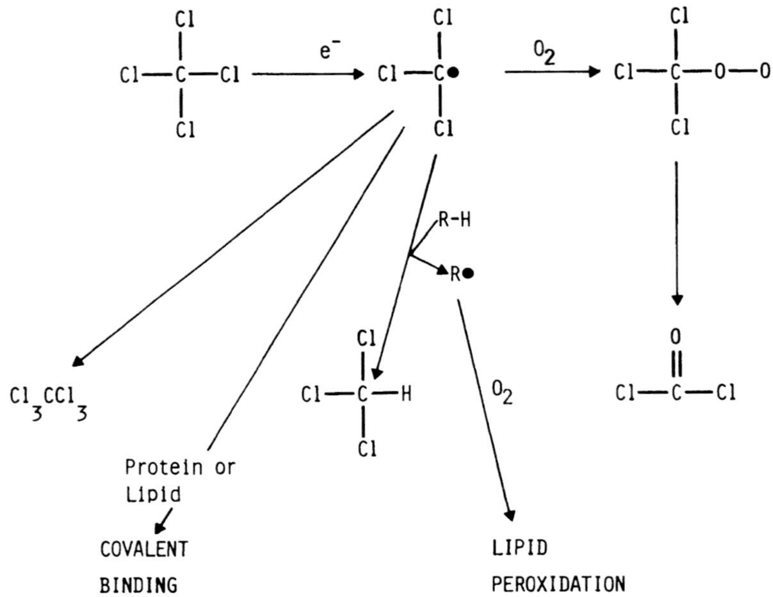

FIGURE 9. Metabolism of carbon tetrachloride. R represents an unsaturated lipid.

with proteins[47] or lipids,[48] with a binding ratio of Cl:C of 3, indicating CCl_3 is the moiety bound to lipids.[48] This may be detected as covalent binding to these macromolecules.

Another alternative is reaction of the trichloromethyl radical with oxygen to form the trichloromethyl peroxy radical.[49] This radical is felt by some to be more reactive and therefore more likely to be involved in the toxicity than the trichloromethyl radical. It may also degrade to phosgene,[50] which is known to be toxic, or the latter could be formed by another route (Figure 9). These initial reactions probably take place in the endoplasmic reticulum and covalent binding in this region and destruction of the cytochrome P-450 probably results from the reductive metabolism.[40] What connects these initial reactions to the ensuing pathological lesions is less clear, however.

Peroxidation of lipids certainly occurs following on from initial free radical attack on unsaturated lipids (Figure 7). The eventual products of this lipid peroxidation, especially 4-hydroxynonenal, a hydroxy alkenal, will cause a reduction in protein synthesis, and inhibit cytochrome P-450 and glucose-6-phosphatase.[40] However, such products of lipid peroxidation are unlikely to be responsible for all the pathological lesions.[40]

More recently, the important role of calcium homeostasis in cellular damage has been recognized.[34,40] The initial attack at the endoplasmic reticulum reduces the capacity of this organelle to sequester Ca^{2+} by 85 to 90%.[40] This is an early event, which follows covalent binding to protein, and occurs at the same time as lipid peroxidation. Immediately after, damage to the endoplasmic reticulum becomes apparent as dispersion of ribosomes, depression of protein synthesis, and accumulation of triglycerides begins.[15,40] The damage to the calcium sequestering capacity of the endoplasmic reticulum may be caused either by lipid peroxidation or by covalent binding of reactive metabolites to proteins and lipids of the endoplasmic reticulum.[45] The result, however, will be an increase in cellular calcium levels, which may cause a variety of intracellular disturbances at some distance from the initial lesion.

The hepatic damage due to carbon tetrachloride is, however, clearly a result of initial microsomal enzyme-mediated metabolic activation which identifies the liver as the major target organ. If small doses of carbon tetrachloride are given, which destroy 75% of the hepatic cytochrome P-450, then subsequent large doses are not hepatotoxic.[15] This clearly illustrates the importance hepatic metabolism may have in target organ toxicity. The biochemical and pathological changes which occur in vivo following toxic doses of carbon tetrachloride can be clearly related to the initial event of metabolic activation. Thus, within 1 min the trichloromethyl moiety is covalently bound to microsomal lipid and protein in the ratio 11:3[15] with three molecules of chlorine bound to one of carbon.[48] This indicates that the CCl_3 moiety is bound.

Within 15 min of dosing, there is electron microscopic evidence of injury to the endoplasmic reticulum with progressive changes which include vacuolation of the cisternae, loss of ribosomes from the rough endoplasmic reticulum, and fragmentation of the smooth endoplasmic reticulum.[14,40,45] Biochemical evidence also indicates that the endoplasmic reticulum is a target with inhibition of glucose-6-phosphatase, and depression of protein synthesis and loss of cytochrome P-450.[14,15] Thus, microsomal enzyme activity is reduced, incorporation of leucine into glycoprotein, glycine into albumin, fibrinogen, and globulins is decreased within 1 hr.[14] By 1 hr after dosing the lipid content of the hepatocyte is elevated,[15] mitochondrial damage occurs but much later, possibly 10 to 12 hr after, when the cisternae disappear, and the organelle enlarges.[14,43] Biochemically, loss of respiratory control and impaired fatty acid oxidation are seen.[43] The Golgi apparatus also shows damage, losing certain dense particles, which may reflect the depressed lipoprotein assembly involved in the pathogenesis of fatty liver.[14,15,43] Changes in fatty acids occur with a decrease in unsaturated fatty acids, and an increase in cross linking[15] probably due in part to the lipid peroxidation initiated by the trichloromethyl free radical.

The liver is clearly the major target for carbon tetrachloride toxicity and in particular, the endoplasmic reticulum of the centrilobular hepatocyte. This is due at least in part to the abilities of this organ to carry out biotransformation and to the wide range of metabolic processes which occur in its cells.

There are numerous other examples of hepatotoxins which owe their toxicity to metabolic activation carried out in the liver. For example, the common analgesic drug paracetamol,[16] the solvent bromobenzene,[51] and the drug metabolite acetylhydrazine[52] will all cause centrilobular hepatic necrosis via reactive metabolites. There will be certain differences from carbon tetrachloride, particularly with regard to the importance of lipid peroxidation and covalent binding of metabolites to cellular macromolecules. Whether the mechanisms of cellular damage for these compounds are similar remains to be seen. However, metabolic activation of the parent compund by hepatic microsomal enzymes followed by reaction of a reactive metabolite with cellular constituents is almost certainly involved in each case.

Therefore, examples of hepatotoxicity in which there are factors other than, or as well as microsomal enzyme-mediated metabolic activation, will now be considered.

2. *Furosemide*

In high doses the diuretic drug furosemide causes centrilobular hepatic necrosis in experimental animals.[53] The mechanism underlying this toxic effect involves initial metabolic activation by the microsomal enzymes followed by covalent binding of a reactive metabolite to hepatic macromolecules (Figure 10).[53] Thus, furosemide is similar to other hepatotoxins such as paracetamol and bromobenzene in requiring metabolic activation.[54] With furosemide-induced hepatic necrosis, however, there is a clear dose threshold which is not due to the depletion of GSH, in contrast to the situation with paracetamol or bromobenzene.[54]

Studies of the distribution of furosemide in vivo revealed that after hepatotoxic doses the distribution of the drug between plasma and liver was markedly altered.[55] Thus, for a 5-fold increase in dose, the liver concentration increased approximately 36-fold, whereas the plasma concentration increased about 5-fold (Figure 11).[55]

This massive increase in liver concentration, which coincides with the threshold for

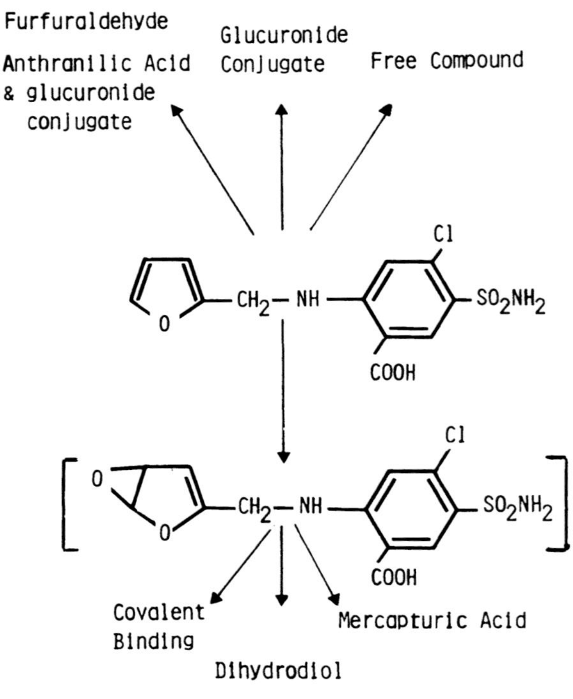

FIGURE 10. Metabolism of furosemide. The epoxide intermediate shown in brackets is the postulated reactive intermediate.

hepatotoxicity, was thought to be due to saturation of plasma protein-binding sites. Thus, the liver is able to extract more free furosemide from the plasma. More recently, it has been found that hepatotoxic doses of furosemide saturate the transport systems involved in biliary excretion.[56] Thus, the normal excretory mechanism for furosemide is dose-dependent and high doses lead to a disproportionate rise in the concentration in the liver. Competitive inhibition of biliary excretion with bromosulfophthalein will also potentiate the hepatic necrosis. Thus, in this example the liver is a target organ both because it is an excretory organ and it is able to metabolically activate furosemide.

Another example of susceptibility of the liver because of its role as an excretory organ is that of proxicromil (Figure 12). This compound is hepatotoxic in dogs after high doses, although this species does not metabolize the compound to any great extent. Indeed, other species such as the rat which do metabolize proxicromil are not susceptible. It seems that hepatotoxic doses of this compound saturate the biliary excretory mechanisms, thereby allowing accumulation of the parent compound to toxic levels.[57]

3. *Allyl Alcohol*

The liver may be a target organ for toxic compounds as a result of metabolism via enzymes other than the microsomal monooxygenases. When this is the case, damage may occur in areas of the lobule other than the centrilobular area. An example of this is allyl alcohol which causes periportal necrosis in rats.[43,58]

It has been demonstrated that allyl alcohol is metabolized by alcohol dehydrogenase to acrolein in vitro and various studies indicate this also occurs in vivo (Figure 13).[58] Thus, the inhibitor of alcohol dehydrogenase, pyrazole, reduces the hepatotoxicity of allyl alcohol. It has been suggested[58,59] that the periportal localization of the damage is due to the presence of higher levels of alcohol dehydrogenase in this part of the lobule.[60] Other sources, however,

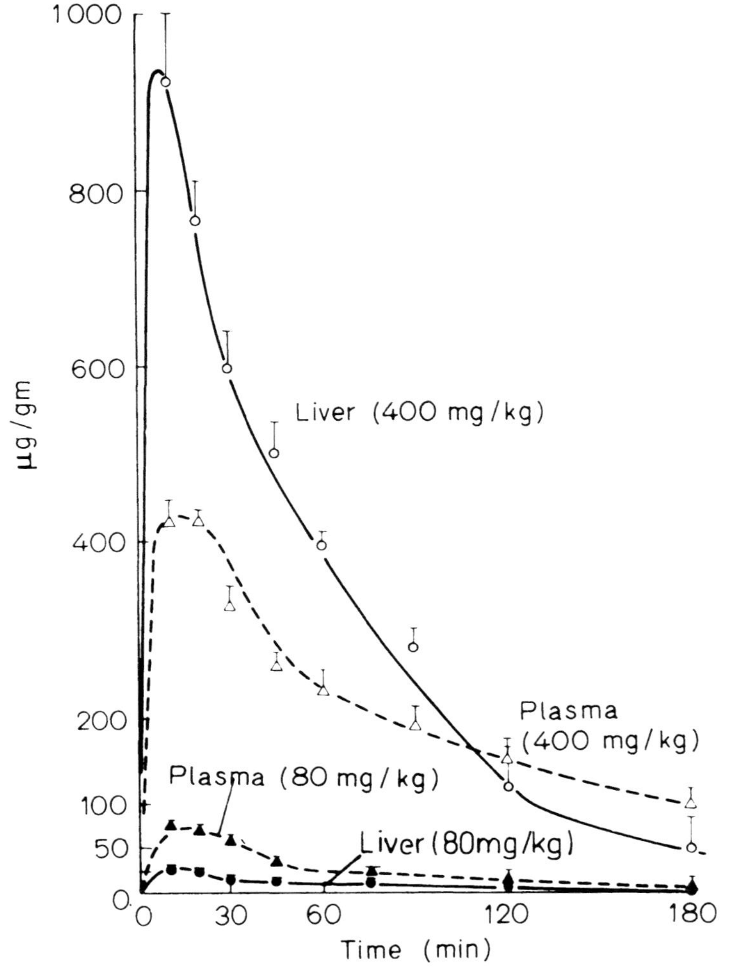

FIGURE 11. Change in distribution of furosemide in plasma and liver after toxic and nontoxic doses were administered to mice. Unpublished data of Mitchell, J. R. et al., in Concepts in Biochemical Phamacology, Part 3, Vol. 28, Gillette, J. R., and Mitchell, J. R., Eds., Springer-Verlag, Berlin, 1975. With permission.

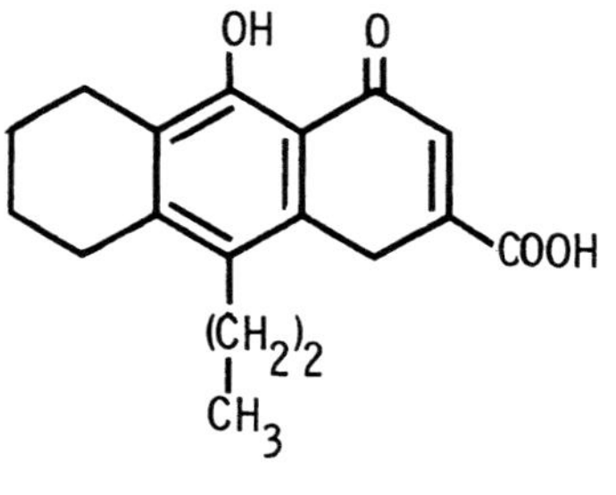

FIGURE 12. Structure of proxicromil.

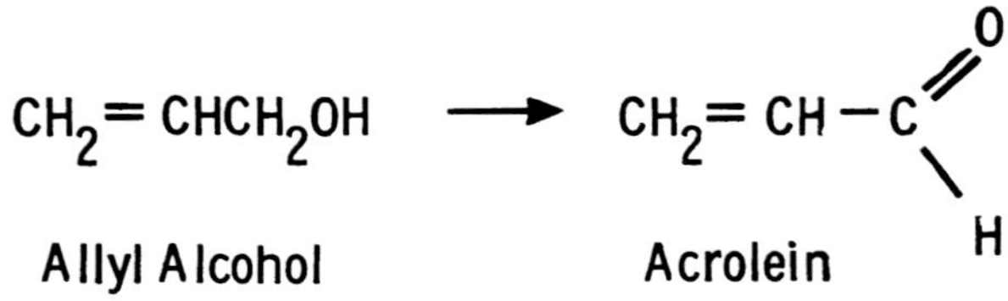

FIGURE 13. Metabolism of allyl alcohol via alcohol dehydrogenase.

have indicated that alcohol dehydrogenase is localized in the centrilobular area.[61] It may be, however, that sufficient levels of alcohol dehdyrogenase are present in the portal area that metabolism of allyl alcohol still takes place.

The presumed reactive metabolite is acrolein (Figure 13). Deuterated allyl alcohol ([1,1-$^2H_2$] allyl alcohol) was less toxic and more slowly metabolized suggesting that alcohol dehydrogenase-mediated metabolism to allyl aldehyde or acrolein is responsible for the hepatic necrosis.[62] Acrolein is a reactive compound and may react directly with cellular macromolecules and bind covalently to them as detected in vivo.[60] The liver is again the target organ in this example, despite the fact that alcohol dehydrogenase is found in many tissues. Presumably, however, the liver is still the site of greatest metabolism and hence suffers the toxic effects.

4. Halothane

The widely used anesthetic agent halothane is an interesting example to consider because the mechanism of its hepatotoxicity seems to be very different from the other hepatotoxins so far discussed.

Thus, the hepatic injury produced by halothane in humans is usually a centrilobular necrosis, which may be accompanied by fever, myalgia, arthralgia, and rashes.[14,63] This is a relatively rare occurrence, however, perhaps following one in several thousand anesthesias, and it seems that a milder form of liver dysfunction with raised serum enzymes and perhaps mild necrosis may occur in up to 20% of recipients of the drug.[64,65] A predisposing factor seems to be multiple exposures[64] which with the symptoms seen in the severe form of hepatotoxicity suggests an immunological involvement.

The hepatic damage is not readily reproducible in experimental animals after a single dose. Under certain conditions halothane may be hepatotoxic to experimental animals.[66] Thus, pretreatment of animals with the microsomal enzyme inducer phenobarbital, use of a low oxygen concentration, use of male rats and exposure to a 1% halothane concentration for 2 hr were all necessary to cause hepatic damage.[66] This and other data suggest that the hepatotoxicity in experimental animals is the result of anaerobic metabolism to a reactive metabolite. The metabolism of halothane does indeed include reductive as well as oxidative pathways (Figure 14).

The major metabolite, trifluoroacetic acid resulting from the oxidative pathway, is not hepatotoxic. The radical intermediate produced via the anaerobic, reductive pathway (Figure 14), may be responsible for the toxicity in experimental animals. The reductive pathway(s) yielding difluorochloroethylene and trifluorochloroethane are known to occur in humans.[67] However, in man severe hepatic damage caused by halothane is a rare event despite the fact that metabolism to potentially toxic metabolites normally occurs. This indicates that other factors may be involved. Studies in human patients have indicated that those with the severe halothane-induced liver damage have features which suggest an immunological basis to the hepatotoxicity.[68] Thus, high titres of liver and kidney antimicrosomal antibodies and antibodies to normal liver cell membrane components may have been detected.[68] Furthermore, such patients had circulating antibodies against hepatocytes from rabbits treated with halothane.[68] These antibodies were specific for the rabbit hepatocyte cell membrane and cy-

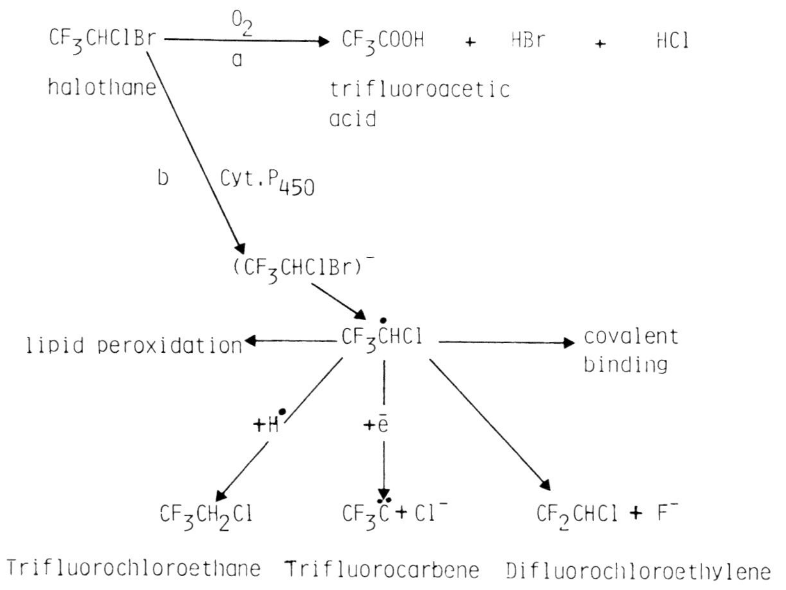

FIGURE 14. Metabolism of halothane and proposed metabolic activation to a reactive metabolite. (From Timbrell, J. A., *Br. J. Clin. Pharmacol.*, 15, 3, 1983. With permission.)

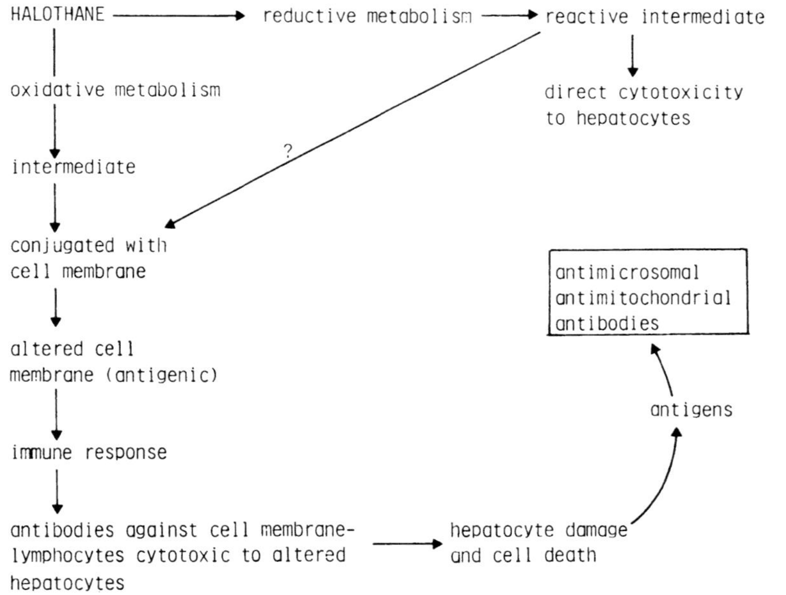

FIGURE 15. Postulated mechanisms for halothane hepatotoxicity. (From Timbrell, J. A., *Br. J. Clin. Pharmacol.*, 15, 3, 1983. With permission.)

totoxic to the hepatocytes in vitro.[68] Manipulation of the metabolism in rabbits indicated that it was the oxidative rather than reductive pathway which gave rise to the supposed antigen.[69] As the oxidative pathway is the major one, there must be other factors involved in the development of the severe, immunologically based hepatic damage. The postulated mechanisms of halothane hepatotoxicity, therefore, embrace the two types of hepatic damage, mild and severe, with the reductive and oxidative pathways being responsible, respectively (Figure 15). It may be that combinations of direct cytotoxicity due to the reactive metabolite

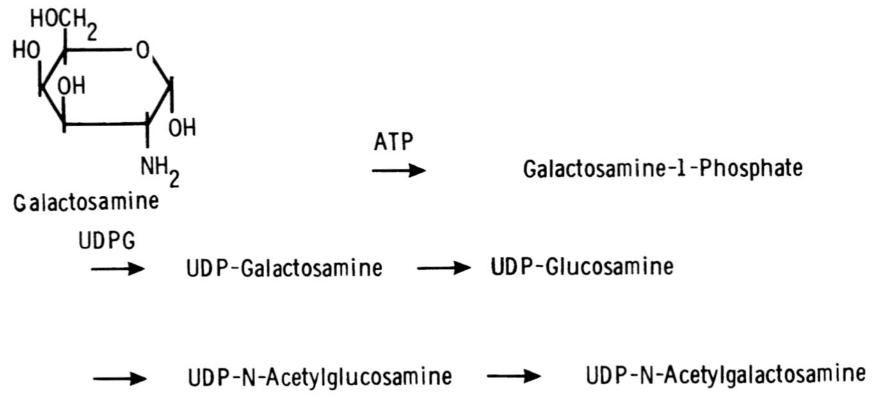

FIGURE 16. Incorporation of galactosamine into carbohydrate metabolism and formation of UDP derivatives. (From Timbrell, J. A., *Principles of Biochemical Toxicology*, Taylor & Francis, London, 1982. With permission.)

produced via the reductive pathway, and immunotoxicity may occur, giving rise to the variable spectrum of clinical features.

Halothane may cause hepatic damage to the liver, therefore, by a straightforward cytotoxic effect of a reactive metabolite, as shown in experimental animals. However, another mechanism at least in man may involve metabolic activation to yield a reactive metabolite which combines with liver cell protein, making it antigenic. This then stimulates antibody production, and these may bind to the altered hepatocyte surface. The binding of the antibody to the hepatocyte then makes the cell a target for cytotoxic lymphocytes.

In this example, therefore, the liver is the target for an immunological attack, after the cell is altered by the reactive metabolite of the toxin.

5. Galactosamine

The amino sugar galactosamine will cause hepatic damage in a variety of species after a single dose.[70] Chronic exposure leads to cirrhosis[70] and hepatocellular carcinoma.[71] The necrosis is not zonal but diffuse and multifocal and may be accompanied by accumulation of fat.[14,70]

Galactosamine is of particular interest as it causes hepatic damage via a mechanism which does not involve metabolic activation. Galactosamine seems to act as an analogue of galactose, but in doing so it traps uridylate in a metabolic "blind alley". This interference with intermediary metabolism leads to a failure in supply of uridylate[70] causing a cellular derangement and necrosis. Thus, galactosamine produces a dose-related liver lesion in a variety of species.[70] Species differences in susceptibility seem to reflect differences in uridylate synthesis. Thus, mice have a high rate of uridylate synthesis and are relatively resistant whereas guinea pigs have a low rate of synthesis and are particularly susceptible.

Cellular necrosis is evident 6 hr after a single dose of galactosamine and progresses over 48 hr.[14,70] Accumulation of fat also occurs and can be detected by 3 hr. Within 2 hr of dosing, ultrastructural changes may be detected. These include hypertrophy of the smooth endoplasmic reticulum and remnants of mitochondria, glycogen, and other debris inside autophagic vacuoles.[14] Nuclei show a condensation of chromatin and nucleoli are fragmented and dispersed by 4 hr.[14] Biochemically, fragmentation and dispersal of RNA takes place along with other nuclear changes. The mechanism of galactosamine-induced hepatic damage involves an initial incorporation into galactose metabolism (Figure 16). This leads to the production of uridine diphosphate glucosamine (UDP-glucosamine). This product, unlike UDP-glucose (the normal product), is not a uridylate donor in reactions catalyzed by uridyltransferase. Uridine, therefore, is trapped as UDP-glucosamine, UDP-*N*-acetylglucosamine, and UDP-*N*-acetylgalactosamine and a deficiency of uridine nucleotides occurs.[70] Although new uridylic acid can be synthesized in the liver, the synthetic ability may be exceeded when toxic doses of galactosamine are given. Thus, the liver soon suffers a depletion of uridine triphosphate (UTP), UDP, uridine monophosphate (UMP), UDP-glucose, UDP-galactose, and UDP-glucuronic acid. It is suggested that the critical level for UDP is 15 to 30% of normal.[70]

Other means of depleting uridylate in vivo such as administering 2-deoxygalactose or D-galactose plus 6-azauridine will also lead to the biochemical and pathological lesions seen after galactosamine treatment.[14,70]

Conversely, administration of uridine or increasing uridylate biosynthesis in vivo will protect against the hepatotoxicity of galactosamine.[70] During periods of increased uridylate synthesis such as after partial hepatectomy and in neonatal rats, susceptibility to galactosamine is decreased.[14]

UTP depletion is an initial biochemical lesion, which clearly develops into a structural derangement. The lack of UTP will depress uracil-dependent RNA synthesis and via thymidine triphosphate (TTP) may also effect DNA synthesis.[70] The depression in RNA synthesis will clearly depress protein synthesis, apparent by 2 hr, and this may be partly responsible for the accumulation of fat, the reduction of synthesis of membrane structures, and other alterations in cellular biochemistry.

Galactosamine specifically damages the liver and is taken up by this organ. Three hours after dosing, 20% of a dose of galactosamine is found in the liver.[70] This presumably reflects the incorporation into galactose metabolism which is occurring in the liver. The liver is again the target organ because of its metabolic capabilities and function.

IV. CHOLESTATIC DAMAGE

A. General Mechanisms

Cholestatic injury ranges from interference with bile flow or formation to overt bile duct damage. These effects may in some cases be accompanied by parenchymal injury.[14] The liver is a target organ in this case because of it secretory function, as well as its metabolic function. The biliary system may be damaged in a variety of ways, leading to cholestasis, bile duct injury, or to a reduction in the uptake and biliary excretion of compounds.

Thus, foreign compounds may affect various parts of the biliary system, although interference with bile secretion, or some form of injury to or alteration of the canaliculus seems to be important in many cases of cholestasis.[14,72] Production and secretion of bile depends on (1) uptake of various constitutents, such as bilirubin, into the hepatocyte from the sinusoidal blood; (2) biotransformation of various constituents, such as the conjugation of bilirubin with glucuronic acid; (3) secretion of bile from the hepatocyte into the canaliculus; and (4) flow of bile down the ductules into the bile ducts, which may also involve modification of the constitutents.[14] Interference with any of these processes by foreign compounds may lead to cholestasis. For example, uptake of bile constituents and precursors into the hepatocyte involves ligandin (GSH transferase) and the Z carrier protein.[14,73] Rifampicin interferes with this process as well as with bile secretion.[74] Conjugation of bilirubin with glucuronic acid prior to biliary excretion takes place in the hepatocyte and this may be inhibited by compounds such as novobiocin.[75]

The major causes of an interference in bile flow which may be due to a hepatotoxic foreign compound are as follows:[14]

1. A decrease in ATP level.[76] This may be involved in cholestasis caused by compounds such as ethionine,[76] but is probably not a major cause in most cholestatic injury.

2. Damage to the canalicular membrane.[77,78] This is certainly a major factor in the cholestasis caused by compounds such as sporidesmin,[78] lithocholate,[79] and cytochalasin B.[78] Injury may involve interference with the fibrillar network of the canaliculus which contains actin.[78] The disruption of this network leads to arrest of bile flow.[78]
3. Precipitation of insoluble substances in the lumen of the canaliculus. This may be a factor in some cases of cholestasis,[77,78] such as that caused by chlorpromazine.[77]
4. Decrease in bile salt-dependent bile flow. Bile flow is due in part to the osmotic pressure created by the secretion of bile acids into the canaliculus,[80] A decrease in bile acid synthesis will therefore lead to a decrease in bile flow.[14] This may be involved in the cholestatic effects of anabolic and contraceptive steroids.[14,81]
5. Decrease in bile salt-independent bile flow. Bile flow is also related to the movement of Na^+ into the canaliculus which involves Na^+K^+ATPase activity.[14,76,80,82] A decrease in the activity of this energy-dependent pump will lead to a decrease in bile flow.[14] Thus, compounds that inhibit Na^+K^+ATPase activity such as the C17 alkylated anabolic steroids[82] and chlorpromazine[83] may cause cholestasis. However, such compounds also have other effects on the biliary system which may also contribute to the cholestatic effects.
6. Alteration in micelle formation.[77] Bile salts, normally present in micelles in the bile, may be excluded by foreign compounds which are structurally similar.[14,84] For example, the C17 alkylated steroids may act in this way[4] as well as interfering in other ways with bile formation.
7. Bile duct damage. Direct damage to the bile ductules and ducts will interfere with bile flow.[77] For example, α-naphthylisothiocyanate, discussed in more detail below, causes cholestasis and bile duct lesions.[14,85]

B. Specific Example

1. α-Naphthylisothiocyanate

This compound has been used extensively as a model hepatotoxin for the production and study of cholestasis.[14]

Acute doses of α-naphthylisothiocyanate (50 to 150 mg/kg) given to rats will cause intrahepatic cholestasis, bile duct damage, and hepatic parenchymal necrosis.[14,86,87] Some species only show parenchymal damage, some such as mice and rats, a mixture of both cholestasis and parenchymal injury.[14,86,87] Chronic doses will lead to marked bile duct proliferation and biliary cirrhosis.[14,88]

After an acute dose of α-naphthylisothiocyanate, intracellular changes may be detected within 3 hr.[89] Thus, mitochondrial injury is apparent at this time as indicated by ultrastructural observation and by enzyme measurements.[14] Also, dilation of the Golgi apparatus can be seen at 3 hr,[14,89] and plasma membrane changes become apparent after 6 hr and are seen as damage at 24 hr.[14,89,90] Between 12 and 24 hr, the epithelial cells of the interlobular portal bile ducts show swelling, vacuolization, and necrosis and these changes become maximal at 48 hr after dosing.[89] This may lead to occlusion of the lumen of the bile ducts which also begin to proliferate.[89] The bile ductular proliferation is maximal at 4 days after dosing.[89]

Other changes detectable 24 hr after dosing are dilation of the canaliculi, inflammation of the membrane, and loss of microvilli.[14,89] Changes within hepatocytes may also be detected such as increases in the smooth endoplasmic reticulum and decreases in the rough endoplasmic reticulum in the periportal region.[14,90] Hepatocytes, as well as necrosis of the bile duct epithelial cells, become necrotic and portal inflammation is seen. The parenchymal injury, focal necrosis, has been variously described as periportal,[91] and mid-zonal to centrilobular.[14,88]

Biochemical changes reflect both the parenchymal and cholestatic lesions. Thus, the serum transaminases SGOT and SGPT are elevated, and serum alkaline phosphatase is increased,

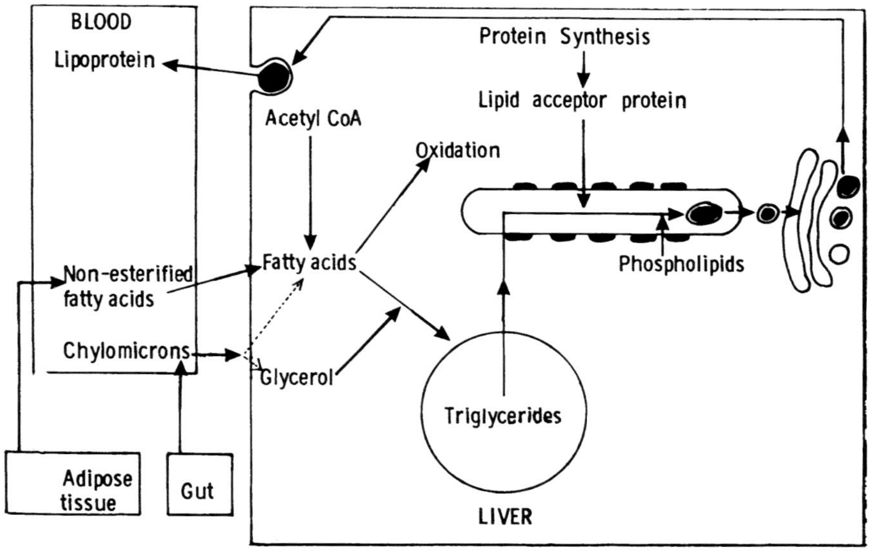

FIGURE 17. Pathways, involved in lipid metabolism and distribution. (From Timbrell, J. A., *Principles of Biochemical Toxicology*, Taylor & Francis, London, 1982. With permission.)

reflecting bile duct involvement.[89] Plasma levels of bilirubin, mainly conjugated, and cholesterol are elevated due to decreased excretion and decreased uptake by the liver.[14] Bromosulfophthalein excretion and drug metabolism are reduced in vivo before there is evidence of cholestasis and hepatic levels of cytochrome P-450 are decreased.[14,85]

The mechanism of α-naphthylisothiocyanate hepatotoxicity is not fully understood but certain evidence indicates that metabolism is necessary.[14,92] Thus, phenobarbital induction of liver enzymes increases the cholestasis and SKF 525A reduces it.[93] Compounds which inhibit protein synthesis also inhibit the cholestasis, presumably by decreasing synthesis of drug-metabolizing enzymes.[92] However, the hepatic parenchymal injury does not seem to be due to metabolism[14,94] and this may explain the variation in effects between different species, some of which show only parenchymal damage.[14] The mechanism by which α-naphthylisothiocyanate or a metabolite damages bile duct epithelial cells is not known, however.[14] This damage results in reduction or cessation of bile flow, one of the effects of α-naphthylisothiocyanate. The bile duct cells are clearly targets for α-naphthylisothiocyanate and the liver is therefore the target organ, both because of its role in bile production and secretion and in the metabolic activation of compounds.

V. FATTY LIVER (STEATOSIS)

A. General Mechanisms

Numerous hepatotoxins cause the accumulation of fat, either as large (macrovesicular) or small (microvesicular) droplets.[14] This may accompany necrosis or it may be the sole lesion. There are several mechanisms underlying the development of steatosis[95] which will be briefly discussed and then one example will be considered. The mechanisms underlying steatosis due to foreign compounds are inhibition of transport of lipids out of the liver, increased uptake of fatty acids or increased synthesis of fatty acids by the liver, and decreased oxidation of fatty acids.[95]

The most important mechanism as far as toxic foreign compounds are concerned is the first. This may be due to an effect on the synthesis of the lipoprotein transport complex or to an effect on the transport of that complex through the plasma membrane (Figure 17).[95]

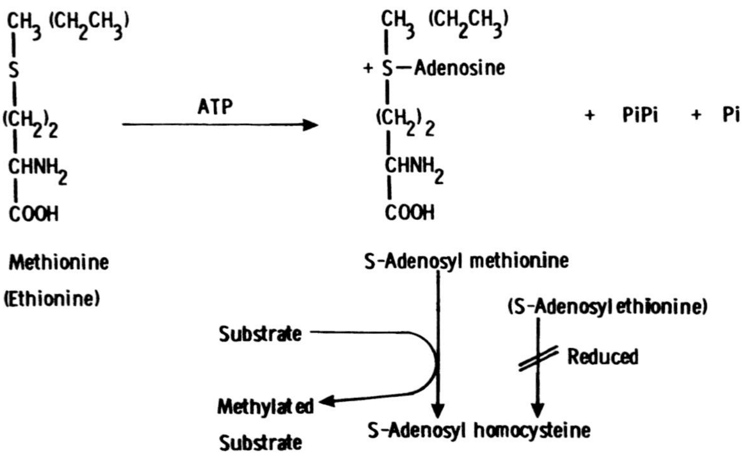

FIGURE 18. Metabolism of methionine and the analogue ethionine. (From Timbrell, J. A., *Principles of Biochemical Toxicology*, Taylor & Francis, London, 1982. With permission.)

Inhibition of lipid transport via interference with protein synthesis is one mechanism by which many hepatotoxins such as ethionine, puromycin, and tetracycline cause steatosis.[14] This may be due to inhibition of the synthesis of the apoprotein part of the very low density lipoprotein (VLDL) involved in lipid transport or due to an inhibition of assembly of the VLDL.

However, not all hepatotoxins giving rise to steatosis inhibit protein synthesis or vice versa. Ethanol in acute doses causes fatty liver by interfering with several aspects of triglyceride disposition.[14,95] As a result of increasing the NADH/NAD ratio, ethanol increases the synthesis of fatty acids, and hence triglycerides, and decreases the oxidation of fatty acids by the mitochondria.[14,95,96]

The liver is a target organ for fatty accumulation in all these cases because of the hepatic localization of triglyceride metabolism although fatty accumulation may occur in other organs such as the kidney and heart following exposure to toxic compounds such as hydrazine.[24,97]

Carbon tetrachloride as already described will cause fatty accumulation along with centrilobular necrosis by inhibiting protein synthesis,[14,15,95] but will not be considered further here.

One example of a xenobiotic-induced fatty liver, due to inhibition of lipid transport out of the liver, will now be considered.

B. Specific Example

1. Ethionine

Ethionine is an analogue of the naturally occurring amino acid methionine. After acute doses to experimental animals of various species, ethionine causes fatty liver,[14,98] with the female rat being more prone to the effect than the male.[99] The accumulation of triglycerides is maximal after 24 hr and the level reaches 15 to 20 times normal.[99] The accumulation starts around the endoplasmic reticulum in periportal hepatocytes and then progresses toward the centrilobular area.[14,99]

Biochemically, the mechanism is not unlike that of galactosamine hepatotoxicity. Thus, in the case of ethionine, it is ATP which is trapped rather than UTP (Figure 18).[99,100] Ethionine combines with adenosine to yield *S*-adenosylethionine in an analogous manner to methionine.[14,100] However, *S*-adenosylmethionine acts as a methyl donor in vivo and the remaining *S*-adenosyl homocysteine may then release adenine for recycling (Figure 18). *S*-

Adenosylethionine, however, does not so readily donate its ethyl group and therefore traps the adenine moiety.[14]

The result of this is a depletion of ATP[14] resulting in a reduction in protein synthesis,[100] which in turn leads to a deficiency in the production of the apolipoprotein complex which transports lipids out of the liver.[14,99,100] There is consequently an accumulation of triglycerides. Plasma levels of triglycerides, cholesterol, lipoprotein, and phospholipid are in turn decreased.[99] Ethionine also has other biochemical effects, such as inhibition of RNA polymerase and synthesis, which will also influence protein synthesis.[100] As with a number of other compounds, inhibition of protein synthesis is not sufficient to kill the cell but is sufficient to lead an accumulation of lipid following reduced lipid transport out of the cell. Thus, the liver is the target organ in this example because it is here that lipid mobilization, transport, and metabolism occur[3,6,95,101] and interference with transport out of the liver leads to accumulation.

VI. VASCULAR DAMAGE

A. Specific Example

1. Pyrrolizidine Alkaloids

Certain toxic substances may cause damage to the vascular system of the liver as well as to the hepatocyte.

An example of this type of liver injury is the veno-occlusive disease caused by pyrrolizidine alkaloids.[14,102,103] This occurs after acute doses, and after chronic dosing cirrhosis[14] and carcinoma develop.[14,104] The overall picture therefore is of necrosis of the sinusoid and endothelial cells of a central vein which results in swelling and closure of the lumen, detectable 6 hr after dosing.[103,105] This leads to a reduction in liver blood flow. Centrilobular hepatic necrosis is detectable 12 hr after dosing in experimental animals.[103] The pathology then progresses to blockage of the venous return from the liver via the central and sublobular veins 2 to 5 days after dosing.[103] Seven to ten days postdosing there is collagenous veno-occlusion of central and sublobular veins.[103] This blockage in the hepatic vascular outflow system leads to the development of venous collaterals.[103]

These events occur in a variety of experimental animals and man.[102] Human exposure normally results either from ingestion of cereal crops contaminated with plants of the *crotolaria, heliotropium* and *senecio* species, or the use of such plants as traditional herbal medicines.[103] The lower level exposure via contaminated cereal crops may be chronic and lead to liver cirrhosis. Larger doses, as ingested in traditional medicines, may lead to veno-occlusive disease.[103] The cirrhosis which may develop later is characteristic as it spares the portal triads and therefore appears as reverse lobulation with the portal triad at the "center" of the lobule.[103]

The biochemical effects of pyrrolizidine alkaloids can be detected very shortly after dosing.[14] Thus, protein and RNA synthesis are decreased by 15 min.[14,103,106,107] Cell division is affected and this leads to the development of abnormally large cells or megalocytes.[102,108]

The mechanism of toxicity involves metabolism of the pyrrolizidine to the corresponding pyrrole.[103] This involves a microsomal enzyme-mediated dehydrogenation (Figure 19). Requirements for toxicity of pyrrolizidine alkaloids are a 1,2 double bond, and esterification of one or more hydroxyl groups, with a branched chain acid involved in the esterification (Figure 19).[103] Various metabolic routes may occur, such as N-oxidation, and possibly epoxidation and dehydrogenation.[103] The levels of pyrrole in the liver correlate with the toxicity,[109] and in vitro only rat liver microsomes are capable of producing significant quantities of the pyrrolic metabolite from the corresponding pyrrolizidine.[103,110] Furthermore, there is a correlation between species variation in susceptibility and ability to produce pyrrolic metabolites.[103,111] The pyrroles themselves when injected are toxic to a variety of organs in

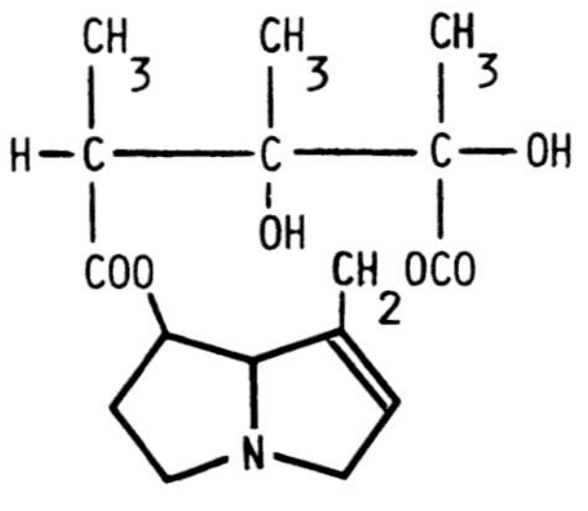

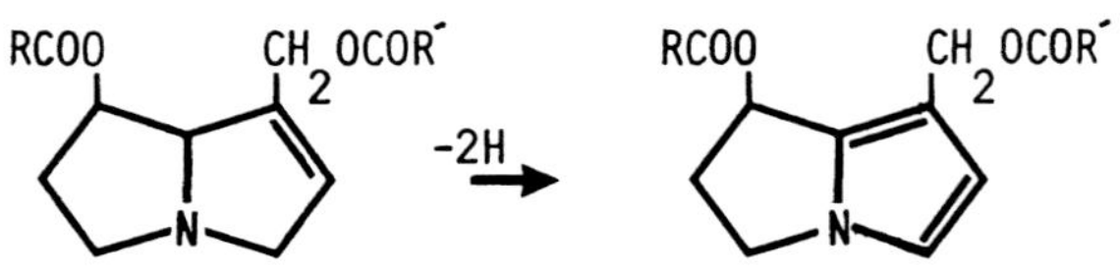

FIGURE 19. Structure of the pyrrolizidine alkaloid monocrotaline and the microsomal enzyme-mediated metabolic activation of the pyrrolizidine alkaloid nucleus.

both species which are susceptible and nonsusceptible to pyrrolizidine alkaloids.[103] Clearly, then, the toxicity of the alkaloids depends on metabolism to pyrrolic metabolites in the liver *in situ*.[103] The pyrrolic compounds react with DNA and possibly proteins[103,112] and are antimitotic.[103]

The liver is a target organ in this example again because of metabolism of the alkaloids to toxic pyrrolic intermediates *in situ*.[103] These pyrrolic metabolites are stable enough to be isolated and therefore also seem able to leave the hepatocyte and damage the hepatic vascular system as well as the hepatocyte. This leads to the sinusoidal and venous endothelial necrosis as well as the centrilobular necrosis. Other organs do not seem to produce significant quantities of the pyrrolic metabolites yet are susceptible to the toxic effects of these pyrroles if they are injected.[103] Thus, intravenous injection of the pyrrolic metabolites may give rise to lung necrosis and necrosis of the walls of small veins.[103] It is clear, therefore, that production of the reactive pyrrolic metabolites in the liver is the reason for this organ being susceptible.

In this example, however, the reactive metabolites although produced in the hepatocyte, damage other cells as well, notably the endothelial cells of the sinusoidal and central vein. This reflects the ability of the metabolites, presumably the pyrrolic products, to diffuse out of the hepatocyte and come into contact with the vascular and sinusoidal epithelial cells.

VII. SUMMARY

The liver is a target organ for toxic compounds because of its position in the organism, metabolic capabilities, and secretory and excretory functions.

The majority of foreign/toxic compounds enter the organism via the oral route, and the liver is therefore exposed to relatively high concentrations before other parts of the body. The liver has many enzymes specifically able to metabolize such compounds, but this metabolism may increase their toxicity.

The structure and function of the liver are partly responsible for its being a target organ and the types of damage reflect this. The large metabolic capability both for foreign compounds and endogenous substances is one factor in toxic liver damage. The sinusoidal structure allows efficient extraction of foreign compounds from the blood. The biliary system may excrete potentially toxic molecules and this function may also be disrupted by foreign substances. The metabolism, synthesis, and transport of lipids which take place in the liver may be interrupted by toxic compounds. The types of hepatic damage which stem from the structure and function such as cytotoxicity, cholestasis, and steatosis have been discussed with regard to examples and mechanisms.

REFERENCES

1. **Davis, M. and Williams, R.,** Hepatic disorders, in *Textbook of Adverse Drug Reactions,* Davis, D. M., Ed., Oxford University Press, London, 1977, 146.
2. **Shand, D. G., Evans, G. H., and Nies, A. A.,** The almost complete hepatic extraction of propranolol during intravenous administration in the dog, *Life Sci.,* 10(1), 1417, 1971.
3. **Jones, A. C. and Spring-Milles, E.,** The liver and gall bladder, in *Histology,* 4th ed., Weiss, L. and Creep, R. O., Ed., McGraw Hill, New York, 1977, 701.
4. **Campra, J. L. and Reynolds, T. B.,** The hepatic circulation, in *The Liver — Biology and Pathobiology,* Arias, I. M., Popper, H., Schacter, D., and Shafritz, D. A., Eds., Raven Press, New York, 1982, 627.
5. **Rappaport, A. M.,** Anatomic considerations, in *Diseases of the Liver,* 3rd ed., Schiff, L., Ed., J. B. Lippincott, Philadelphia, 1979, 1.
6. **Rappaport, A. M.,** Physioanatomical basis of toxic liver injury, in *Toxic Injury of the Liver,* Part A, Farber, E. and Fisher, M. M., Eds., Marcel Dekker, New York, 1979, 1.
7. **Arias, I. M., Popper, H., Schacter, D., and Shafritz, D. A.,** *Sinusoidal Cells in the Liver — Biology and Pathobiology,* Arias, I., Popper, H., Schacter, D., and Shafritz, D. A., Eds., Raven Press, New York, 1982, 493.
8. **Jungermann, K. and Katz, N.,** Metabolic heterogeneity of liver parenchyma, in *Metabolic Compartmentation,* Sies, H., Ed., Academic Press, London, 1982, 411.
9. **Yashimura, S., Komatsu, N., and Watanabe, K.,** Purification and immunohistochemical localization of rat-liver glutathione peroxidase, *Biochem. Biophys. Acta,* 621, 130, 1980.
10. **Smith, M. T., Loveridge, N., Wills, E. D., and Chayen, J.,** The distribution of glutathione in the rat liver lobule, *Biochem. J.,* 182, 103, 1979.
11. **van Berkel, T. J. C.,** Functions of hepatic non-parenchymal cells, in *Metabolic Compartmentation,* Sies, H., Ed., Academic Press, London, 1982, 437.
12. **Govier, W. C.,** Reticuloendothelial cells as the site of sulfanilamide acetylation in the rabbit, *J. Pharmacol. Exp. Ther.,* 150, 305, 1965.
13. **Suolinna, E.-M.,** Metabolism of sulphanilamide and related drugs in isolated rat and rabbit liver cells, *Drug Metab. Dispos.,* 8, 205, 1980.
14. **Zimmerman, H. J.,** *Hepatotoxicity: The Adverse Effects of Drugs and Other Chemicals on the Liver,* Appleton Century Crofts, New York. 1978.
15. **Recknagel, R. O. and Glende, E. A.,** Carbon tetrachloride hepatotoxicity: an example of lethal cleavage, *CRC Crit. Rev. Toxicol.,* 2, 263, 1973.
16. **Mitchell, J. R., Jollow, D. J., Potter, W. Z., Davis, D. E., Gillette, J. R., and Brodie, B. B.,** Acetaminophen-induced hepatic necrosis. I. Role of drug metabolism, *J. Pharmacol. Exp. Ther.,* 187, 185, 1973.
17. **Piazza, J. G.,** Zur kenntnis der wirkung der allylverbindungen, *Z. Exp. Pathol. Ther.,* 17, 318, 1915.
18. **Evans, M. A. and Harbison, R. D.,** Cocaine induced hepatic necrosis in mice, *Toxicol. Appl. Pharmacol.,* 45. 739, 1978.
19. **Keppler, D., Lesch, R., Reutter, W., and Dekker, K.,** Experimental hepatitis induced by D-galactosomine, *Exp. Mol. Pathol.,* 9, 279, 1968.
20. **Capelle, P., Dhumeaux, D., Mora, M., Feldman, G., and Berthelot, P.,** Effect of rifampicin on liver function in man, *Gut,* 13, 366, 1972.
21. **Kenright, S. and Levi, A. J.,** Sites of competition in the selective uptake of rifampcin-SV, flavaspidic acid, bilirubin and bromosulphathalein, *Gut,* 15, 220, 1974.
22. **Heikel, T. A., Knight, B. C., Rimington, C., Richie, H. D., and Williams, E. J.,** Studies on biliary excretion in the rabbit. I. The effect of icterogenin and rehmannic acid on bile flow and the excretion of

bilirubin, phylloerythrin, coprophorophyrin, alkaline phosphatase and bromosulphalein, *Proc. R. Soc. London Ser. B:*, 153, 47, 1960.
23. **Wheeler, H. O.,** Secretion of bile, in *Diseases of the Liver,* 4th ed., Schiff, L., Ed., J. B. Lippincott, Philadelphia, 1975, 87.
24. **Scales, M. D. C. and Timbrell, J. A.,** Hydrazine hepatotoxicity. I. Pathological findings, *J. Toxicol. Environ. Health,* 10, 941, 1982.
25. **Lullman, H., Lullman-Rauch, R., and Wasserman, O.,** Drug-induced phospholipidoses, *CRC Crit. Rev. Toxicol.,* 4, 185, 1975.
26. **Neuberger, J. M., Dais, M., and Williams, R.,** Clinical aspects of oral contraceptive associated liver tumours, in *Drug Reactions and the Liver,* Davis, M., Tredger, J. M., and Williams, R., Eds., Pitman, London, 1981, 271.
27. **Wogan, G. N.,** Aflatoxins and their relationship to hepatocellular carcinoma, in *Hepatocellular Carcinoma,* Okuda, K. and Peters, R. L., Eds., J. Wiley & Sons, New York, 1976, 25.
28. **Heath, C. W., Falk, H., and Creech, J. L.,** Characteristics of cases of angiosarcoma of the liver among vinyl chloride workers in the United States, *Ann. N.Y. Acad. Sci.,* 246, 231, 1975.
29. **McLean, E. K. and Mattocks, A. R.,** Environmental liver injury: plant toxins, in *Toxic Injury of the Liver,* Farber, E. and Fisher, M. M., Eds., Marcel Dekker, New York, 1980, 517.
30. **Sterup, K. and Mosbech, J.,** Budd-Chiari syndrome after taking oral contraceptives, *Br. Med. J.,* 4, 660, 1967.
31. **Robinson, D. S. and Seakins, A.,** The development in the rat of fatty livers associated with reduced plasma lipoprotein synthesis, *Biochim. Biophys. Acta,* 62, 163, 1962.
32. **Farber, E.,** Ethionine fatty liver, *Adv. Lipid Res.,* 5, 119, 1967.
33. **Judah, J. D., Ahmed, K., McLean, A. E. M., and Christie, G. S.,** Ion transport in ethionine intoxication, *Lab. Invest.,* 15, 167, 1966.
34. **Farber, J. L.,** Reactions of the liver to injury: necrosis, in *Toxic Injury to the Liver,* Farber, E. and Fisher, M. M., Eds., Marcel Dekker, New York, 1980, 215.
35. **Farber, J. L. and Abrams, J.,** Unpublished observations in chapter, "Reactions of the liver to injury: necrosis", in *Toxic Injury to the Liver,* Farber, E. and Fisher, M. M., Eds., Marcel Dekker, New York, 1980, 215.
36. **Schanne, F. A. and Farber, J. L.,** Unpublished observations in chapter, "Reactions of the Liver to Injury: Necrosis", in *Toxic Injury to the Liver,* Farber, E. and Fisher, M. M., Eds., Marcel Dekker, New York, 1980, 215.
37. **Slater, T. F.,** *Free Radicals in Tissue Injury,* Pion, London, 1972.
38. **Gillette, J. R., Mitchell, J. R., and Brodie, B. B.,** Biochemical mechanisms of drug toxicity, *Ann. Rev. Pharmacol.,* 14, 271, 1974.
39. **Slater, T. F.,** Biochemical studies on liver injury, in *Biochemical Mechanisms of Liver Injury,* Slater, T. F., Ed., Academic Press, London, 1978, 1.
40. **Recknagel, R. O.,** Carbon tetrachloride hepatotoxicity: status quo and future prospects, *Trends Pharm. Sci.,* March 1983, 129.
41. **Woodward, K. N. and Timbrell, J. A.,** Acetylhydrazine hepatotoxicity: the role of covalent binding, *Toxicology,* 30, 65, 1984.
42. **Boyd, M., Statham, C. N., and Longo, N. S.,** The pulmonary Clara cell as a target for toxic chemicals requiring metabolic activation: studies with carbon tetrachloride, *J. Pharmacol. Exp. Ther.,* 212, 109, 1980.
43. **Rouiller, Ch.,** Experimental toxic injury of the liver, in *The Liver,* Vol. 2, Rouiller, Ch., Ed., Academic Press, New York, 1964, 335.
44. **Cameron, G. R. and Karunaratne, W. A. E.,** Carbon tetrachloride cirrhosis in relation to liver degeneration, *J. Pathol. Bacteriol.,* 42, 1, 1936.
45. **Reynolds, E. S. and Moslen, M. T.,** Environmental liver injury: halogenated hydrocarbons, in *Toxic Injury to the Liver,* Farber, E. and Fisher, M. M., Eds., Marcel Dekker, New York, 1980, 541.
46. **Gooding, P. E., Chayen, J., Sawyer, B., and Slater, T. F.,** Cytochrome P-450 distribution in rat liver and the effect of sodium phenobarbitone administration, *Chem. Biol. Interact.,* 20, 299, 1978.
47. **Reynolds, E. S. and Ree, H. J.,** Liver parenchymal cell injury. VII. Membrane denaturation following carbon tetrachloride, *Lab. Invest.,* 25, 269, 1971.
48. **Gordis, E.,** Lipid metabolites of carbon tetrachloride, *J. Clin. Invest.,* 48, 203, 1969.
49. **Slater, T. F.,** Free radicals as reactive intermediates in tissue injury, in *Biological Reactive Intermediates. II. Chemical Mechanisms and Biological Effects, Part A,* Snyder, R., Parke, D. V., Kocsis, J. J., Jollow, D. J., Gibson, G. G., and Witmer, C. M., Eds., Plenum Press, New York, 1982, 575.
50. **Kubic, V. L. and Anders, M. W.,** Metabolism of carbon tetrachloride to phosgene, *Life Sci.,* 26, 2151, 1980.

51. **Jollow, D. J., Mitchell, J. R., Zampaglione, W., and Gillette, J. R.,** Bromobenzene-induced hepatic necrosis. Protective role of glutathione and evidence for 3,4-bromobenzene oxide as the hepatotoxic metabolite, *Pharmacology,* 11, 151, 1974.
52. **Timbrell, J. A., Mitchell, J. R., Snodgrass, W. R., and Nelson, S. D.,** Isoniazid hepatotoxicity: the relationship between covalent binding and metabolism in vivo, *J. Pharmacol. Exp. Ther.,* 213, 364, 1980.
53. **Mitchell, J. R., Nelson, W. L., Potter, W. Z., Sasame, H. A., and Jollow, D. J.,** Metabolic activation of furosemide to a chemically reactive hepatotoxic metabolite, *J. Pharmacol. Exp. Ther.,* 199, 41, 1976.
54. **Mitchell, J. R., Potter, W. Z., Hinson, J. A., Snodgrass, W. R., Timbrell, J. A., and Gillette, J. R.,** Toxic drug reactions, in *Concepts in Biochemical Pharmacology, Part 3,* Vol. 28, Gillette, J. R. and Mitchell, J. R., Eds., Springer-Verlag, Berlin, 1975, 383.
55. **Potter, W. Z., Jollow, D. J., and Mitchell, J. R.,** Unpublished data, in *Concepts in Biochemical Pharmacology, Part 3,* Vol. 28, Gillette, J. R. and Mitchell, J. R., Eds., Springer-Verlag, Berlin, 1975, 394.
56. **Spitznagle, L. A., Wirth, P. J., Boobis, S. W., Thorgiersson, S. S., and Nelson, W. L.,** The role of biliary excretion in the hepatotoxicity of furosemide in the mouse, *Toxicol. Appl. Pharmacol.,* 39, 283, 1977.
57. **Smith, D. A.,** Differences in toxicity due to species variation in the metabolism of an oral anti-allergy agent, *Br. J. Pharmacol.,* 66, 422P, 1979.
58. **Reid, W. D.,** Mechanism of allyl alcohol-induced hepatic necrosis, *Experientia,* 28, 1058, 1972.
59. **Rees, R. K. and Tarlow, M. J.,** The hepatotoxic action of allyl formate, *Biochem. J.,* 104, 757, 1967.
60. **Greenberger, N. J., Cohen, R. B., and Isselbacher, K. J.,** The effect of chronic ethanol administration on liver alcohol dehydrogenase activity in the rat, *Lab. Invest.,* 14, 264, 1965.
61. **Morrison, G. R. and Brock, F. E.,** Quantitative measurement of alcohol dehydrogenase in the lobule of normal livers, *J. Lab. Clin. Med.,* 70, 116, 1967.
62. **Patel, J. M., Gordon, W. P., Nelson, S. D., and Liebman, K. C.,** Comparison of hepatic biotransformation and toxicity of allyl alcohol and [1,1-$^2H_2$] allyl alcohol in rats, *Drug Metab. Dispos.,* 11, 164, 1983.
63. **Strunin, L.,** *The Liver and Anaesthesia,* W. B. Saunders, London, 1977.
64. **Trowell, J., Peto, R., and Smith, A. C.,** Controlled trial of repeated halothane anaesthesias in patients with carcinoma of the uterine cervix treated with radium, *Lancet,* 2, 821, 1975.
65. **Timbrell, J. A.,** Drug hepatotoxicity, *Br. J. Clin. Pharmacol.,* 15, 3, 1983.
66. **Sipes, I. G., Gandolfi, A. J., and Brown, B. R.,** Halothane associated hepatitis: evidence of direct toxicity, in *Drug Reactions and the Liver,* Davis, M., Tredger, J. M., and Williams, R., Eds., Pitman, London, 1981, 219.
67. **Maiorino, R. M., Sipes, I. G., Gandolfi, A. J., and Brown, B. R.,** Quantitative analysis of volatile halothane metabolites in biological tissues by gas chromatography, *J. Chromatogr.,* 164, 63, 1979.
68. **Neuberger, J. and Davis, M.,** Advances in understanding halothane hepatitis, *Trends Pharm. Sci.,* January 1983, 19.
69. **Neuberger, J. M., Tredger, J. M., and Davis, M.,** Oxidative metabolism of halothane in the production of hepatocyte neoantigens, in *Drug Reactions and the Liver,* Davis M., Tredger, J. M., and Williams, R., Eds., Pitman, London, 1981, 245.
70. **Decker, K. and Keppler, D.,** Galactosamine-induced liver injury, in *Progress in Liver Disease,* Vol. 4, Popper, H. and Schaffner, F., Eds., Grune & Stratton, New York, 1972, 183.
71. **Miller, E. C. and Miller, J. A.,** Hepatocarcinogenesis by chemicals, in *Progress in Liver Disease,* Vol. 5, Popper, H. and Schaffner, F., Eds., Grune & Stratton, New York, 1972, 699.
72. **Phillips, M. J., Oda, M., and Furatsu, K.,** Reactions of the liver to injury: cholestasis, its ultrastructural aspects, in *Toxic Injury of the Liver,* Farber, E. and Fisher, M. M., Eds., Marcel Dekker, New York, 1979, 333.
73. **Arias, I. M., Fleischner, G., and Listowsky, I.,** On the structure and function of ligandin and Z protein, in *The Hepatobiliary System: Fundamental and Pathological Mechanisms,* Taylor, W., Ed., Plenum Press, New York, 1975, 81.
74. **Capelle, P., Dhumeaux, D., Mora, M., Feldman, G., and Berthelot, P.,** Effect of rifampicin on liver function in man, *Gut,* 13, 366, 1972.
75. **Hargreaves, T. and Lathe, G. H.,** Inhibitory aspects of bile secretion, *Nature (London),* 200, 1172, 1963.
76. **Slater, T. F. and Eakins, M. N.,** Biochemical studies on bile secretion, in *The Hepatobiliary System: Fundamental and Pathological Mechanisms,* Taylor, W., Ed., Plenum Press, New York, 1975, 61.
77. **Popper, H., Schaffner, F., and Denk, H.,** Molecular pathology of cholestasis, in *The Hepatobiliary System: Fundamental and Pathological Mechanisms,* Taylor, W., Ed., Plenum Press, New York, 1975, 605.
78. **Phillips, M. J., Oda, M., Mak, E., and Steiner, J. W.,** Fine structure of the biliary free, in *The Hepatobiliary System: Fundamental and Pathological Mechanisms,* Taylor, W., Ed., Plenum Press, New York, 1976, 245.

79. **Palmer, R.,** Toxic effects of lithocholate on the liver and biliary tree, in *The Hepatobiliary System: Fundamental and Pathological Mechanisms,* Taylor, W., Ed., Plenum Press, New York, 1975, 227.
80. **Prandi, D., Erlinger, S., Glasinovic, J.-C., and Dumont, M.,** Canalicular bile production in man, *Eur. J. Clin. Invest.,* 5, 1, 1975.
81. **Gumucio, J. J. and Validivieso, V. D.,** Studies on the mechanism of the ethynylestradiol impairment of bile flow and bile salt excretion in the rat, *Gastroenterology,* 61, 339, 1971.
82. **Wheeler, H. O.,** Secretion of bile, in *Diseases of the Liver,* 4th ed., Schiff, L., Ed., J. B. Lippincott, Philadelphia, 1975, 87.
83. **Samuels, A. M. and Caney, M. C.,** Effects of chlorpromazine (CPZ) and its metabolites on Na^+-K^+-ATPase activity of canalicular enriched liver plasma membranes, *Gastroenterology,* 70, 933, 1976.
84. **Schaffner, F. and Raisfeld, I. H.,** Drugs and the liver: a review of metabolism and adverse reactions, *Adv. Intern. Med.,* 15, 221, 1969.
85. **Plaa, G. L.,** Hyperbilirubinemia and cholestases, a different form of liver injury produced in animals, *Essays Toxicol.,* 2, 137, 1970.
86. **Capizzo, F. and Roberts, R. J.,** α-Naphthylisothiocyanate (ANIT)-induced hepatotoxicity and disposition in various species, *Toxicol. Appl. Pharmacol.,* 19, 176, 1971.
87. **Ungar, H., Moran, E., Eisner, M., and Eliakim, M.,** Rat intrahepatic biliary tract lesions from alpha-naphthylisothiocyanate, *Arch. Pathol.,* 73, 427, 1962.
88. **McLean, M. R. and Rees, K. R.,** Hyperplasia of bile ducts induced by alpha-naphthyl-iso-thiocyanate: experimental biliary cirrhosis free from biliary obstruction, *J. Pathol. Bacteriol.,* 76, 175, 1958.
89. **Zimmerman, H. J.,** Experimental hepatotoxicity, in *Experimental Production of Disease, Part 5, Liver: Handbook of Experimental Pharmacology,* Vol. 16, Eichler, O., Ed., Springer-Verlag, Berlin, 1976, 1.
90. **Steiner, J. W. and Baglione, M.,** Electron microscopy of the cytoplasm of parenchymal liver cells in α-naphthylisothiocyanate induced cirrhosis, *Lab. Invest.,* 12, 765, 1963.
91. **Goldfarb, S., Singer, K., and Popper, H.,** Experimental cholangitis due to alpha-naphthylisothiocyanate (ANIT), *Am. J. Pathol.,* 40, 685, 1962.
92. **Plaa, G. L.,** Toxic responses of the liver, in *Cassarett and Doull's Toxicology: The Basic Science of Poisons,* 2nd ed., Doull, J., Klassen, C. D., and Amdur, M. O., Eds., Macmillan, New York, 1980, 206.
93. **Indacochea-Redmond, N., Witschi, H. P., and Plaa, G. L.,** Effects of inhibitors of protein and ribonucleic acid synthesis on α-naphthylisothiocyanate induced hyperbilirubinaemia, sulphobromophthalein retention and prolongation of pentobarbital hypnosis, *J. Pharmacol. Exp. Ther.,* 189, 278, 1974.
94. **Indacochea-Redmond, N., Witschi, H. P., and Plaa, G. L.,** Effect of inhibitors of protein and ribonucleic acid synthesis on the hyperbilirubinemia and cholestasis produced by α-naphthylisothiocyanate, *J. Pharmacol. Exp. Ther.,* 184, 780, 1973.
95. **Dianzani, M. U.,** Reactions of the liver to injury: fatty liver, in *Toxic Injury to the Liver,* Farber, E. and Fisher, M., Eds., Marcel Dekker, New York, 1979, 281.
96. **Lieber, C. S.,** Effects of ethanol on lipid metabolism, *Lipids,* 9, 103, 1974.
97. **Patrick, R. L. and Back, K. C.,** Pathology and toxicology of repeated doses of hydrazine and 1,1-dimethylhydrazine in monkeys and rats, *Ind. Med. Surg.,* 34, 430, 1965.
98. **Farber, E., Simpson, M. V., and Tarver, H.,** Studies on ethionine. II. The interference with lipid metabolism, *J. Biol. Chem.,* 182, 91, 1950.
99. **Farber, E.,** Ethionine fatty liver, *Adv. Lipid Res.,* 5, 119, 1967.
100. **Farber, E.,** The pathology of translation, in *Biochemistry of Disease,* Vol. 2, *The Pathology of Transcription and Translation,* Farber, E., Ed., Marcel Dekker, New York, 1972, 123.
101. **Farber, E.,** The liver in protein and lipid metabolism, in *The Liver,* Gall, E. A. and Mostofi, F. K., Eds., Williams & Wilkins, Baltimore, 1973, 121.
102. **McLean, E. K.,** The toxic actions of pyrrolidizine (Senecio) alkaloids, *Pharmacol. Rev.,* 22, 429, 1970.
103. **McLean, E. K. and Mattocks, A. R.,** Environmental liver injury: plant toxins, in *Toxic Injury to the Liver,* Farber, E. and Fisher, M. M., Eds., Marcel Dekker, New York, 1980, 517.
104. **Schoental, R.,** Toxicology and carcinogenic action of pyrrolizidine alkaloids, *Cancer Res.,* 28, 2237, 1968.
105. **Allen, J. R., Carstens, L. A., and Katagiri, G.,** Hepatic veins of monkeys with veno-occlusive disease, *Arch. Pathol.,* 87, 279, 1969.
106. **Harris, C., Reddy, J., Chiga, M., and Svoboda, D.,** Polyribosomal disaggregation in rat liver by lasiocarpine, *Biochem. Biophys. Acta,* 182, 587, 1969.
107. **Warwick, G. P.,** Metabolism of liver carcinogens and other factors influencing liver cancer induction, in *Liver Cancer,* WHO/IARC Publ. No. 1, International Agency for Research on Cancer, Lyon, 1971, 121.
108. **Jago, M. V.,** The development of the hepatic megalocytosis of chronic pyrrolizidine poisoning, *Am. J. Pathol.,* 56, 405, 1969.
109. **Mattocks, A. R. and White, I. N. H.,** The conversion of pyrrolizidine alkaloids to N-oxides and to dihydropyrrolizine derivatives by rat liver microsomes *in vitro, Chem. Biol. Interact.,* 3, 383, 1971.

110. **Mattocks, A. R.,** Acute hepatotoxicity and pyrrolic metabolites in rats dosed with pyrrolidizine alkaloids, *Chem. Biol. Interact.*, 5, 277, 1972.
111. **White, I. N. H., Mattocks, A. R., and Butler, W. H.,** The conversion of the pyrrolizidine alkaloid retrorsine to pyrrolic derivatives *in vivo* and *in vitro* and its acute toxicity to various animal species, *Chem. Biol. Interact.*, 6, 207, 1973.
112. **Hsu, J. C., Robertson, R. C., Shumaker, R. C., and Allen, J. R.,** Binding of tritiated dehydroretronecine, *Res. Commun. Chem. Pathol. Pharmacol.*, 11, 99, 1975.

Chapter 9

THE NERVOUS SYSTEM AS A TARGET ORGAN FOR TOXICITY

R. J. Anderson, R. J. Richardson, and B. W. Schwab

TABLE OF CONTENTS

I. INTRODUCTION

Neurotoxicology is the study of adverse effects of chemicals on the nervous system. As with any other organ system, there are three basic types of changes that can occur following chemical insult: biochemical, structural (morphological), and functional. These three levels of change in the tissue provide three different approaches to the assessment of toxicity in the system. A flow chart indicating possible sequences and interrelationships of events in the neurotoxic process is shown in Figure 1. It should be noted that the same fundamental principles of toxicity to any organ apply to this description of neurotoxicity. However, there are pecularities of biochemistry, geometry, and function of nervous tissue, rendering this system uniquely susceptible to certain toxic agents and requiring specialized techniques for the measurement and assessment of toxicity[25,101,118] as indicated below:[101]

1. Neurotoxicity can produce a wide range of signs and symptoms covering the spectrum from such things as subtle behavioral or psychiatric changes to profound movement disorders, sensory loss, or paralysis. The probability of detecting a toxic effect in the nervous system is, therefore, often greater than that for detecting toxicity in other organ systems, because many of the signs or symptoms will be apparent either to an external observer or to the affected individual.
2. The central nervous system (CNS) has a high metabolic requirement for oxygen and utilizes glucose almost exclusively as its fuel source. Therefore, the CNS will be uniquely sensitive to any chemical that alters oxygen or glucose availability or utilization, e.g., carbon monoxide, cyanide, or insulin.
3. The nervous system has specific targets either not found in other organ systems or not as vital to maintenance of function. These include neurotransmitter systems (synthetic and degradative enzymes, storage vesicles, uptake and release mechanisms, and receptors), ion channels, and axonal transport systems.
4. The elongated structure of individual neurons renders them sensitive to toxic insult in that repair of structures distant from the cell body of the neuron will be retarded due to the delay incurred by the requirement to transport necessary materials to the injured site.
5. Generally, neurons do not undergo cell division. Consequently, killed neurons are not replaced by functional units. This is in marked contrast to toxicity in regeneratable organ systems, e.g., liver, kidney, skin, or gut.
6. Regeneration of damaged axons in the CNS arising from surviving neuronal cell bodies is seldom effective. In the peripheral nervous system (PNS), regeneration of axons is slow and often imperfect.
7. Due to the integrative function of the nervous system, it is called upon to coordinate actions and sensory inputs throughout the entire body, and to perform associative functions as well as learning and memory within itself. Consequently, a range of approaches is needed to detect and properly evaluate neurotoxicity. These approaches range across studies of behavior, neuromuscular function, electrophysiological parameters, neurochemistry, and morphology. As we will see, it is sometimes possible to integrate findings from diverse areas into a coherent picture, but, more often than not, we are left with isolated pieces of information that do not allow us to say why the nervous system or one of its components was the selective target of a given toxicant.

Referring to Figure 1, we see that exposure to a xenobiotic chemical is followed by its absorption, distribution, and possible biotransformation. Some of the material may be kept clear of the nervous system through excretion, storage, or metabolic inactivation. The xenobiotic and/or possible metabolites may then enter the nervous system; the extent of entry

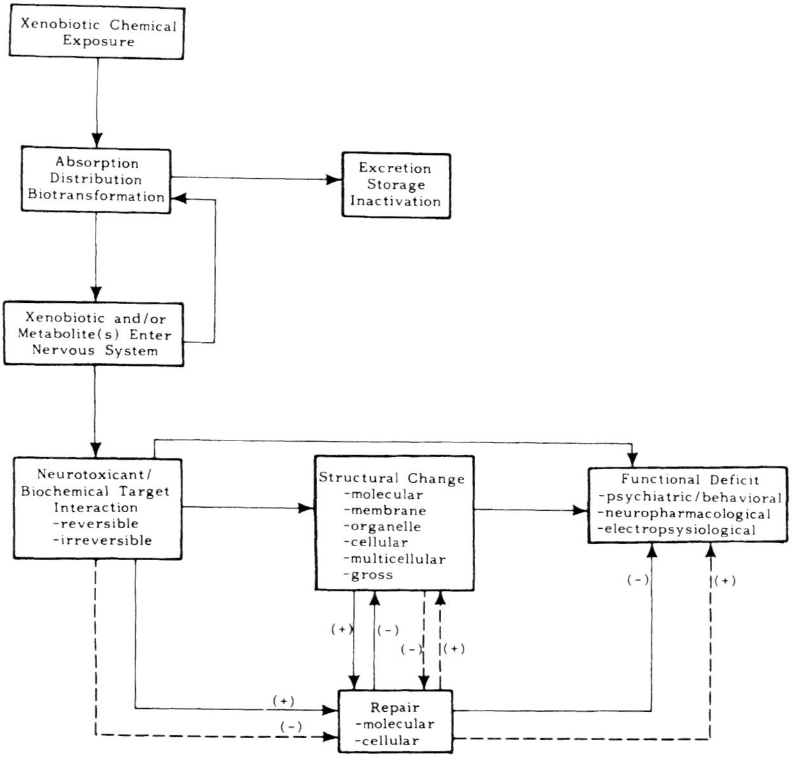

FIGURE 1. Flow chart depicting the major flow of events in neurotoxicity. Dashed or continuous lines indicate, respectively, distinct pathways to be followed in the diagram. (+) Indicates an increase or stimulation; (−) indicates a decrease or depression.

will be dependent upon the effectiveness of the molecules of concern in penetrating the blood-brain and blood-nerve barriers. The initial event that leads to eventual observable toxicity must necessarily be biochemical, although we are often unable to detect the earliest relevant interaction between the neurotoxicant and its biochemical target. Exceptions to this include inhibition of acetylcholinesterase leading to the acute manifestations of organophosphorus ester toxicity and the phosphorylation and aging of neurotoxic esterase leading to delayed organophosphorus neuropathy.[32] Other promising lines of biochemical investigation include the formation of pyrrole rings and protein crosslinks in γ-diketone neuropathies.[5] The initial molecular perturbation of the system may be reversible or irreversible, but in either event the molecular-level change can lead directly to functional deficits or to structural changes revealed to the pathologist at the electron or light microscopic, or gross level. Some morphological changes at the organelle or membrane level may be discernible by electron microscopy, but other subtle structural alterations and changes at the molecular level require biophysical and biochemical techniques for their detection.[70,101] The flow chart includes the possibility of bypassing structural changes on the way to functional deficits, but if we include structural changes at the molecular level this direct pathway does not really exist.

What we are saying here is that any toxic interaction involves a structural change, even if only at a molecular level. One distinction that can be made is one of the detectability of the change (e.g., a conformational change in an enzyme or an ion-channel protein). It is important to realize that even if it were possible in a given case to detect a molecular-level structural change, we are not necessarily obliged to designate the change as a neurotoxic event. We may decide, for example, that if the molecular perturbation is repaired at such a rate that higher-order structural changes or functional deficits are averted, then the chemical has produced an adaptive response rather than a toxic response. On the other hand, a molecular change can be spontaneously reversible, yet the animal will suffer severe functional deficits or even death. The reactivatable inhibition of acetylcholinesterase by a carbamate insecticide is a familiar example.

The flow chart in Figure 1 has implicit differences from a similar diagram drawn for other organ systems, but the only immediately apparent explicit difference lies within the area of functional deficits — both in the nature of the functions and the methods used in their measurement. Thus, processes affecting mentation and mood might be evaluated by psychiatric studies in humans; some counterparts of these and other nervous system functions may be assessed using behavioral techniques in experimental animals.[46,97] Alterations in neuropharmacological responsiveness can occur as a result of changes in receptor number or ligand binding, which are biochemical effects. Nevertheless, the effects of such changes are often observable at the whole-animal level using behavioral methods; moreover, the in vitro assays for characterization of ligand binding to receptors often reflect the whole-animal response to well-studied neuropharmacological agents and so may be included as functional indices.[19,31,101] Therefore, to understand how and why the nervous system responds as a target organ to toxicants, Figure 1 must be placed within the context of points 1 to 7 listed above.

Electrophysiological techniques offer one of the best-suited approaches to the objective and specific assessment of nervous system function. Even though the nervous system is made up of cells containing metabolic machinery in common with cells in other organ systems, the distinguishing characteristic of the nervous system is its specialization for the transmission of electrical impulses from one cell to another, sometimes over relatively large distances.[69]

Whether a biochemical event or a structural change leads to clinical expression as a functional deficit depends upon the capacity of the organism to repair damage at either the molecular or cellular level. Figure 1 indicates various conceivable outcomes when repair processes are considered. (Solid lines going in the same direction in the figure are meant to be followed through, as are dotted lines. A (+) sign indicates a stimulation or increase; a (−) sign indicates a depression or decrease.) For example, it is conceivable that the initial biochemical interaction would have a stimulatory effect on repair (solid arrow; +): this would result in a decrease in functional deficit (solid arrow; −). Conversely, it is also possible that an overwhelming biochemical insult would depress repair mechanisms (dotted arrow; −), resulting in an increase in functional deficits (dotted arrow; +). The same arguments apply for the effects of structural changes on repair and vice-versa. The details of molecular repair mechanisms in the nervous system are only beginning to be approached with sufficiently sophisticated techniques to eventually lead to their elucidation. Repair on the cellular level, including such processes as axonal sprouting and regeneration has been described quite well, but much remains to be discovered about the factors governing these processes.[28,51,89] Moreover, as pointed out above, repair and regeneration are limited, particularly within the CNS. This chapter focuses on morphological and electrophysiological approaches to the assessment of neurotoxicity, although it is appreciated that mechanistic understanding will depend on molecular approaches, especially those that are currently under development.[109]

II. MORPHOLOGICAL ASPECTS

The current value of morphological examination is in providing an end-point in screening and mechanistic studies of neurotoxicity. Morphological examination of the nervous system can range in sophistication from gross examination of postmortem or biopsy specimens to stereologic quantitation of fine structure. Thus, while the "sensitivity" of morphological assessment in detecting neurotoxicity has been debated, the degree of detection obtained from morphological study is, in fact, variant according to the etiology of the toxic effect and the particular type of methodology the investigator chooses to utilize. There are many effects of chemicals on the nervous system not associated with alteration in structure. The highly specific and transient actions of local anesthetics and certain pesticides on ion channels are in this category.[86] In contrast, some chemicals may cause structural damage so florid that even gross examination of specimens is adequate. Such was the case in experimental production of nervous system tumors with methyl- and ethylnitrosourea.[30] In the situations in between if possible, methods must match the subtlety or specificity of the lesion. Morphological techniques must also be weighed against biochemical, electrophysiological,[6,83,129] and behavioral[91] end-points, and an appropriate test must be chosen according to the specific needs of the problem.

In this section, morphological techniques which have been used in neurotoxicology will be presented in brief. The appearance of normal nerve in the various preparations will be described for the purpose of contrast with examples of structural alterations caused by toxic chemicals. Should the reader require more detail than is possible in this chapter, there are excellent volumes on neurotoxicology and neuropathology that may be consulted.[8,36,119,134,135]

A. Conventions in Terminology

The nerve cell, or neuron, is typically an elongated structure, and it has become common in structural neurotoxicology to describe alterations according to the portion of the cell affected. Additionally, terms have evolved that refer to the neuron plus its associated cells. A brief review of the conventions of terminology will be provided here.

The enlarged portion of the neuron which contains the nucleus and protein synthetic apparatus is variously referred to as the perikaryon, cell body, or soma. There is a great deal of rough endoplasmic reticulum in the cell body, and neuromorphologists often refer to this structure as Nissl substance, in reference to its staining characteristics.

Dendritic processes radiate from the cell body in varying patterns, depending on the cell type. In some preparations, dendritic processes called spines can be appreciated and are believed to be one of the sites of synaptic contact between nerve cells. Dendrites carry electrical signals, but are not specialized for the propagation of an impulse without loss of amplitude, as is the axon, described below.

The axon is another elongated process that exits from the cell body. At times this structure can be enormously long (on the order of a meter in the motor neurons of the human adult sciatic nerve) and can be distinguished morphologically from dendrites by its tendency to travel in tracts (i.e., in a parallel arrangement with other axons) and its association with supporting cells. The distinguishing electrophysiological feature of the axon is its ability to generate and carry an action potential.

The supporting cell of the peripheral nervous system (PNS: the nervous system outside the brain and spinal cord) is the Schwann cell. All peripheral axons are closely associated with Schwann cells and in certain instances the Schwann cell elaborates a multilamellar set of membranes around the axon. This structure is myelin. The comparable cell of the CNS is the oligodendroglial cell. Differences in the PNS and CNS supporting cells are obvious in the electron microscope, where it can be seen that Schwann cells have a basal lamina surrounding their plasma membrane, a structure lacking in the oligodendroglial cell. Oligodendroglial cells tend to myelinate several axons whereas the Schwann cell myelinates only one. When myelin is not elaborated, both the Schwann cell and the oligodendroglial cell are associated with multiple axons. The axon plus its supporting cell is known as a fiber. Thus there are myelinated and unmyelinated fibers. In the CNS only, astroglial cells are dispersed among the fibers. Fibers are bundled in an endoneurial space delimited by a connective tissue sheath known as the perineurium. The endoneurial space contains fluid, amorphous connective tissue, and fibroblasts. In the PNS but not the CNS, longitudinally arranged collagen fibers are also contained in the endoneurial space. The bundle is called a fascicle, and a nerve, or nerve trunk, is comprised of a highly variant number of fascicles bound by a loose connective tissue sheath called the epineurium. The fascicles in a nerve are varied in size and may contain both myelinated and unmyelinated fibers.

The end of an axon is a specialized structure, and many different types exist. When an axon is carrying electrical information away from the cell body (somatofugally), the ending is distinguished by close juxtaposition to another cell. This structure is the synapse. The axonal (presynaptic) side of the synapse is generally enlarged and synaptic vesicles, presumably containing transmitter substances, can be seen with the electron microscope. A postsynaptic density can be seen when the axon synapses on another nerve cell. A specialized synapse, the motor end plate, exists at the junction between axons and muscle (see Figures 4 and 5). An electrical signal is transferred to a synapse via a motor fiber. The signal is efferent (i.e., conducting outward to an effector). When an axon carries information toward the cell body (somatopetally), the axon ending is one of a number of structures designed to transduce some sensation (e.g., taste, odor, pressure, pain, vibration) into an electrical signal. The signal is transferred from a sensory ending via a sensory fiber and is afferent (i.e., conducting inward, usually from the PNS to the CNS).

From this discussion it can be seen that a toxic neuropathy can refer to toxic effects on any of a number of cell types or their component parts. Spencer and Schaumburg[120] suggested using the terms neuronopathy (affecting the cell body and therefore the entire cell), axonopathy, and myelinopathy to confer more precision on the description of neurotoxic effects. Neuropathy can be central or peripheral or both. The terms myelopathy (neuropathy in the spinal cord), radiculopathy (neuropathy of the nerve roots entering and exiting the spinal cord), and encephalopathy (brain) may be encountered. Finally, most toxic neuropathies are polyneuropathies, inferring that the effect is on a number of nerves, as opposed to a mononeuropathy which might occur, for instance, when an individual nerve is traumatized.

B. Neuropathology Methods

1. "Thick sections": Cryostat and Paraffin Sections for the Light Microscope

Preparations of fresh tissue may be made by quick-freezing specimens in ultra-cold (approximately −150°C) isopentane and sectioning in a cryostat. Formaldehyde-fixed tissue (fixation is a general term for procedures which preserve structure; in the case of aldehydes, fixation is achieved by the cross-linking of proteins) may be embedded in paraffin and sectioned with a microtome. These techniques are quite different, but have been grouped here because they are used to produce sections which are substantially thicker (5 to 100 μm) than those generated using the procedures listed below. This has the effect of diminishing the resolving power of the light microscope so that nerve cell structure is less easily appreciated in frozen and paraffin sections. With classic hematoxylin and eosin (H & E) staining of paraffin sections, the nuclei of perikarya and glial cells are blue, and fiber tracts are present as pink fibrous material. Bickford and Sprague[15] have used H & E to observe axonal degeneration in hens following treatment with high doses of neurotoxic organophosphorus compounds. The degeneration appears as clear spots of vacuolization, and glial proliferation is obvious as inordinate numbers of nuclei are present. In longitudinal section, the clear axon tracts, sometimes with amorphous debris contained, have been called "digestion chambers" in this and other neuropathies.

Enhanced appearance of thick sections can be obtained with stains for specific structures. Of particular note are the stains for myelin (e.g., Luxol fast blue and Weigert's stain), silver impregnation for axons,[81] and histochemical stains for acetylcholinesterase, which is useful for the visualization of the neuromuscular junction.[56,63,66] The reader is directed to detailed methodologies contained in monographs on the subject, such as that by Kiernan.[64] The degeneration of myelin results in the esterification of cholesterol with fatty acids,[139] and that alteration is exploited in several special stains for degenerating myelin, clearly a useful technique for demyelination of a primary or secondary nature. For instance, oil red O stains cholesterol esters. A selective osmium staining method for degenerating myelin was developed by Marchi[80] in the 19th century and is still used. The method requires osmication in the presence of an oxidizing agent. The reduction of osmium tetroxide in the presence of fatty acids normally blackens structures such as myelin which have high lipid content. Esterified cholesterol is quite hydrophobic, while the lipids of intact myelin are relatively hydrophilic. Therefore, treatment with osmium tetroxide plus oxidizing agent blackens degenerating myelin, but the penetration of water-soluble oxidizing agent prevents the reaction in the more hydrophilic intact myelin. α-Naphthylamine can be used following the osmication to stain osmium-containing structures orange (the OTAN method).[1] An advantage of sections from frozen and paraffin-embedded tissue is that their surface area is far in excess of that which can be achieved with other preparations. Sections of the brain of small animals and transverse sections of spinal cord of large animals and humans can be obtained.

This expedites screening procedures for structural damage, as when Crapper and Tomko[29] sectioned an entire hemisphere of cat brain following intracranial injection of aluminum chloride. With a silver stain for neurons they saw swelling of cell bodies which they interpreted as neurofibrillary degeneration. The technique is also beneficial for determining patterns of damage. For example, it is now well known that trimethyltin chloride is a compound that produces signs of CNS toxicity accompanied by structural damage in several areas of the cortex.[22] However, one of the most striking findings about trimethyltin toxicity in the rat is that CNS damage is relatively restricted to specific neurons of the hippocampus. In Plates 1 and 2* this selectivity of damage is apparent because the sections are large enough to encompass the architecture of an entire cross-section of the hippocampus and appropriate stains for damaged structures are used.

Golgi preparations are brain samples fixed in potassium dichromate or potassium dichromate plus osmium tetroxide, stained with silver nitrate and cut in 100-μm thick sections. The procedure results in an apparently random staining of only a small number of nerve cells, but with the exception of myelinated axons, all portions of the cell (cell body, dendrites, axon endings) are stained dark brown. The mechanism of this selective, yet total staining of neurons is unknown.[65] Because of the small number of cells stained, the complexity of the CNS is avoided and the complete structure of a single nerve cell can be studied. Using a Golgi preparation, Norton and Culver[90] noted that the development of dendritic spines of caudate cells in carbon monoxide-treated 5-day-old rats were significantly reduced, but the ultimate number of caudate cell dendritic spines was higher in treated animals. They postulated that these findings were the morphological correlate of transient behavioral effects in these animals.

2. "Semi-Thin Sections": Plastic Embedding for the Light Microscope

Nervous tissue may be embedded in epoxy and other types of resins, colloquially known as "plastic". The increased support offered by these media allows thinner sections to be cut. Tissue for plastic embedding must be better fixed than specimens prepared for thick section. Failing this, the advantage of increased resolving power gained with the use of thinner sections is lost due to poor preservation of structure. Fixation is generally accomplished using a combination of paraformaldehyde and glutaraldehyde. (Glutaraldehyde is the superior protein fixative, but is a poor penentrator, making it necessary to use the rapidly penetrating paraformaldehyde to protect against postmortem structural changes while complete fixation is accomplished.) The preferable method, where possible, is to fix tissue by

* Color plates follow page 182.

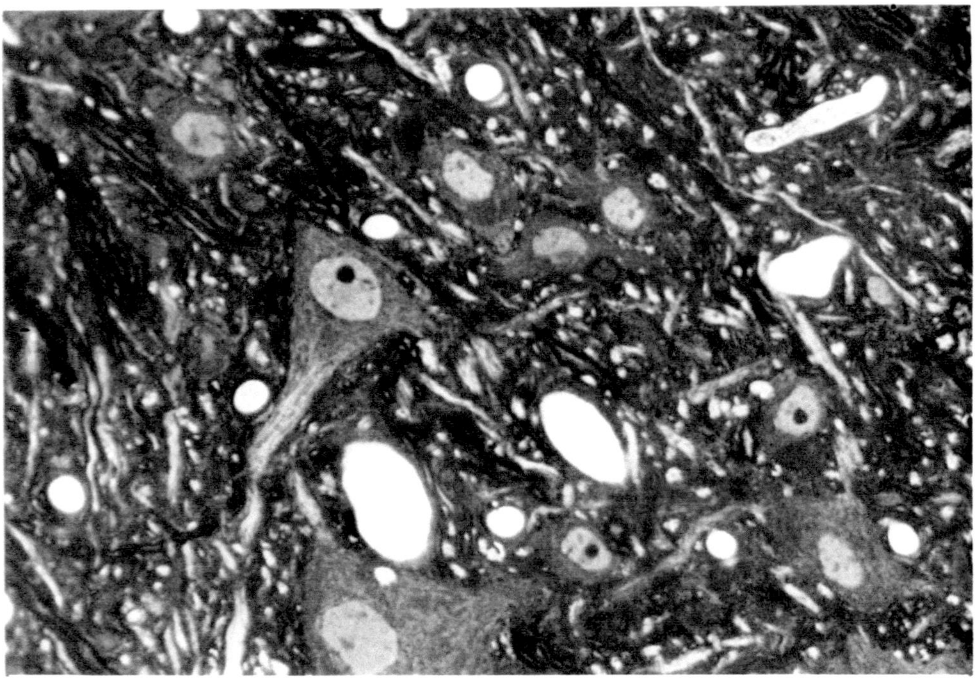

FIGURE 2. Cell bodies: perikarya in the ventral horn of the rat lumbar spinal cord; 1-μm section embedded in Spurr's medium stained with toluidine blue. Note the light staining nucleus and granular Nissl substance.

vascular perfusion of the experimental animal. Tissue samples are obtained carefully to prevent handling artifacts and postfixed in osmium tetroxide to crosslink lipids. The samples are then dehydrated stepwise in increasing concentrations of ethanol, "cleared" of alcohol with a solvent such as acetone or propylene oxide, and infiltrated with resin. The tissue samples are oriented in embedding molds and resins are polymerized with low heat (60°C) for 24 to 48 hr. Sections 1-μm thick are obtained from the hardened blocks with an ultramicrotome equipped with a glass knife.

With a 1-μm section, the resolution of the light microscope is limited only by the optical characteristics of the investigator's instrument and the wavelength of its light source. Resolution can approach 200 nm, so that in a section stained with toluidine blue, the cytoarchitecture of most nerve cells can be easily observed. In a toluidine blue stained semi-thin section the nucleus of a nerve cell is light blue and centered in the cell body. The cytoplasmic space is inhomogeneously stained a darker blue due to Nissl substance and often has a stellate appearance. An example of the appearance of cell bodies (from the anterior horn of rat spinal cord) in tissue prepared with this technique is shown in Figure 2. As exemplified in Figure 3, the nerve trunk in cross section reveals a darkly stained perineurium. Myelinated fibers possess darkly stained myelin surrounding lightly stained axons. Axonal organelles are at the limits of resolution of the light microscope, as are unmyelinated fibers which appear as light areas within blue-stained glial cytoplasm. The terminations of axons are not easily visualized with this technique.

The ease with which structural alteration can be determined in this preparation has made it extremely popular with neurotoxicologists concerned with the actions of chemicals on myelinated fibers of the spinal cord and peripheral nervous system. For instance, using this technique, clear spots have been noted in the nuclei of dorsal root ganglion cell bodies following the administration of doxorubicin: it has been postulated that this is the result of

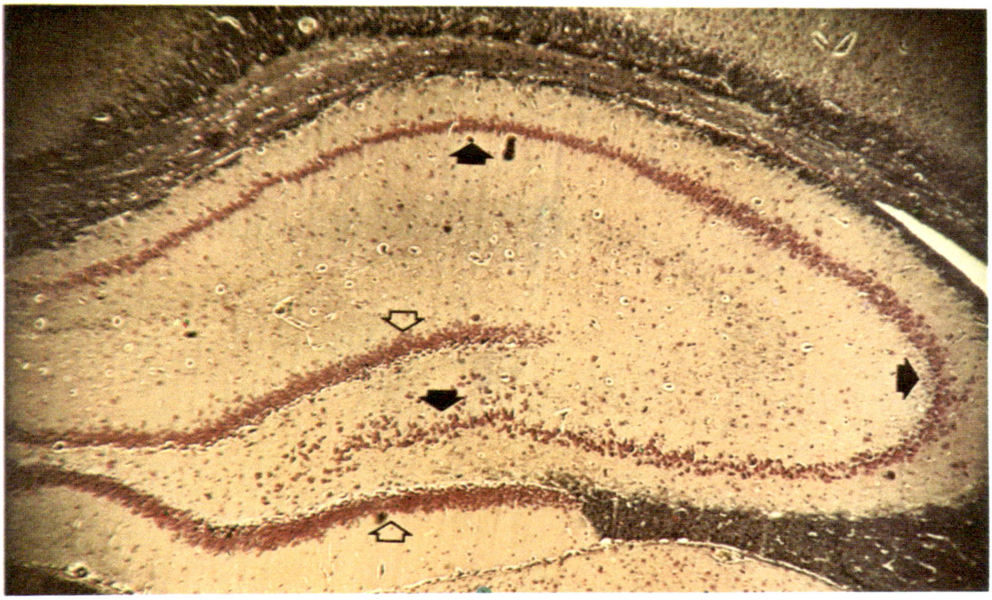

PLATE 1. Cell bodies: rat hippocampal formation; paraffin section stained with iron-hematoxylin and counter-stained with cresyl violet. Iron blackens myelinated fibers and hematoxylin gives pink appearance to the dendrites, nerve terminals, and supporting cells referred to as neuropil. Cresyl violet stains Nissl substance, revealing layers of pyramidal cells of the hippocampus (closed arrows) and granular cells of the dentate gyrus (open arrows). (Courtesy of Donald Robertson, University of Michigan, Ann Arbor.)

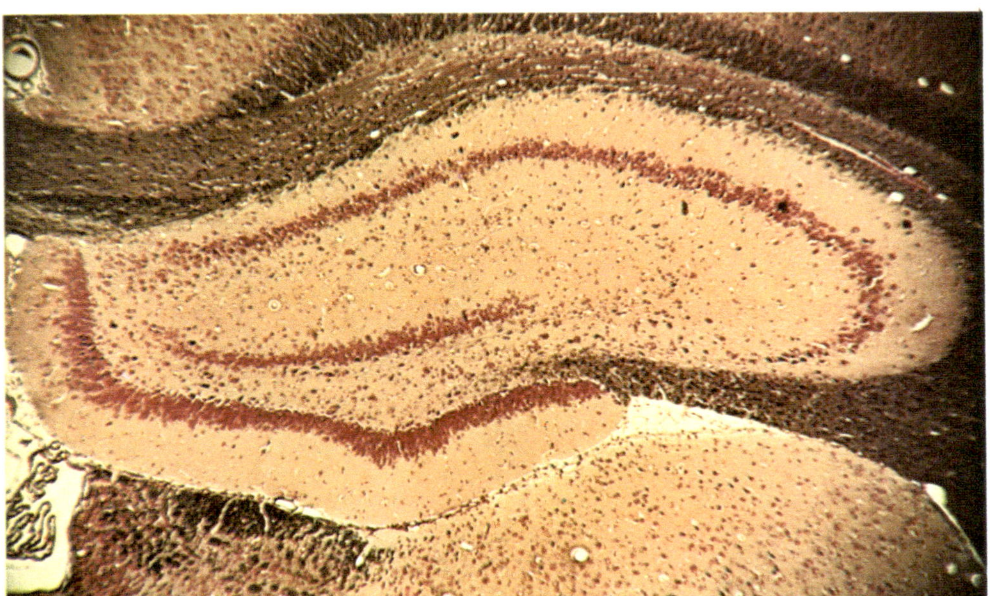

PLATE 2. Cell bodies: as in Plate 1 from rat treated with trimethyltin. Note the loss of pyramidal cells in the region between the dentate cell layers. (Courtesy of Donald Robertson, University of Michigan, Ann Arbor.)

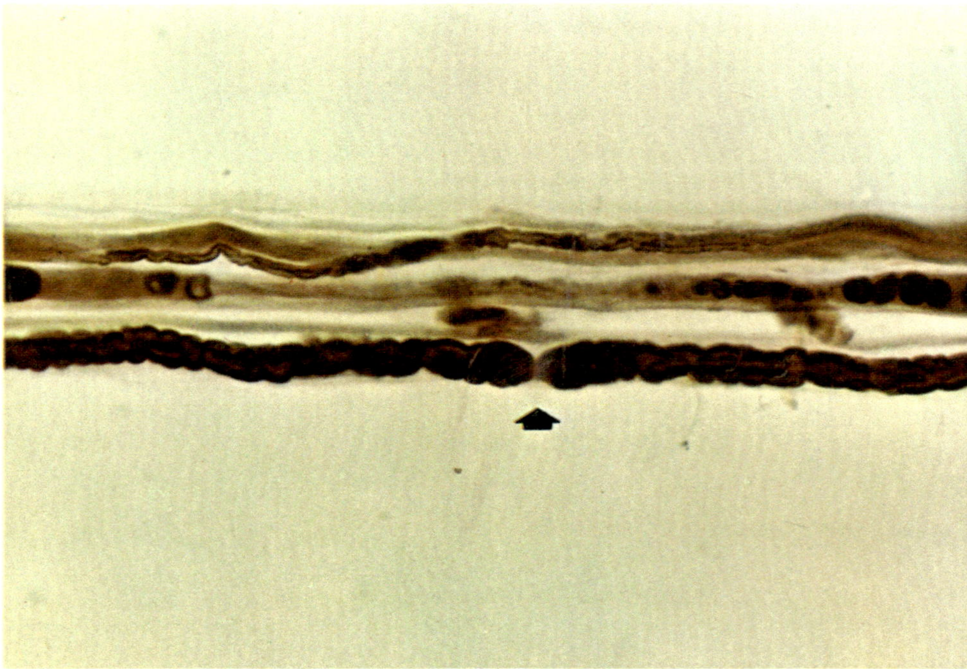

PLATE 3. Axons: rat plantar nerve; teased fibers post fixed in osmium. A degenerated fiber is seen as myelin debris within a Schwann cell tube. A node of Ranvier is seen in the normal fiber (arrow). This preparation is from a "normal" rat, pointing out the fact that one must be aware of the background pathological change in control animals. (Preparation courtesy of Richard DeGrandchamp, University of Michigan, Ann Arbor.)

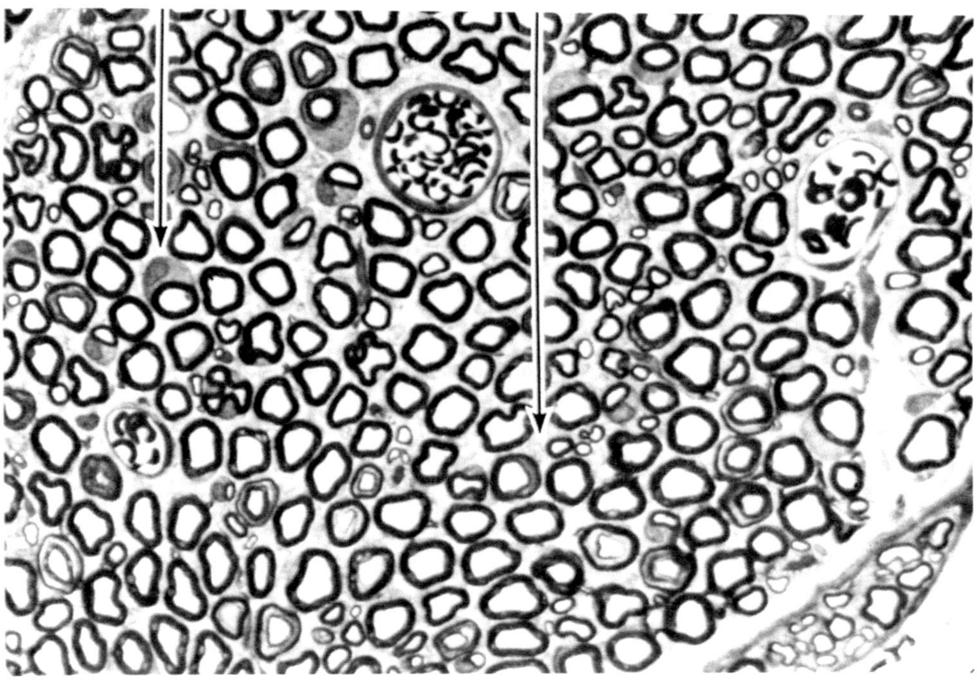

FIGURE 3. Axons: rat tibial nerve; 1-μm sections embedded in Spurr's medium stained with toluidine blue. Myelin stains dark blue and surrounds a lightly stained axon. Unmyelinated axons (right arrow) are at the limits of resolution of the light microscope. Nuclei and cytoplasm of Schwann cells are distinguishable (left arrow). The tissue, which was fixed by immersion, still has blood cells in the vasculature.

chromatin breakdown following intercalation into DNA.[27] In many neuronopathies, including that due to doxorubicin, an early finding which is easily appreciated using plastic semi-thin sections is eccentricity of the nucleus. Eccentricity of the nucleus is one of a constellation of structural alterations of the cell body known as chromatolysis. The cell body becomes rounded and Nissl substance is dispersed or marginated to the region of the plasma membrane. Chromatolysis originally referred to alterations in the cell body following transection of its axon.[53] The observed changes are presumed to indicate that the cell is preparing for a regenerative process. Recently, authors have used the word to describe changes following exposure to neurotoxic chemicals such as acrylamide.[26,126,127] The significance of the finding in intoxications is currently a subject of debate.

Demyelination due to chemicals such as acetyl ethyl tetramethyl tetralin is a spectacular finding in semi-thin sections.[125] Hugh myelin bubbles appear with preserved or shrunken axons in their centers. This is primary demyelination, which should be contrasted with demyelination secondary to loss of myelin-axon contact when the axon degenerates. Axonal degeneration due to chemicals such as 2,5-hexanedione[117] or zinc pyridinethione[105] appears as dark blue spots, but is often preceded by axonal swellings in which axonal diameter is excessive in relation to myelin thickness. The ultrastructure of these swellings will be covered below. Spatial analysis suggests that the effect of these chemicals is to cause an axonopathy distal from the cell body in long, large-diameter myelinated fibers. Long-term peripheral neuropathy results in complete clearance of axons and proliferation of Schwann cells still within the basal lamina (not visible). These tubes of Schwann cells devoid of axons are known as bands of Bungner, and may aid in the regeneration of axons in the PNS by providing a guide to the target cell. Why certain chemicals selectively affect myelin or different regions of the axon is far from clear, but it is thought that careful classification of the pathology can provide clues about pathogenesis.[120]

A further advantage of plastic embedding is that sections for electron microscopy can be obtained from the same tissue blocks, although many authors claim that the need to do so is rare.[7,121] The disadvantage of the method lies in the extremely small surface area of the prepared section. While complete transverse sections of peripheral nerve and spinal cord of small animals are possible, sampling of the brain requires considerable dissection. Many laboratories are currently experimenting with softer resins (e.g., methacrylates) and broad-faced (Ralph) glass knives which make it possible to obtain 1-μm sections of larger surface area.

3. Teased Fibers

Specimens of peripheral nerve fixed and processed for plastic embedding can instead be placed in a low viscocity resin such as Spurr's medium and individual myelinated fibers dissected from a fascicle with fine needles. The dissection is executed on a glass slide and, when the fibers are oriented properly, the preparation can be hardened like other epoxy resins. Because osmium postfixation of the tissue blackens structures with high lipid content (reduction of osmium tetroxide in the presence of unsaturated fatty acids is responsible for this effect), myelin sheaths are clearly visible under the light microscope.

This method represents one of the most rapid techniques of assessing toxicity to peripheral nerves. Segmental demyelination is evident as clear spaces in the fiber. Degenerating fibers appear as a ''string of beads'' of myelin debris. Structural change beyond these findings cannot be seen in teased fibers, but the preparation can then be re-embedded in conventional resins and sectioned for the light or electron microscope.

An example of normal and degenerated teased fibers is shown in Plate 3*. In addition to demonstrating the morphology seen with this technique, it brings up the important point that there is ''background'' damage in supposedly normal animals. The sample was obtained from an untreated rat. Often, neurotoxic damage is a matter of degree rather than unique structural findings.

4. ''Thin Sections'': TEM

Transmission electron microscopy (TEM) requires thinner sections (50 nm) than those used with the light microscope. They may be obtained from plastic embedded tissue blocks with the aid of an ultramicrotome equipped with a glass or, preferably, diamond knife. Sections are collected on copper grids and stained with lead and uranyl acetate.

With well-prepared specimens, a resolving power of a few nanometers and high magnifications can be attained with the transmission electron microscope. Thus, organelles including endoplasmic reticulum, Golgi apparatus, microtubules, neurofilaments, and synaptic vesicles can be observed, as well as plasma and nuclear membranes and myelin lamellae. Structures in intracellular spaces are also easily appreciated.

Electron microscopic investigation of neurotoxic effects is extremely labor-intensive, and the size of the area examined is small. For these reasons TEM has been used more selectively than other methods of structural neurotoxicology. Electron microscopy has been used to divulge the ultrastructural correlates of damage previously observed with the light microscope. Organophosphorus acetylcholinesterase inhibitors produce a transient myopathy visible with the light microscope. When inspected with the electron microscope, it was found that the toxic effect was actually a neuromyopathy. The limitation of the early structural change to the neuromuscular junction suggested that the etiology of the lesion might be an

* Color plates follow page 182.

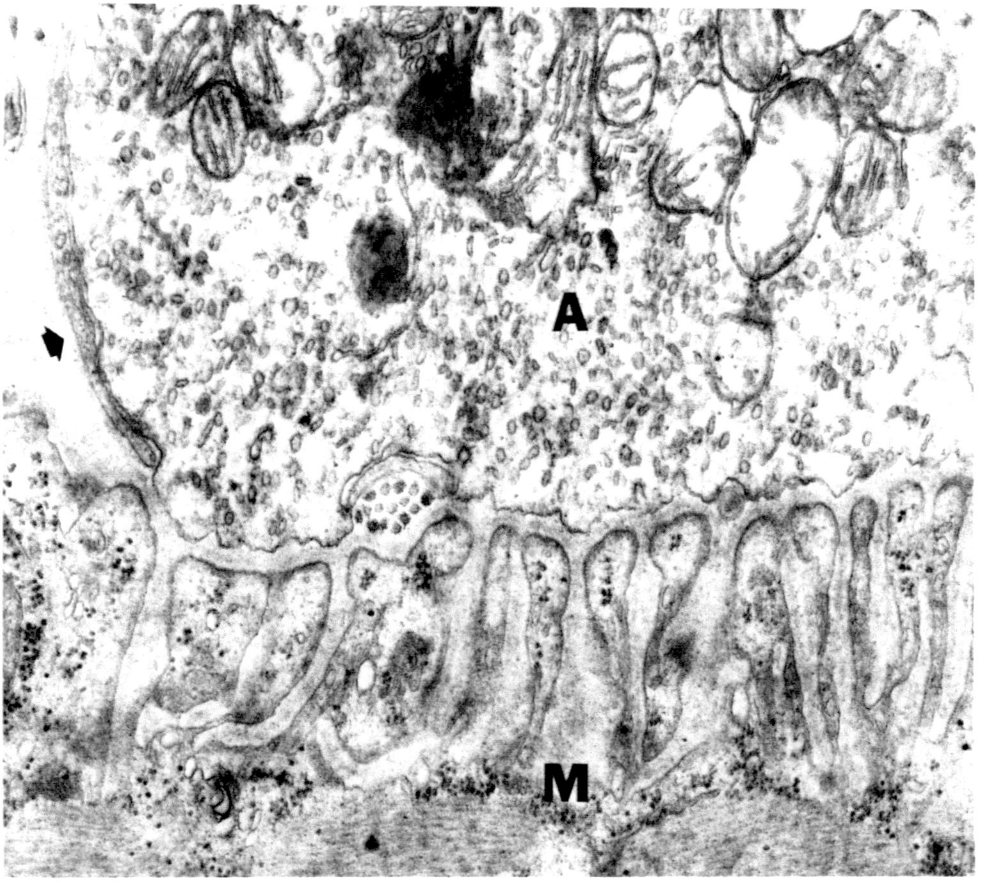

FIGURE 4. Axon terminals: motor endplate in rat diaphragm. 60-nm section of Epon/Araldite®-embedded specimen stained with lead and uranyl acetate. In the electron microscope, the mitochondria and synaptic vesicles of the axon terminal (A) are evident, as are the postsynaptic clefts and contractile appratus of the muscle cell (M). The ensheathing cell of the axon terminal can also be seen (arrow).

excitotoxic effect of excess acetylcholine.[72-74] The effect of organophorphorus anticholinesterases on the neuromuscular junction is shown in Figures 4 and 5.

The axonal swellings associated with distal axonopathy have been analyzed ultrastructurally. In hexacarbon,[117] zinc pyridinethione,[105] and organophosphorus neuropathies,[20] swellings were found to be adjacent to a node of Ranvier (paranodal). In the case of hexacarbon neuropathy, the paranodal swellings contain high concentrations of neurofilaments, whereas in organophosphorus and zinc pyridinethione neuropathy, the swellings are reported to be filled with membrano-vesicular elements. It is worth noting that the neuropathologist must be cautious of overinterpreting such findings. While it might be assumed that the neurofilamentous swellings seen in many neuropathies are somehow linked to ultimate axonal degeneration, it has been found that 2,2′-iminodipropionitrile(β,β′-iminodipropionitrile; IDPN) causes a comparable accumulation of neurofilaments in a more proximal location, but axonal degeneration does not occur.[54] The compound 3,4-dimethyl-2,5-hexanedione is an interesting hybrid in that at the right dose in rats, it produces proximal axonal swellings and distal axonal degeneration.[4] Thus, the link between neurofilamentous neuropathy and axonal degeneration remains to be explained.

Electron microscope examination has been used to examine subtle changes unaccompanied

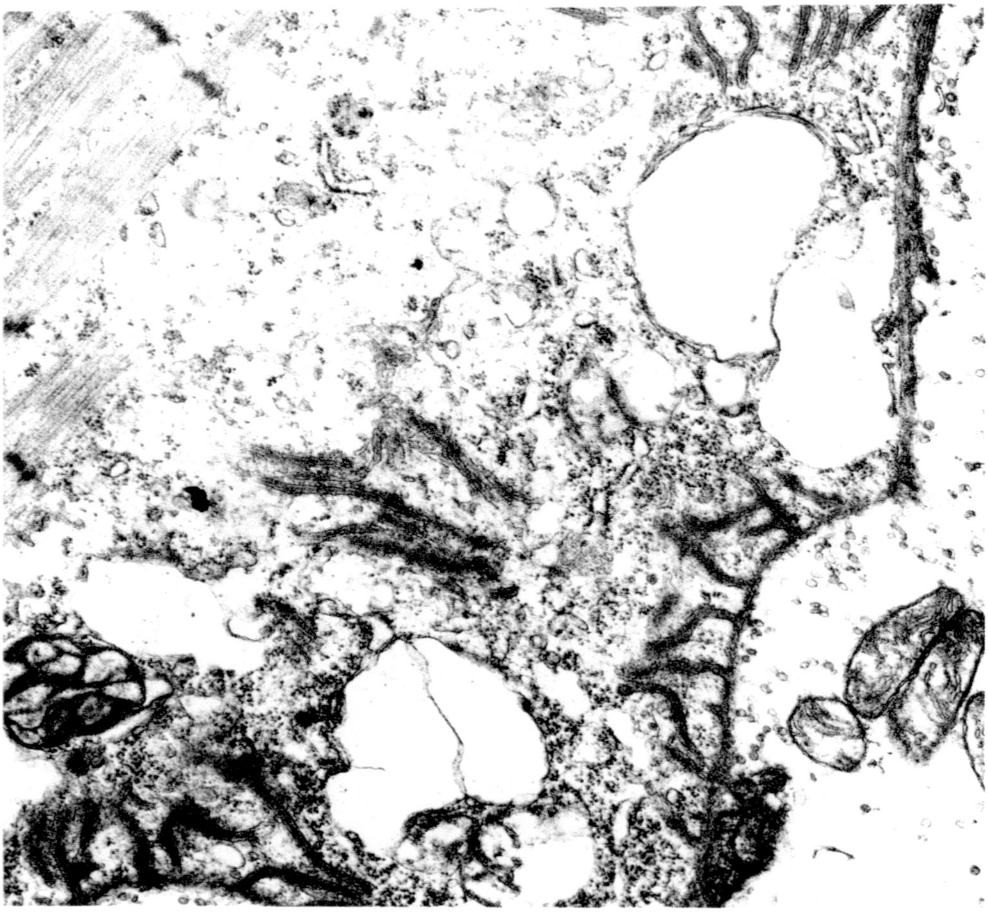

FIGURE 5. Axon terminals: motor endplate in rat diaphragm. 60-nm section of Epon-embedded specimen stained as above. The animal was treated with the anticholinesterse compound paraoxon. There is a paucity of synaptic vesicles in the axon terminals, and vacuoles have replaced many mitochondria of the axon terminal and the muscle cell. The contractile apparatus of the muscle is fragmented beneath the endplate.

by overt alterations, as in Spencer's study of mitochondrial changes in thallium toxicity,[116] and in viewing structures which cannot be seen with the light microscope, as in Schaumburg's observation of Pacinian corpuscle damage in acrylamide neuropathy.[108]

5. Morphometry

Rather than being a separate technique, morphometry is the quantitative measurement of structures viewed with any of the above techniques. Morphometry is labor-intensive, but the objective nature of the data makes the techniques valuable. Computer assistance greatly facilitates morphometric techniques and is becoming commonplace.[37] The simplest morphometry is a single-number description of some aspect of structure. The spine count in Golgi preparations described above is an example. Measurement of fascicular or endoneurial area is in the same category. Yount et al.[141] reported that the area of fascicles in the tibial nerve of rats is decreased following treatment with dichloroacetate.

One of the most common morphometric techniques is to analyze axonal or fiber diameter as a frequency histogram. Myelinated fibers, which may be measured from light microscope images or photographs, are bimodally distributed with respect to diameter. Thus, relation of fiber size to neurotoxic sensitivity can be determined. In this way, it was shown that thalidomide treatment in humans caused a large-fiber sensory axonopathy,[45] and the *p*-bromophenylacetylurea caused distal large-fiber loss in the sensory tracts of the CNS and PNS.[96] Unmyelinated fibers have a unimodal distribution of fiber diameters.

Measurements of teased fibers have been extremely useful in the study of structural neurotoxicity. In normal nerves, there is a linear relationship between the diameter of a myelinated fiber and the length of its internode. Linear regression of these two features yields a line of high correlation coefficient and characteristic slope. When a myelinated axon degenerates and then regenerates through a band of Bungner, it is remyelinated by all the Schwann cells which have proliferated and as a consequence the internodes are short and of constant length regardless of axon diameter. Linear regression of length vs. diameter in this case reveals a good correlation coefficient with a slope near zero. This is to be contrasted with segmental de- and remyelination where internodal lengths vary greatly. The correlation coefficient in this case, as Fullerton et al.[44] have demonstrated with lead demyelination, is extremely poor.

If morphometry is to be meaningful, the pathologist must be fastidious in his fixation and tissue processing procedures. Ohnishi et al.[92-94] published a series of papers on optimizing fixation of human peripheral nerves for morphometry. Many of the findings are probably applicable to experimental animals and may also be used in the CNS.

III. ELECTROPHYSIOLOGICAL ASPECTS

With few exceptions, electrophysiological techniques are better at determining the site rather than the mechanism of action of a neurotoxic substance. By comparing functional measures of various components of the nervous system, electrophysiological techniques can detect the first component of the nervous system that has been altered by the substance. This often points to the most vulnerable part of the nervous system and also confirms an effect at the onset of the toxicity, rather than being an index of cumulative damage. This is valuable information, but it does not necessarily demonstrate, or even suggest, the underlying mechanisms responsible for the change in function. Biochemical techniques are usually the only way of establishing a mechanism of action. However, electrophysiological techniques can often provide corroborating data.

The value of electrophysiological techniques is that they are usually a reflection of altered function in the nervous system. This means, especially in the peripheral nervous system where there is a small reserve for compensatory processes, that the observed functional change is biologically meaningful and probably reflects a real, or at least immediate, loss to the organism. In contrast, there are instances where considerable pathological lesions or greatly altered biochemical processes can be induced by a substance without any apparent altered biological function or awareness on the part of the subject. These latter changes, even though they may be dramatic, may not have immediate consequences but may affect the reserve capacity of the system to respond to additional stressors or to the ravages of time.

However, if sensory perception is distorted, if nerves are not conducting, if reflexes are disrupted, if individual neurons are firing too much or not enough, then it is likely that the subject as a whole is not able to function properly and these specific deficits in neuronal function may be used to describe, if not to explain, the neurotoxicity that has obviously occurred. It is at this point that biochemical and electrophysiological techniques must work together to sort out the underlying mechanisms which are responsible for the functional deficits.

The detection of early changes in physiological function may have another valuable purpose. Subtle changes in electrophysiological activity often preceed (or are early indications) of a pathological insult. Since repair in the nervous system is often difficult or even impossible, electrophysiological methods may be the harbinger of a permanent deficit. If damage is not corrected early, the subject may have no chance of compensating for the loss. Conversely, in cases where recovery of the nervous system is possible, these same techniques can be and have been used with great success in following recovery from the neurotoxic episode by using measures of improvement of nervous system function.[113]

A. Peripheral Nervous System

Peripheral neuropathy is among the most common types of neurotoxicity seen clinically. In severe cases, there is histopathological change of the nerves and clearly indicated neurologic signs of motor and sensory deficits, particularly in the distal limbs. Clinical electrophysiological measurements, which have generally been restricted to electromyography and peripheral nerve conduction velocity, also show alterations indicating neurotoxicity.[13] However, the noninvasive measures taken clinically cannot usually distinguish between selective losses of axons vs. myelin, large vs. small diameter axons, sensory vs. motor axons, or proximal vs. distal axonal segments.

1. Axonal Function

Many studies of neurotoxic substances have used conduction velocity as a measure of peripheral nerve function. The measurement simply indicates the speed with which a nerve action potential travels along the nerve axon and is calculated in m/sec. When conduction velocity is measured in a peripheral nerve bundle, as is usually the case, the resultant nerve compound action potential is the summation of many events occurring synchronously. There are a number of considerations which make interpretation of conduction velocity difficult, and it should not be surprising that little in the way of useful information about the action of neurotoxic agents has been dervied from it.[130] First, most peripheral nerves are mixed motor and sensory nerves. Therefore, the measurement always reflects the function of some axons propagating orthodromically and the rest propagating antidromically. Both sensory and motor nerves conduct impulses slightly faster in their natural direction, so distortion of the "true" conduction velocity is inevitable. Second, conduction velocity is usually calculated using the onset of the compound potential and thus is a measure of only the fastest axons in the nerve bundle. If the fastest axons are the most susceptible, as in the case of acrylamide intoxication, a reduced conduction velocity is an early sign of neurotoxicity.[57] However, alterations in the slower conducting axons are buried in other components of the waveform and marked changes in their function can be easily overlooked. Finally, since the individual action potentials which make up the measurements are all-or-none events, they are not easily altered in speed of conduction. A more likely and common event is axonal conduction blockade, either by disruption of ion conductances across the membrane, disruption of Na^+/K^+ ATPase, or pathologic nerve damage. Blockade of conduction eliminates the axon from the population altogether and its contribution to the compound action potential is lost. However, unless the damaged axon was one of the most rapidly conducting in the nerve bundle the loss would not be reflected as a change in conduction velocity. Rather, the loss would be more apparent as diminution in the action potential amplitude. DeJesus et al.[34] showed that zinc pyridinethione, which selectively degenerates distal axons, induced a selective loss in the amplitude of the compound action potential, but no change in conduction velocity. Similarly, the earliest electrophysiological change after methylmercury chloride intoxication (which also produces an axonopathy) is a reduced action potential amplitude.[84]

The most useful information from measurements of conduction velocity have come from studies of single axons. If a representative sample of fibers in the nerve bundle (100 or more) is taken, the distribution of axons, based upon their individual conduction velocities, can be compared before and after exposure to neurotoxic substances. Using this type of analysis, no loss in conduction velocities of sensory nor motor axons was induced with acrylamide in doses which produced overt neurotoxicity in the animal,[74,76-78,128,129] whereas IDPN induced a progressive loss of signal from faster conducting motor axons during chronic exposure.[47] The reason for this difference is that acrylamide appears to attack nerve endings initially and spare axonal function, whereas, IDPN produces large swellings at the proximal end of the axon, causing a decrement in conduction due to the increased surface area of the axon. Calculating conduction velocity from single axons is a tedious process and is somewhat biased in that the faster (larger diameter) axons tend to be preferentially selected and are thus overrepresented. In addition, even in single fiber studies, the measurement of conduction velocity is averaged over the length of the axon. In fact, the conduction velocity of an axon changes over its course and depends on the changing axonal diameter, terminal branching, and degree of myelination. Therefore, even under the best circumstances, it is almost impossible to obtain an unbiased assessment of the effects of a neurotoxic agent on axonal conduction.

A somewhat related measurement of peripheral nerve function appears to be of greater advantage in assessing peripheral axonopathies. Friedli and Meyer[42] have proposed using the strength-duration curve as a sensitive electrophysiological endpoint. Using several percutaneous stimulation sites, the stimulus intensity required for minimally perceptible sensation was determined at 12 different stimulus durations. The stimulus strength vs. duration was then plotted and compared between normal subjects and a patient group. The stimulus strength at sensory threshold was increased as the degree of axonal failure increased. The authors suggested that rheobase (the amount of current necessary for a threshold sensory response at infinite stimulus duration) was a valuable and less extensive method of measuring this deficit and that the strength-duration curve is a useful complement to other measures of peripheral neuropathy.

Another measure of axonal function which has been proposed to have value in determining the actions of neurotoxic agents is the relative refractory period. Applying paired stimuli to the nerve at an interstimulus interval greater than 10 msec results in two equal compound action potentials. However, when the nerve is stimulated to elicit a second action potential during its refractory phase, there is a diminished response. Three factors contribute to the diminished response in a whole nerve: (1) incomplete recovery of the single axonal action potential amplitude, (2) increased dispersion of the single axon action potential arrival times, and (3) propagation failure of individual axons.[124] Changes in refractoriness of the individual axons can result from demyelination or from changes in membrane-mediated ion conductances, particularly potassium conductance during the repolarization phase of the membrane response.

Various methods have been proposed for quantitating the refractory period of peripheral axons. Roberts and Trollope[103] proposed the calculation of the minimal refractory period point, which roughly correlates to the longest interstimulus interval just necessary to achieve the absolute refractory period. However, the refractory period is an exponential function and defining one point along the course of the response is deceptive. Smith[112] has presented a graphic representation of the relative refractory period which shows in visual detail the changes in the complete relative refractory period (see Figure 6). Using this technique Smith and Hall[114] showed a correlation between the histopathology and changes in refractory period induced by the demyelinating agent, lysophosphatidyl choline. Quantitative changes can also be assessed by performing a log-linear conversion of the relative refractory period data and then comparing the slopes and intercepts of the linear functions. Using this technique[3] it has been possible to demonstrate that hexane and 2,5-hexandione increase the relative refractory period prior to overt neurotoxicity in rats. Phenol, which is known to be a rather selective demyelinating agent,[107] also produced a large increase in the relative refractory period several weeks after a single exposure.[2]

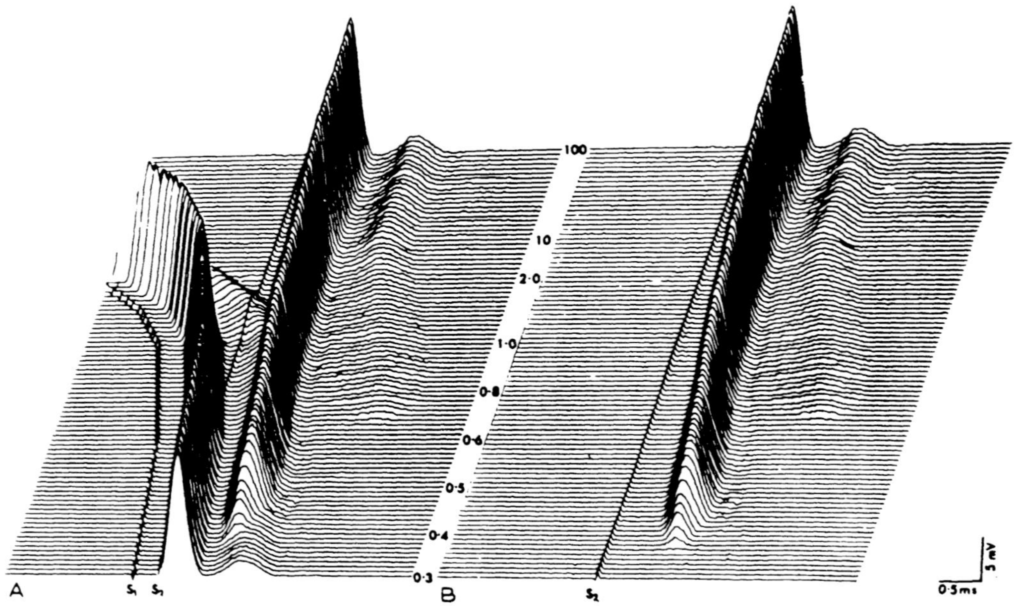

FIGURE 6. (A) Three-dimensional array of monophasic compound nerve action potentials recorded from cat sural nerve in response to pairs of supramaximal stimuli (20 × threshold); the stimulus separation (shown in msec) is increased between each record. The faster fibers have shorter RPTs than the slower fibers. The increased latency of the fibers during the relative refractory period is especially noticeable among the slower fibers. (B) The compound action potentials resulting from the second stimulus alone; the response to the first stimulus has been subtracted. The three-dimensional appearance of these figures is due to computer plotting of the data with a program incorporating "hidden line suppression". (From Smith, K. J., *J. Neurol. Sci.*, 48, 191, 1980. With permission.)

The reason for the increase in relative refractory period with lysophosphatidyl choline and phenol is the primary demyelination produced by these agents (see the section on myelin below). The reason for the early change in refractory period with hexane and its active metabolite, 2,5-hexanedione (compounds which produce primary axonal degeneration) is not known, but may be indicative of a change in axonal membrane repolarization.

Even though it is possible to record the resting membrane potential from the cell soma, there are instances when considerable axonal damage has occurred, yet the cell soma remains intact. This is particularly common in peripheral nerve axons, which are separated by a great distance from the cell body. It is possible that marked changes in peripheral axonal function and membrane responsiveness can occur prior to deficits in the cell body. Unfortunately, it is difficult to record intracellularly from mammalian peripheral nerve axons without causing more damage through physical trauma than the neurotoxic agent under study may be causing. As an alternative, the sucrose gap technique has been used as a measure of the averaged resting membrane potential and membrane responsiveness in the peripheral nerve bundle. By depolarizing one end of the peripheral nerve with a potassium solution and recording between the depolarized and nondepolarized ends of the nerve across a gap which is electrically insulted with a perfusing sucrose solution, an electrical potential can be measured. In addition, test solutions of varying concentrations of potassium can be used to challenge the responsiveness of the nerve membrane to potassium. This technique has been used in studies of nerves taken from animals exposed to organophosphorus agents (OPs). Chronic exposure to OPs for several weeks induces progressive changes in the membrane responsiveness to potassium, which may be a compensatory mechanism in the rat to resist the threat of peripheral neuropathy. Similarly, hexane and 2,5-hexanedione induce a change in the nerve membrane following chronic exposure, which mimics the effect

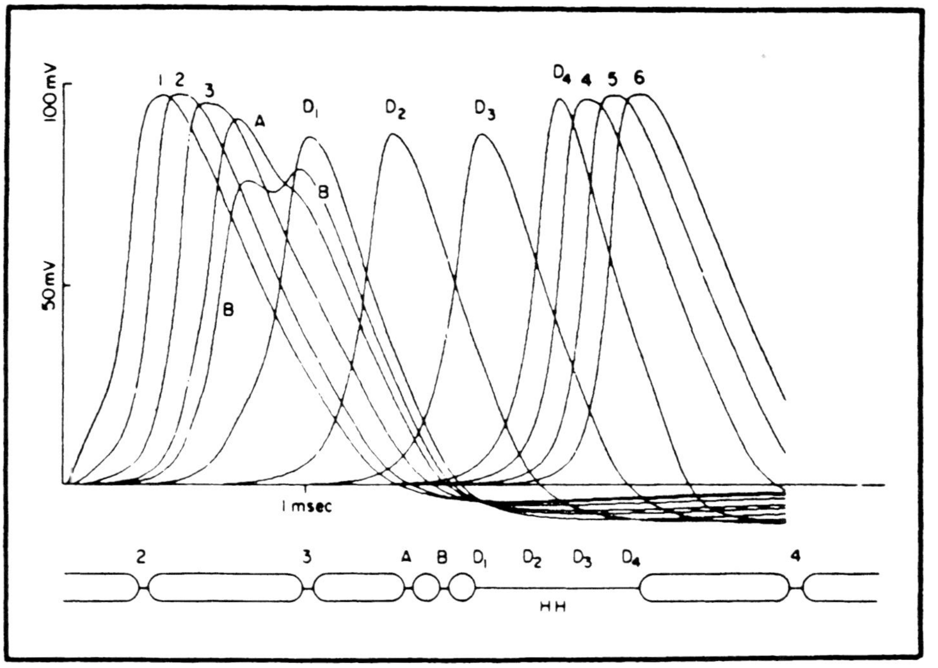

FIGURE 7. Computed action potentials in a fiber in which the axon membrane in the demyelinated (D_1 through D_4) region has the characteristics of a Hodgkin-Huxley membrane. Despite the lower density of sodium channels than at normal nodes, conduction proceeds through the demyelinated zone. Unimpeded conduction requires the presence of closely spaced nodes of Ranvier (400 μm internode distance in this situation) proximal to the demyelinated region. (From Waxman, S. G., *Neurology*, 28, 27, 1978. With permission.)

of acutely applied ouabain to untreated nerves.[3] This suggests that the hexane neurotoxicity may be preceded by an inhibition of the Na^+K^+ATPase in the nerve or otherwise a change in the membrane handling of potassium fluxes.

2. Myelin Function

The physical and bioenergetic constraints which determine conduction velocity and action potential propagation in the nervous system has recently been reexamined.[138] In myelinated nerves the geometry of the axon is such that the ratio of axon diameter to fiber diameter and of internodal length to fiber diameter are within the ranges which are most efficient for electrical conduction.[136] Thus, any alteration of the myelin will have an adverse effect on conduction.

The effects of demyelination on conduction of peripheral nerve axons has been extensively studied and provide a large body of evidence upon which to draw conclusions about the deficits induced by novel neurotoxic agents.[130] Demyelination occurs in steps, and thus results in a continuum of electrophysiological changes. Initial thinning of myelin at heterogeneous sites along the axon results initially in a mixture of saltatory and continuous conduction. Thus, from one end of the axon to the other, conduction velocity may slow down as the impulse passes through a demyelinated region and then resume normal conduction in sections of the nerve which remain intact (see Figure 7). Conduction velocity is slowed in proportion to the average extent of demyelination along the length of the fiber, each transition from saltatory to continuous conduction contributing to the slowed conduction velocity of the nerve.[99] If the average conduction velocity is calculated (as it often is experimentally), there is a slowing of the calculated nerve conduction velocity very early in the intoxication process. There are many examples of neurotoxic agents in which demyelination is accompanied by early decreases in conduction velocity, including hexachlorophene,[33,82] lead,[43,95] and phenylmercuric acetate.[111]

Concomitant with the slowed conduction through the demyelinated region is a dimunution in the safety margin for impulse transmission. This enhances the refractoriness of the damaged section of the nerve axon. Electrophysiologically, it is detected as a longer refractory period during paired or high frequency stimulation. The severity of demyelination resulting in a decreased safety factor is directly correlated with increases in the refractory period. Thus, an increased refractory period can be used as a sensitive indicator of minimal and possibly very patchy demyelination even prior to changes in conduction velocity.[99] Marked increases in the relative refractory period of peripheral nerve have been reported for two demyelinating agents, lysophosphatidyl choline[114] and phenol.[2]

The demands for depolarization of the excitable membrane increase as the surface area of the demyelinated region increases, due to increased capacitative and resistive shunting.[137] This is compounded by the fact that there is a complementary distribution of ion channels in mammalian nerves, with sodium channels more dense at the nodes of Ranvier and potassium channels located under the myelin.[39,137] Thus, as a result of demyelination, conduction is not only attenuated because of a shift to continuous conduction in the demyelinated region, but also due to the low density of sodium channels which will not support impulse conduction and the high density of potassium channels which clamp the membrane close to the resting potential.[102] If the current provided by the invading action potential is insufficient to raise the demyelinated region to threshold, conduction fails. Thus, in severe cases of demyelination there is not only a slowing of conduction velocity, but also a decrease in the size of the compound action potential.

3. Nerve Terminal Studies

For many substances which produce a peripheral neuropathy, the most vulnerable component of the neuron is the nerve terminal. For sensory neurons, which are often more vulnerable than motor neurons to neuropathic agents, direct measurements of nerve terminal activity can be recorded. Natural stimulation of the sensory receptors (e.g., muscle stretch for muscle spindles, increased muscle tension for Golgi tendon organs, cutaneous stimulation of Pacinian corpuscles) activate the transduction process in the nerve ending and can be recorded from single afferent fibers in the dorsal root of the spinal cord. Concurrent measurement of axonal conduction velocity can establish whether changes in sensory responsiveness are due to toxicity of the nerve axon or to a selective loss of the nerve terminal.

Deficits in muscle spindle afferent activity in cats have been reported for diisopropylphosphorofluoridate (DFP) which produces a peripheral neuropathy in cats as well as in other species. Loss of muscle spindle responsiveness after DFP exposure occurs prior to changes in axonal conduction, with a greater deficit seen in the smaller diameter secondary afferent nerve terminals.[9] Detecting alterations in motor nerve terminals is more difficult electrophysiologically. Physical constraints at the neuromuscular junction make it almost impossible to record the activity of motor nerve terminals directly. However, the neuromuscular junction has been widely studied as an example of a cholinergic synapse and many of the techniques are applicable to studies of neurotoxicology. Two types of experiments can be used to study the motor nerve terminal. Both are indirect. The most commonly used technique is to record the intracellular activity of the muscle cell in the region of the motor nerve terminal. Changes in the muscle membrane responsiveness to nerve stimulation is taken as evidence for altered nerve terminal function. Since the measurements are an indirect measure of the motor nerve terminal, the data must be interpreted with caution and direct effects of a toxic agent on the muscle must first be ruled out. However, some changes indicate selective effects on the nerve terminal. For instance, changes in miniature end plate potential (mepp) frequency and amplitude suggest changes in spontaneous neurotransmitter release. Changes in end plate potential (epp) suggest changes in nerve-stimulated transmitter release.

Cangiano et al.[24] reported that 2,5-hexandione produced changes in motor nerve terminals at the time of the early clinical signs but prior to ultrastructural changes in nerve or muscle. The observed increase in mepp frequency and decrease in epp quantal content were suggested to be due to an impaired neuronal metabolism and consequent depolarization of the nerve terminal. Satoh et al.[106] showed that the inhibition of neuromuscular transmission induced by cadmium is due to a blockade of the endplate potential. Long-term exposure to methylmercuric chloride produced a presynaptic localized loss of spontaneous and evoked transmitter release, whereas mercuric chloride had both pre- and postsynaptic actions.[62] Finally, the muscle weakness seen with triethyl tin has been attributed to a direct muscle effect, since no change in neurotransmitter release was observed.[17]

The second way of assessing motor nerve terminal function is measuring the antidromic potentials propagated from peripheral nerve stimulation. Single pulse, low frequency stimulation of the motor nerve results in an orthodromic action potential and eventually a muscle twitch. Concurrently, an antidromic action potential is propagated centrally and can be recorded from the ventral root. After a high frequency train of stimulation, subsequent single stimuli evoke a short burst of action potentials which are initiated at the motor nerve terminal. These multiple action potentials result in an increased twitch tension in the muscle (post-tetanic potentiation) and can also be recorded centrally. If nerve terminal function is altered, the development of this post-tetanic potentiation response may also be altered.

For example, although DFP does not alter neurotransmitter release[16] or affect axonal conduction velocity, there is a progressive loss of post-tetanic muscle potentiation and post-tetanic nerve repetitive activity which correlates with the development of neurological deficits.[79] Ultrastructural changes in the motor nerve terminal have been reported to occur concurrently with these neurological and electrophysiological changes.[35] Furthermore, phenylmethanesulfonyl fluoride, which inhibits neurotoxic esterase but does not undergo aging and consequently not only does not produce a peripheral neuropathy but protects against delayed neurotoxic OPs in hens,[60] prevents the neurological, electrophysiological, and ultrastructural deficits induced by DFP in cats.[10,35] A similar loss in post-tetanic repetitive nerve activity occurs after acrylamide administration.[74,76-78] However, the motor nerve terminal deficit is less severe and probably occurs after acrylamide-induced disruption of sensory nerve endings.

4. *Membrane Selective Studies*

Ultimately, the best electrophysiological evidence is a direct measurement of the effect of a neurotoxic agent on the excitable membrane. Most other electrophysiological measurements (EEG, axonal conduction velocity, resting membrane potential, reflex responsiveness, refractory period) are only expressions of changes in ion flux, either while the nerve is at rest or as a result of neurotransmitter-mediated synaptic activation. Measurement of ion conductances and related effects such as the active transport of ions via Na^+/K^+ ATPase are the closest that electrophysiological studies come to biochemical studies in determining the mechanism of action of the neurotoxic agent.

Intracellular recording techniques are now widely used to examine the effects of compounds on excitable membranes. Careful manipulation of the perfusing media and experimental conditions have made it possible to deduce specific molecular actions of some neurotoxic substances. For instance, the use of selective ion channel blocking agents has provided evidence that Hg^{2+} exerts its neurotoxic effect at an intracellular site and gains entry to the cell via both the sodium and calcium channels.[85] DFP has been shown to bind to the ion conductance complex of the nicotinic receptor and modifies neurotransmission directly, in addition to inhibiting acetylcholinesterase.[68] Soman, another organophosphorus agent, activates the sodium current in electroplaques.[23]

Until recently, the best way of determining sodium, potassium, or other ion conductances was by using the voltage clamp technique. Most applications of this technique to neurotoxicology have centered around the effects of insecticides and other selective neurotoxins such as tetrodotoxin. Using voltage clamp techniques, Narahashi[87] found that the hyperexcitatory activity of DDT was the result of specific binding of DDT to the sodium channel, resulting in a prolongation of the sodium current.

More recently, patch clamp techniques have provided the most direct measure yet of ion channel behavior. In studying the effects of various pyrethroids, Narahashi[88] found that the excitatory activity of the pyrethroids was the result of a modification of some, but not all, sodium channels. In those sodium channels which are modified by tetramethrin, for instance, the modified channel remains open longer than normal, resulting in increased neuronal excitability.[88]

Voltage clamp and patch clamp techniques have been restricted thus far to neuronal membranes from either invertebrates or cultured cell lines, and the effects of acute neurotoxic exposure. When it is possible to use these or similar techniques to examine the chronic effects of neurotoxic agents in mammalian tissue, these methods will provide invaluable information about the molecular mechanisms of action of toxic substances.

B. Central Nervous System

1. Electroencephalography

Initial studies of toxicity to the CNS used electroencephalographic (EEG) measurements similar to those used clinically. Due to the complexity of the EEG signal, assessment of neurotoxicity was limited to gross assessment of changes in excitability or depression. This technique was used, for example, to study the CNS effects of soman and showed that the cortex usually exhibited the first sign of excitation.[75] These techniques were also used by Van Meter et al.[133] to demonstrate that the CNS toxicity of a variety of anticholinesterases was due to a direct effect, since lethality occurred despite prior inhibition of cholinesterase and protection with atropine.

In an effort to quantitate the EEG signal, investigators have turned to Fourier analysis, which transforms the EEG from the time to the frequency domain. Quantitative changes in the power of the various frequency components can more easily be compared. Frequency bands corresponding to the alpha, delta, and theta components of the EEG are altered during changes in mental state (due to such variables as sleep/wakefulness, anesthetic depth, or blood gas balance), and compressed spectral arrays graphically show changes as a result of various experimental manipulations. These techniques have been widely exploited by anesthesiologists and psychophysiologists but have not been used in studies of neurotoxicity. Thus, the value of this technique as a means of evaluating neurotoxicity remains to be determined. However, the potential value of this type of analysis should not be overlooked.

Intra- and extracellular recordings of brain neuronal activity have been used, either alone or in conjunction with EEG measurements, in some studies of neurotoxic substances and have provided useful information about selective sites of toxic action in the brain. For example, studies by Bay et al.[11] and Beers et al.[12] have established that the central respiratory neurons in the medulla and brain stem are the most sensitive to the acute effects of the potent organophorphorus agents, and that despite other CNS actions (e.g., convulsions), lethality is almost certainly due to central respiratory depression.

In a study of the acute effects of divalent metal ions, Rozear et al.[104] found that cadmium, cobalt, nickel, manganese, zinc, and iron all had a depressant effect on cortical and brain stem neurons, both spontaneous activity and that evoked by glutamate activation. Since the metal salts were administered iontophoretically, the toxicity of these ions in vivo may be much less, depending upon their accessibility to the brain.

Bjorklund et al.[18] found that postnatal exposure to lead for 3 weeks to newborn rats resulted in a mean decrease in spontaneous firing rate of cerebellar Purkinje cells, compared to weight-matched controls. Since the neuronal recordings were made in adult rats after neonatal exposure, these data were interpreted to indicate that lead exposure during development produced a permanent alteration in CNS function.

These examples show that selective studies of neuronal activity in response to exposure to neurotoxic agents can provide valuable information about toxic actions at the cellular level. Observations of changes in neuronal activity are an important intermediate step in guiding further specific studies to determine the molecular mechanism of the toxic action.

2. Special Senses

Behavioral methods have been described for evaluating the ototoxicity of various agents in experimental animal models.[123] Using operant conditioning techniques, a variety of species (e.g., guinea pig, chinchilla, cat, monkey) can be trained to respond to sound. Tone-on-tone masking behavioral techniques have been used to obtain psychophysical tuning curves which correlate to the tuning curves obtained from electrophysiologic recording of single auditory neurons.[110] From these, measures of auditory threshold, differential frequency sensitivity, frequency selectivity, and loudness discrimination can be made.

Using these techniques, several aminoglycosidic antibiotics (which are well-known ototoxic agents) have been shown to produce a selective loss of high frequency sound perception.[122] Histologic examination has revealed that losses of hearing at various frequencies directly correlated with losses of inner and outer hair cells along the basilar membrane.[55] High frequency hearing losses correlate with damage at the basal end of the membrane and low frequency hearing losses with damage near the apex.[123] Interestingly, similar hearing losses and hair cell damage are observed following exposure to industrial noise.[123]

For studies of toxicity in the visual system, visual evoked potentials are most often used. Monopolar recording from the visual cortex permits the measurement of a variety of highly processed visual information. These techniques have recently been reviewed.[41] Briefly, a visual stimulus is selected to evoke a rather well-defined degree of sensory processing. A diffuse, unpatterned flash of light evokes a visual potential which contains largely temporal information. The latency from stimulus presentation to processing in the visual cortex can be used as a measure of conduction along the visual tract, which includes not only axonal transmission but also synaptic activity. Patterned stimuli contain not only temporal information but also spatial information such as contours, edges, and shapes. These subtle but specific functions of the visual system can provide evidence of the damage induced by neurotoxic agents. If specific types of data or interpretation are required, many other stimulus parameters such as luminance, wavelength of the light, frequency and duration of the stimulus, background illumination, and the condition of the eye (pupil diameter and accommodation) must be controlled.

Using these techniques, Fox et al.[40] showed that neonatal exposure to lead resulted in an increased latency of the visual evoked potential and also alterations in the response to paired stimuli, which were interpreted as a decreased conduction velocity in the visual projection system. Similarly, Johnson et al.[59] showed that methyl *n*-butyl ketone increased the latency of the visual evoked potential and concluded that this deficit correlated with pathologic damage in the optic tract and hypothalamus. On the other hand, Joy[61] found that dieldrin increased the excitability of visual cortex neurons in response to lateral geniculate stimulation.

3. Somatosensory Evoked Potentials

Selective vulnerability of the nervous system to neurotoxic agents has been recently reviewed,[131] pointing out that many primary sensory axons are not only affected at their peripheral terminals but also at their central terminals. Clioquinol appears to be one of only a few agents which are preferentially toxic to the central sensory nerve endings,[67] producing a selective disruption of long latency spinal reflexes.[52]

To measure changes in the central projection of sensory neurons, the somatosensory evoked potential is most commonly used.[14,38,100] Peripheral electrical stimulation at the tail or limb extremities is recorded centrally at points along the spinal cord, brain stem, and/or cerebral cortex. Signal averaging is usually necessary to increase the resolution of this small, multicomponent signal. Using this technique, Arezzo et al.[6] in the monkey and Boyes and Cooper[21] in the rat demonstrated that the earliest signs of acrylamide neurotoxicity could be measured as an increased latency of the somatosensory evoked potential. These observations correlated with early morphological changes in the gracile fasciculus[6] and confirmed earlier observations by Prineas[98] that the central projections of sensory nerves are especially vulnerable to acrylamide.

4. Spinal Cord Studies

Although spinal cord reflex responsiveness has rarely been used as a means of studying neurotoxic agents, there are a number of useful features to this type of investigation. Stimulation of the dorsal root and recording of the evoked compound action potential from the corresponding ventral root provides a means of studying the simplest and most accessible monosynaptic connection in the CNS. This segmental monosynaptic connection is responsible for mediating the stretch reflex, receiving input from large Ia afferents from muscle and projecting the large alpha motor neuron efferents to modulate muscle tone. Since there is now a greater awareness that agents which are toxic to peripheral sensory nerves also affect their central processes, the afferent limb of the monosynaptic reflex arc is another example of a central sensory process which can be studied. Neuropathic agents which affect sensory nerve endings might logically be expected to affect the presynaptic nerve terminal innervating the monosynaptic reflex. Indeed, Goldstein and Lowndes[50] have shown that acrylamide decreases the amplitude of the monosynaptic reflex in cats prior to the development of other neurotoxic effects. These results were interpreted as an effect of acrylamide on the central nerve terminal of the sensory afferents.

The monosynaptic reflex is also suppressed following tris(2-methylphenyl)phosphate (tri-*o*-cresyl phosphate; TOCP) administration.[71] Since histological damage to the spinal cord has been reported following TOCP,[58] these data suggest that the diminished reflex response is due to loss of sensory input. However, both the dorsal and ventral root compound action potentials were also decreased in amplitude,[71] suggesting that TOCP neuropathy was occurring concurrently in both sensory and motor axons.

A more direct measure of afferent and efferent activity in the spinal reflex arc is obtained by intracellular recording from these neurons. The large alpha motor neurons are easily recorded intracellularly and can provide useful information about the effects of neurotoxic agents on motor output from the spinal cord. In addition to features of the neuronal cell membrane such as resting membrane potential and membrane resistance, various features of membrane excitability in response to stimulation can also be recorded. These measurement have been quite informative in cases where the neurotoxic agent affects not only the peripheral axon, but also the cell soma. One such example is IDPN which induces large swellings of the proximal axon in the region of the initial segment. Lowndes has shown that his enlargement of the axon results in changes in the propagation of action potentials from the neuron and also changes the characteristics of threshold excitability of the initial segment.[41]

The sensory cell bodies in the dorsal root ganglia are also large and permit intracellular recording of sensory afferent activity. In addition to recording membrane characteristics, the sensory neuron can be evoked by activation of the afferent nerve terminals. Since many peripheral neurotoxic substances have their earliest effects on sensory nerves, intracellular recording from the cell soma can provide information about the specific types of changes occurring in impulse conduction during intoxication. For example, Somjen et al.[115] showed that methylmercury decreased the ability of the dorsal root ganglion cells to discharge repetitively, an effect which correlated with the overt neurotoxicity in the animals. The deficits appeared to occur in the cell body prior to changes in the sensory axons.

C. Special Studies: Cultured Neurons

The use of cultured cell lines for examining neurotoxic effects in vitro has recently been reviewed.[48] There is a great appeal to this technique for the study of neurotoxic substances. Many of the confounding factors which make data interpretation difficult in vivo are eliminated and direct effects of the toxic substance on the neuron can be studied, using accurately known toxicant concentrations. In addition, since neurons and glia can be cultured separately, selective effects on these two important components of the nervous system can be studied individually. This is particularly important for the neurotoxicologist, since neurotoxic effects may involve either or both of these cell types.

Most neurotoxicology studies using tissue culture thus far have concentrated either on biochemical changes[49,132] or on morphological changes during the development of neurons and myelination.[140] Electrophysiological studies, unfortunately, have not been performed in culture by neurotoxicologists, although tissue culture has been widely exploited by electrophysiologists and neuropharmacologists. The number and variety of discrete studies using tissue culture systems are almost endless and provide many opportunities for studying specific changes in neuronal and glial function during exposure to neurotoxic substances. Tissue culture thus provides one of the most fertile lines of research for electrophysiological studies of neurotoxic substances.

REFERENCES

1. **Adams, C. W. M. and Bayliss, O. B.,** Reappraisal of osmium and OTAN histochemical reactions, *Histochemie*, 16, 162, 1968.
2. **Anderson, R. J.,** Relative refractory period as a measure of peripheral nerve neurotoxicity, *Toxicol. Appl. Pharmacol.*, 71, 391, 1983.
3. **Anderson, R. J. and Dunham, C. B.,** Electrophysiologic deficits in peripheral nerve as a discriminator of early hexacarbon neurotoxicity, *J. Toxicol. Environ. Health*, 13, 835, 1984.
4. **Anthony, D. C., Boekelheide, K., and Graham, D. G.,** The effect of 3,4-dimethyl substitution on the neurotoxicity of 2,5-hexanedione. I. Accelerated clinical neuropathy is accompanied by more proximal axonal swellings, *Toxicol. Appl. Pharmacol.*, 71, 362, 1983.
5. **Anthony, D. C., Boekelheide, K., Anderson, C. W., and Graham, D. G.,** The effect of 3,4-dimethyl substitution on the neurotoxicity of 2,5-hexanedione. II. Dimethyl substitution accelerates pyrrole formation and protein crosslinking, *Toxicol. Appl. Pharmacol.*, 71, 371, 1983.
6. **Arezzo, J. C., Schaumburg, H. H., Vaughan, H. G., Spencer, P. S., and Barna, J.,** Hindlimb somatosensory evoked potentials in the monkey: the effects of distal axonopathy, *Ann. Neurol.*, 12, 24, 1982.
7. **Asbury, A. K. and Johnson, P. C.,** Standard histologic preparations, in *Pathology of Peripheral Nerve*, Asbury, A. K. and Johnson, P. C., Eds., W. B. Saunders, Philadelphia, 1978, 34.
8. **Asbury, A. K. and Johnson, P. C., Eds.,** *Pathology of Peripheral Nerve*, W. B. Saunders, Philadelphia, 1978.
9. **Baker, T. and Lowndes, H. E.,** Muscle spindle function in organophorphorus neuropathy, *Brain Res.*, 185, 77, 1980.
10. **Baker, T., Lowndes, H. E., Johnson, M. K., and Sandborg, I. C.,** The effects of phenylmethanesulfonyl fluoride on delayed organophosphorus neuropathy, *Arch. Toxicol.*, 46, 305, 1980.
11. **Bay, E., Adams, N. L., Von Bredow, J. K. D., and Nelson, J. D.,** Respiratory phase shift of pattern in the medullary reticular formation after soman in the cat, *Brain Res.*, 60, 526, 1973.

12. **Beers, E. T., Foster, R. E., Glenn, J. F., Adams, N. L., Finger, A. V., Randolph, T. C., and Rickett, D. L.,** Comparison of central nervous system and neuromuscular effects of acute infusion of tabun or VX, *Neurosci. Abstr.*, 9, 1164, 1983.
13. **Bergmans, J. A.,** Neurophysiological features of experimental and human neuropathies, in *Clinical and Biological Aspects of Peripheral Nerve Diseases*, Battistin, L. et al., Eds., Alan R. Liss, New York, 1983, 73.
14. **Bhargava, V.,** Effect of diisopropylphosphorofluoridate (DFP) on the somatosensory evoked potentials in rats, *Psychopharmacology*, 25, 376, 1972.
15. **Bickford, A. A. and Sprague, G. L.,** The significance of background neurologic lesions in acute delayed neurotoxicity studies: a comparison of neurohistopathologic lesions induced in commercial hens by tri-o-tolyl phosphate (TOCP) with those observed in negative control hens, *Neurotoxicology*, 4, 283, 1983.
16. **Bierkamper, G. G.,** Electrophysiological effects of diisopropylfluorophosphate on neuromuscular transmission, *Eur. J. Pharmacol.*, 73, 343, 1981.
17. **Bierkamper, G. G.,** *In vitro* assessment of neuromuscular toxicity, *Neurobehav. Toxicol. Teratol.*, 4, 597, 1982.
18. **Bjorklund, H., Palmer, M. R., Lind, B., Hoffer, B. J., and Olson, L.,** Postnatal lead exposure alters spontaneous cerebellar Purkinje neuron discharge, *Environ. Res.*, 31, 448, 1983.
19. **Bondy, S. C.,** Neurotransmitter binding interactions as a screen for neurotoxicity, in *Mechanisms of Actions of Neurotoxic Substances*, Prasad, K. and Vernadakis, A., Eds., Raven Press, New York, 1982, 25.
20. **Bouldin, T. W. and Cavanagh, J. B.,** Organophosphorus neuropathy. II. A fine structural study of the early stages of axonal degeneration, *Am. J. Pathol.*, 94, 253, 1979.
21. **Boyes, W. K. and Cooper, G. P.,** Acrylamide neurotoxicity: effects on farfield somatosensory evoked potentials in rats, *Neurobehav. Toxicol. Teratol.*, 3, 487, 1981.
22. **Brown, A. W., Aldridge, W. N., Street, B. W., and Vershoyle, R. D.,** The behavioral and neuropathologic sequelae of intoxication by trimethyltin in the rat, *Am. J. Pathol.*, 97, 59, 1979.
23. **Bullock, J. O., Farquharson, D. A., and Hoskin, F. C. G.,** Soman and receptor-ligand interaction in electrophorus electroplaques, *Biochem. Pharmacol.*, 26, 337, 1977.
24. **Cangiano, A., Lutzemberger, L., Rizzuto, N., Simonati, A., Rossi, A., and Toschi, G.,** Neurotoxic effects of 2,5-hexandeione in rats: early morphological and functional changes in nerve fibers and neuromuscular junctions, *Neurotoxicology*, 2, 25, 1981.
25. **Cavanagh, J. B.,** The significance of the "dying back" process in experimental and human neurological disease, *Int. Rev. Exp. Pathol.*, 3, 219, 1964.
26. **Cavanagh, J. B.,** The pathokinetics of acrylamide intoxication: a reassessment of the problem, *Neuropathol. Appl. Neurobiol.*, 8, 315, 1982.
27. **Chu, E.-S., Jortner, B. S., Schaumburg, H. H., and Spencer, P. S.,** A single intravenous injection of doxorubicin (Adriamycin) induces sensory neuronopathy in rats, *Neurotoxicology*, 1, 583, 1980.
28. **Cotman, C. W.,** *Neuronal Plasticity*, Raven Press, New York, 1978.
29. **Crapper, D. R. and Tomko, G. J.,** Neuronal correlates of an encephalopathy associated with aluminum neurofibrillary degeneration, *Brain Res.*, 97, 253, 1975.
30. **Craviato, H. M., Weiss, J. F., Goebel, H. H., Weiss, E. deC., and Ranshoff, J. F.,** Preferential induction of central or peripheral nervous system tumors in rats by nitrosourea derivatives, *J. Neuropathol. Exp. Neurol.*, 33, 595, 1974.
31. **Cresse, I.,** Receptor binding as a primary drug screening device, in *Neurotransmitter Receptor Binding*, Yamamura, H. I., Enna, S. J., and Kuhar, M. J., Eds., Raven Press, New York, 1978, 141.
32. **Davis, C. S. and Richardson, R. J.,** Organophorphorus compounds, in *Experimental and Clinical Neurotoxicology*, Spencer, P. S. and Schaumburg, H. H., Eds., Williams & Wilikins, Baltimore, 1980, 527.
33. **DeJesus, C. P. V. and Pleasure, D. E.,** Hexachlorophene neuropathy, *Arch. Neurol.*, 29, 180, 1973.
34. **DeJesus, C. P. V., Towfighi, J., and Snyder, D. R.,** Sural nerve conduction study in the rat: a new technique for studying experimental neuropathies, *Muscle Nerve*, 1, 162, 1978.
35. **Drakontides, A. B. and Baker, T.,** An electrophysiologic and ultrastructural study of the phenylmethanesulfonyl fluoride protection against a delayed organophosphorus neuropathy, *Toxicol. Appl. Pharmacol.*, 70, 411, 1983.
36. **Dyck, P. J., Thomas, P. K., and Lambert, E. H., Eds.,** *Peripheral Neuropathy*, Vol. 1 and 2, W. B. Saunders, Philadelphia, 1975.
37. **Dyck, P. J. and Karnes, J. L.,** Morphometry of neuron columns and fiber tracts in neurobiology and pathology using computer imaging, *Trends Neurosci.*, June, 138, 1981.
38. **Dyer, R. S. and Boyes, W. K.,** Use of neurophysiological challenges for the detection of toxicity, *Fed. Proc. Fed. Am. Soc. Exp. Biol.*, 42, 3201, 1983.
39. **Foster, R. E., Whalen, C. C., and Waxman, S. G.,** Reorganization of the axon membrane in demyelinated peripheral nerve fibers: morphological evidence, *Science*, 210, 661, 1980.

40. **Fox, D. A., Lewkowski, J. P., and Cooper, G. P.,** Acute and chronic effects of neonatal lead exposure on development of the visual evoked response in rats, *Toxicol. Appl. Pharmacol.*, 40, 449, 1977.
41. **Fox, D. A., Lowndes, H. E., and Bierkamper, G. G.,** Electrophysiological techniques in neurotoxicology, in *Nervous System Toxicology,* Mitchell, C. L., Ed., Raven Press, New York, 1982, 299.
42. **Friedli, W. G., and Meyer, M.,** Strength-duration curve: a measure for assessing sensory deficit in peripheral neuropathy, *J. Neurol. Neurosurg. Psychiatr.*, 47, 184, 1984.
43. **Fullerton, P. M.,** Chronic peripheral neuropathy produced by lead poisoning in guinea pigs, *Neuropathol. Exp. Neurol.*, 25, 214, 1966.
44. **Fullerton, P. M., Gilliatt, R. W., Lascelles, R. G., and Morgan-Hughes, J. A.,** Relation between fiber diameter and internodal length in chronic neuropathy, *J. Physiol. (London),* 178, 26P, 1965.
45. **Fullerton, P. M. and O'Sullivan D. J.,** Thalidomide neuropathy: a clinical, electrophysiological and histological follow-up, *J. Neurol. Neurosurg. Psychiatr.*, 31, 543, 1968.
46. **Geller, I., Stevvins, W. C., and Wagner, M. J.,** *Test Methods for Definition of Effects of Toxic Substances on Behavior and Neuromotor Function,* Ankho International, Fayetteville, N.C., 1979.
47. **Gold, B. G., Griffin, J. W., Price, D. L., Cork, L. C., and Lowndes, H. E.,** Iminodipropionitrile neuropathy in cats: physiological consequences of giant axon formations, *Soc. Neurosci. Abstr.*, 6, 804, 1980.
48. **Goldberg, A. M.,** Mechanisms of neurotoxicity as studied in tissue culture systems, *Toxicology,* 17, 201, 1980.
49. **Goldberg, A. M., Brookes, N., and Burt, D. R.,** The use of spinal cord cell cultures in the study of neurotoxicological agents, *Toxicology,* 17, 233, 1980.
50. **Goldstein, B. D. and Lowndes, H. E.,** Spinal cord defect in the peripheral neuropathy resulting from acrylamide, *Neurotoxicology,* 1, 75, 1979.
51. **Gorio, A., Millesi, H., and Mingrino, S.,** *Posttraumatic Peripheral Nerve Regeneration,* Raven Press, New York, 1981.
52. **Goto, M., Akahane, K., Ono, H., Kaneko, T., Fukuda, H., and Tamura, Z.,** Deterioration of spinal reflex in beagles orally ingesting clioquinol, *J. Toxicol. Sci.*, 7, 1, 1982.
53. **Grafstein, B.,** The nerve cell body response to axotomy, *Exp. Neurol.*, 48, 32, 1975.
54. **Griffin, J. W., Hoffman, R. M., Clark, A. W., Carrol, P. T., and Price, D. L.,** Slow axonal transport of neurofilament proteins: impairment by iminodipropionitrile, *Science,* 202, 633, 1978.
55. **Hawkins, J. E., Stebbins, W. C., Johnsson, L. G., Moody, D. B., and Muraski, A.,** The patas monkey as a model for dihydrostreptomycin ototoxicity, *Acta Otolaryngol.*, 83, 123, 1977.
56. **Holt, S. J. and Withers, R. F. J.,** Cytochemical localization of esterases using indoxyl derivatives, *Nature (London),* 170, 1012, 1952.
57. **Hopkins, A. P. and Gilliatt, R. W.,** Motor and sensory conduction velocity in the baboon: normal values and changes during acrylamide neuropathy, *J. Neurosurg. Psychiatr.*, 34, 415, 1971.
58. **Illis, L., Pantagia, G. N., and Cavanagh, J. B.,** Boutons terminaux and tri-ortho-cresyl phosphate neurotoxicity, *Exp. Neurol.*, 14, 160, 1966.
59. **Johnson, B. L., Setzer, J. V., Lewis, T. R., and Anger, W. K.,** Effects of methyl *n*-butyl ketone on behavior and the nervous system, *Am. Ind. Hyg. Assoc. J.*, 38, 567, 1977.
60. **Johnson, M. K.,** Organophosphorus and other inhibitors of brain neurotoxic esterase and the development of delayed neurotoxicity in hens, *Biochem. J.*, 120, 523, 1970.
61. **Joy, R. M.,** The alteration by dieldrin of cortical excitability conditioned by sensory stimuli, *Toxicol. Appl. Pharamcol.*, 38, 357, 1976.
62. **Juang, M. S.,** An electrophysiological study of the action of methylmercuric chloride and mercuric chloride on the sciatic nerve-sartorius muscle preparation of the frog, *Toxicol. Appl. Pharmacol.*, 37, 339, 1976.
63. **Kiernan, J. A. and Berry, M.,** Neuroanatomical methods, in *Methods in Brain Research,* Bradley, P. B., Ed., Wiley Interscience, London, 1975, 1.
64. **Kiernan, J. A.,** Neurohistochemical techniques, in *Histological and Histochemical Methods: Theory and Practice,* Kiernan, J. A., Ed., Pergamon Press, Oxford, 1981, 256.
65. **Kiernan, J. A. and Berry, M.,** Neuroanatomical methods, in *Methods in Brain Research,* Bradley, P. B., Ed., Wiley Interscience, London, 1975, 1.
66. **Koelle, G. B. and Friedenwald, J. S.,** A histochemical method for localizing cholinesterase activity, *Proc. Soc. Exp. Biol. Med.*, 70, 617, 1949.
67. **Krinke, G., Schaumburg, H. H., Spencer, P. S., Thomann, P., and Hess, R.,** Clioquinol and 2,5-hexanedione produce different types of distal axonopathy: a comparative experimental study in the dog, *Acta Neuropathol.*, 47, 213, 1979.
68. **Kuba, K., Albuquerque, E. X., and Barnard, E. A.,** Diisopropylfluorophosphate: suppression of ionic conductance of the cholinergic receptor, *Science,* 181, 853, 1973.
69. **Kuffler, S. W. and Nicholls, J. G.,** *From Neuron to Brain,* Sinauer Associates, Sunderland, Mass., 1976.
70. **Landis, D. M. D.,** Membrane structure and neurological disorders, in *Membrane Abnormalities and Disease,* Vol. 2, Toa, M., Ed., CRC Press, Boca Raton, Fla., 1982, 1.

71. **Lapadula, D. M., Kinnes, C. G., Somjen, G. G., and Abou-Donia, M. B.,** Monosynaptic reflex depression in cats with organophosphorus neuropathy: effects of tri-o-cresyl phosphate, *Neurotoxicology*, 3, 51, 1982.
72. **Laskowski, M. B., Olson, W. H., and Dettbarn, W.-D.,** Ultrastructural changes at the muscle end-plate produced by an irreversible cholinesterase inhibitor, *Exp. Neurol.*, 47, 290, 1975.
73. **Laskowski, M. B., Olson, W. H., and Dettbarn, W.-D.,** Initial ultrastructural changes at the muscle end-plate produced by cholinesterase inhibition, *Exp. Neurol.*, 57, 13, 1977.
74. **Leonard, J. P. and Saltpeter, M. M.,** Agonist-induced myopathy at the neuromuscular junction is mediated by calcium, *J. Cell. Biol.*, 82, 811, 1979.
75. **Lipp, J. A.,** Cerebral electrical activity following soman administration, *Arch. Int. Pharmacodyn.*, 175, 161, 1968.
76. **Lowndes, H. E. and Baker, T.,** Studies on drug-induced neuropathies. III. Motor nerve deficit in cats with experimental acrylamide neuropathy, *Eur. J. Pharmacol.*, 35, 177, 1976.
77. **Lowndes, H. E., Baker, T., Cho, E. S., and Jortner, B. S.,** Position sensitivity of deefferented muscle spindles in experimental acrylamide neuropathy, *J. Pharmacol. Exp. Ther.*, 205, 40, 1978.
78. **Lowndes, H. E., Baker, T., Michelson, L. P., and Vincent-Ablazey, M.,** Attenuated dynamic responses of primary endings of muscle spindles: a basis for depressed tendon responses in acrylamide neuropathy, *Ann. Neurol.*, 3, 433, 1978.
79. **Lowndes, H. E., Baker, T., and Riker, W. F.,** Motor nerve dysfunction in delayed DFP neuropathy, *Eur. J. Pharmacol.*, 29, 66, 1974.
80. **Marchi, V.,** Sur l'origine et la cours des pedoncles cerebelleux et sur leurs rapports avec les autres centres nerveux, *Arch. Ital. Biol.*, 17, 190, 1892.
81. **Marsland, T. A., Glees, P., and Erikson, L. B.,** Modification of the Glees silver impregnation method for paraffin sections, *J. Neuropathol. Exp. Neurol.*, 13, 587, 1954.
82. **Maxwell, I. C. and LeQuesne, P. M.,** Conduction velocity in hexachlorophene neuropathy: correlation between electrophysiological and histological findings, *J. Neurol. Sci.*, 43, 95, 1979.
83. **McLeod, J. G., Prineas, J. W., and Walsh, J. C.,** The relationship of conduction velocity to pathology in peripheral nerves, in *New Developments in Electromyography and Clinical Neurophysiology*, Vol. 2, Desmedt, J. E., Ed., S. Karger, Basel, 1973, 248.
84. **Misumi, J.,** Electrophysiological studies *in vivo* on peripheral nerve function and their application to peripheral neuropathy produced by organic mercury in rats. III. Effects of methylmercuric chloride on compound action potentials in the sciatic and tail nerve in rats, *Kumamoto Med. J.*, 32, 15, 1979.
85. **Miyamoto, M. D.,** Hg^{2+} causes neurotoxicity at an intracellular site following entry through Na and Ca channels, *Brain Res.*, 267, 375, 1983.
86. **Narahashi, T.,** Chemicals as tools in the study of excitable membranes, *Physiol. Rev.*, 54, 813, 1972.
87. **Narahashi, T.,** Nerve membrane as a target of environmental toxicants, in *Experimental and Clinical Neurotoxicology*, Spencer, P. S. and Schaumburg, H. H., Eds., Williams & Wilkins, Baltimore, 1980, 225.
88. **Narahashi, T.,** Cellular and molecular mechanisms of action of insecticides: neurophysiological approach, *Neurobehav. Toxicol. Teratol.*, 4, 753, 1982.
89. **Nicholls, J. G.,** *Repair and Regeneration of the Nervous System*, Springer-Verlag, Berlin, 1982.
90. **Norton, S. and Culver, B.,** A Golgi analysis of caudate neurons in rats exposed to carbon monoxide, *Brain Res.*, 132, 455, 1977.
91. **Norton, S.,** Behavior versus morphology as an indicator of central nervous system toxicity, in *Nervous System Toxicology*, Mitchell, C. L., Ed., Raven Press, New York, 1982, 247.
92. **Ohnishi, A., Offord, K., and Dyck, P. J.,** Studies to improve fixation of human nerves. I. Effect of duration of glutaraldehyde fixation on peripheral nerve morphometry, *J. Neurol. Sci.*, 23, 223, 1974.
93. **Ohnishi, A., O'Brien, P. C., and Dyck, P. J.,** Studies to improve fixation of human nerves. II. Effect of time elapsed between death and glutaraldehyde fixation on relationship of axonal area to number of myelin lamellae, *J. Neurol. Sci.*, 23, 387, 1974.
94. **Ohnishi, A., O'Brien, P. C., and Dyck, P. J.,** Studies to improve fixation of human nerves. III. Effect of osmolality of glutaraldehyde solutions on relationship of axonal area to number of myelin lamellae, *J. Neurol. Sci.*, 27, 193, 1976.
95. **Ohnishi, A., Schilling, K., Brimijoin, W. S., Lambert, E. H., Fairbanks, V. F., and Dyck, P. J.,** Lead neuropathy. I. Morphometry, nerve conduction and choline acetyltransferase transport: new finding of endoneurial edema associated with segmental demyelination, *J. Neuropathol. Exp. Neurol.*, 36, 499, 1977.
96. **Ohnishi, A. and Ikeda, M.,** Morphometric evaluation of primary sensory neurons in experimental *p*-bromophenylacetylurea intoxication, *Acta Neuropathol.*, 52, 111, 1980.
97. **Pincus, J. H. and Tucker, G. J.,** *Behavioral Neurology*, Oxford University Press, New York, 1974.

98. **Prineas, J.,** The pathogenesis of dying-back polyneuropathies. II. An ultrastructural study of experimental acrylamide intoxication in the cat, *J. Neuropathol. Exp. Neurol.*, 28, 598, 1969.
99. **Rasminsky, M.,** Physiological consequences of demyelination, in *Experimental and Clinical Neurotoxicology*, Spencer, P. S. and Schaumburg, H. H., Eds., Williams & Wilkins, Baltimore, 1980, 257.
100. **Rebert, C. S. and Sorenson, S. S.,** Concentration-related effects of hexane on evoked responses from brain and peripheral nerve of the rat, *Neurobehav. Toxicol. Teratol.*, 5, 69, 1983.
101. **Richardson, R. J.,** Toxicology of the nervous system, in *Toxicology: Principles and Practice*, Vol. 1, Reeves, A. L., Ed., John Wiley & Sons, New York, 1981, 107.
102. **Ritchie, J. M. and Chiu, S. Y.,** Distribution of sodium and potassium channels in mammalian myelinated nerve, in *Demyelinating Disease: Basic and Clinical Electrophysiology*, Waxman, S. G. and Ritchie, J. M., Eds., Raven Press, New York, 1981, 329.
103. **Roberts, D. V. and Trollope, I. E.,** Nerve conduction velocity and refractory period as parameters of neurotoxicity, *Electroencephalog. Clin. Neurophysiol.*, 46, 351, 1979.
104. **Rozear, M., DeGroff, R., and Somjen, G.,** Effects of microiontophoretic administration of divalent metal ions on neurons of the central nervous system of cats, *J. Pharmacol. Exp. Ther.*, 176, 109, 1971.
105. **Sahenk, Z. and Mendell, J. R.,** Ultrastructural study of zinc-pyridinethione-induced peripheral neuropathy, *J. Neuropath. Exp. Neurol.*, 38, 532, 1979.
106. **Satoh, E., Asai, F., Itoh, K., Nishimura, M., and Urakawa, N.,** Mechanism of cadmium-induced blockade of neuromuscular transmission, *Eur. J. Pharmacol.*, 77, 251, 1982.
107. **Schaumburg, H. H., Dyck, P., and Weller, R. O.,** The effect of phenol on peripheral nerve, *J. Neuropathol. Exp. Neurol.*, 29, 615, 1970.
108. **Schaumburg, H. H., Wisniewski, H. M., and Spencer, P. S.,** Ultrastructural studies of the dying back process. I. Peripheral nerve terminal and axon degeneration in systemic acrylamide intoxication, *J. Neuropathol. Exp. Neurol.*, 33, 260, 1974.
109. **Schmitt, F. O., Bird, S. J., and Bloom, F. E.,** *Molecular Genetic Neuroscience*, Raven Press, New York, 1982.
110. **Serafin, J. V., Moody, D. B., and Stebbins, W. C.,** Psychophysical tuning curves in monkeys, *J. Acoust. Soc. Am.*, 63, S30, 1978.
111. **Slizewski, M.,** Influence of chronic administration of phenylmercuric acetate on the peripheral nerve system of rat, *Neuropathol. Pol.*, 13, 71, 1975.
112. **Smith, K. J.,** A sensitive method for the detection and quantification of conduction deficits in nerve, *J. Neurol. Sci.*, 48, 191, 1980.
113. **Smith, K. J., Blakemore, W. F., and McDonald, W. I.,** Central remyelination restores secure conduction, *Nature (London)*, 280, 395, 1979.
114. **Smith, K. J. and Hall, S. M.,** Nerve conduction during peripheral demyelination and remyelination, *J. Neurol. Sci.*, 48, 201, 1980.
115. **Somjen, G. G., Herman, S. P., and Klein, R.,** Electrophysiology of methyl mercury poisoning, *J. Pharmacol. Exp. Ther.*, 186, 579, 1973.
116. **Spencer, P. S., Peterson, E. R., Madrid, R., and Raine, C. S.,** Effects of thallium salts on neuronal mitochondria in organotypic cord-ganglia-muscle combination cultures, *J. Cell. Biol.*, 58, 79, 1973.
117. **Spencer, P. S. and Schaumburg, H. H.,** Experimental neuropathy produced by 2,5-hexanedione — a major metabolite of the industrial solvent methyl n-butyl ketone, *J. Neurol. Neurosurg. Psychiatr.*, 38, 771, 1975.
118. **Spencer, P. S., Sabri, M. I., Schaumburg, H. H., and Moore, C.,** Does a defect in energy metabolism in the nerve fiber cause axonal degeneration in polyneuropathies?, *Ann. Neurol.*, 5, 501, 1979.
119. **Spencer, P. S. and Schaumburg, H. H.,** *Experimental and Clinical Neurotoxicology*, Williams & Wilkins, Baltimore, 1980.
120. **Spencer, P. S. and Schaumburg, H. H.,** Classification of neurotoxic disease: a morphological approach, in *Experimental and Clinical Neurotoxicology*, Spencer, P. S. and Schaumburg, H. H., Eds., Williams & Wilkins, Baltimore, 1980, 92.
121. **Spencer, P. S., Bischoff, M. C., and Schaumburg, H. H.,** Neuropathological methods for the detection of neurotoxic disease, in *Experimental and Clinical Neurotoxicology*, Spencer P. S. and Schaumburg, H. H., Eds., Williams & Wilkins, Baltimore, 1980, 743.
122. **Stebbins, W. C., Miller, J. M., Johnsson, L. G., and Hawkins, J. E.,** Ototoxic hearing loss and cochlear pathology in the monkey, *Trans. Am. Otol. Soc.*, 57, 110, 1969.
123. **Stebbins, W. C. and Moody, D. B.,** Comparative behavioral toxicology, *Neurobehav. Toxicol.*, 1, 33, 1979.
124. **Stegeman, D. F., DeWeerd, J. P. C., and Notermans, S. L. H.,** Modelling compound action potentials of peripheral nerves in situ. III. Nerve propagation in the refractory period, *Electroencephalog. Clin. Neurophysiol.*, 55, 668, 1983.

125. **Sterman, A. B. and Spencer, P. S.,** The pathogenesis of primary internodal demyelination produced by acetyl ethyl tetramethyl tetralin: evidence for preserved Schwann cell somal function, *J. Neuropathol. Exp. Neurol.*, 40, 112, 1981.
126. **Sterman, A. B., Sposito, N., and Shirey, L.,** The evolution and specificity of neuronal cell body changes in hexacarbon-induced axonopathy: ultrastructural stereologic analysis, *Ann. Neurol.*, 14, 119, 1983.
127. **Sterman, A. B.,** Altered sensory ganglia in acrylamide neuropathy, *J. Neuropathol. Exp. Neurol.*, 42, 166, 1983.
128. **Sumner, A. J.,** Axonal degenerations, in *Physiology of Peripheral Nerve Disease*, Sumner, A. J., Ed., W. B. Saunders, Philadelphia, 1980, 342.
129. **Sumner, A. J. and Asbury, A. K.,** Physiological studies of the dying-back phenomenon. Muscle stretch afferents in acrylamide neuropathy, *Brain*, 98, 91, 1975.
130. **Thomas, P. K.,** The morphological basis for alterations in nerve conduction in peripheral neuropathy, *Proc. R. Soc. Med. London*, 64, 295, 1971.
131. **Thomas, P. K.,** Selective vulnerability of the centrifugal and centripetal axons of primary sensory neurons, *Muscle Nerve*, 5, S117, 1982.
132. **Trapp, B. D. and Richelson, E.,** Usefulness for neurotoxicology of rotation-mediated aggregating cell cultures, in *Experimental and Clinical Neurotoxicology*, Spencer, P. S. and Schaumburg, H. H., Eds., Williams & Wilkins, Baltimore, 1980, 803.
133. **Van Meter, W. G., Karczmar, A. G., and Fiscus, R. R.,** CNS effects of anticholinesterases in the presence of inhibited cholinesterases, *Arch. Int. Pharmacodyn.*, 231, 249, 1978.
134. **Vinken, R. J. and Bruyn, G. W., Eds.,** Intoxications of the Nervous System. I., II., in *Handbook of Clinical Neurology*, Vol. 36 and 37, North Holland, Amsterdam, 1979.
135. **Waxman, S. G., Ed.,** *Physiology and Pathobiology of Axons*, Raven Press, New York, 1978.
136. **Waxman, S. G.,** Prerequisites for conduction in demyelinated fibers, *Neurology*, 28, 27, 1978.
137. **Waxman, S. G.,** Membranes, myelin and the pathophysiology of multiple sclerosis, *N. Engl. J. Med.*, 306, 1529, 1982.
138. **Waxman, S. G.,** Action potential propagation and conduction velocity — new perspectives and questions, *Trends Neurosci.*, 6, 157, 1983.
139. **Weller, R. O. and Cervos-Navarro, J.,** Techniques of peripheral nerve biopsy and histological preparation, in *Pathology of Peripheral Nerves*, Weller, R. O. and Cervos-Navarro, J., Eds., Butterworths, London, 1977, 18.
140. **Yonezawa, T., Bornstein, M. B., and Peterson, E. F.,** Organotypic cultures of nerve tissue as a model system for neurotoxicity investigation and screening, in *Experimental and Clinical Neurotoxicology*, Spencer, P. S. and Schaumburg, H. H., Eds., Williams & Wilkins, Baltimore, 1980, 788.
141. **Yount, E. A., Felten, S. Y., O'Conner, B. L., Petersen, R. G., Powell, R. S., Yum, M. N., and Harris, R. A.,** Comparison of the metabolic and toxic effects of 2-chloroproprionate and dichloroacetate, *J. Pharmacol. Exp. Ther.*, 222, 501, 1982.

INDEX

A

B

C

D

E

F

G

H

I

K

L

M

N

O

P

Q

R

S

T

U

V

W

X

Z

For Product Safety Concerns and Information please contact our EU representative GPSR@taylorandfrancis.com Taylor & Francis Verlag GmbH, Kaufingerstraße 24, 80331 München, Germany

Batch number: 10733641

Printed by Printforce, the Netherlands